Mathematical Sciences Research Institute
Publications

14

Coxeter Graphs and Towers of Algebras

Frederick M. Goodman Pierre de la Harpe
Vaughan F. R. Jones

Springer-Verlag
New York Berlin Heidelberg
London Paris Tokyo

Frederick M. Goodman
Department of Mathematics
University of Iowa
Iowa City, Iowa 52240
USA

Pierre de la Harpe
Section de Mathématiques
Université de Genève
CH-1211 Genève 24
Switzerland

Vaughan F.R. Jones
Department of Mathematics
University of California - Berkeley
Berkeley, California 94720
USA

Mathematical Sciences Research Institute
1000 Centennial Drive
Berkeley, California 94720
USA

Mathematical Subject Classification (1980): 46L10, 05C50, 16A40

Library of Congress Cataloging-in-Publication Data
Goodman, Frederick M.
Coxeter graphs and towers of algebras.
(Mathematical Sciences Research Institute publications ; 14)
Bibliography: p.
1. Class field towers. 2. Coxeter graphs.
I. La Harpe, Pierre de. II. Jones, Vaughan F.R.,
1952– . III. Title. IV. Series.
QA247.G68 1989 512'.55 89-5991

Printed on acid-free paper.

Camera-ready copy prepared by the authors.
Printed and bound by R. R. Donnelley & Sons, Harrisonburg, Virginia.
Printed in the United States of America.

9 8 7 6 5 4 3 2 1

ISBN 0-387-96979-9 Springer-Verlag New York Berlin Heidelberg
ISBN 3-540-96979-9 Springer-Verlag Berlin Heidelberg New York

PREFACE

The paper [Jo1] on subfactors of von Neumann factors has stimulated much research in von Neumann algebras. Quite unexpectedly, it was discovered soon after the appearance of [Jo1] that certain algebras which are used there for the analysis of subfactors could also be used to define a new polynomial invariant for links [Jo3]. The period of activity following this discovery saw the creation of a number of related invariants as well as the successful use of these invariants in knot theory. Furthermore, recent effort to understand the fundamental nature of the the new link invariants has led to connections with invariant theory, statistical mechanics and quantum field theory. In turn the link invariants, the notion of a quantum group, and the quantum Yang–Baxter equation have had great impact on the study of subfactors.

It is not yet the time to give a comprehensive account of these developments, and we make no attempt to do so here. Our subject is certain algebraic and von Neumann algebraic topics closely related to the original paper [Jo1]. However, in order to promote, in a modest way, the contact between diverse fields of mathematics, we have tried to make this work accessible to the broadest audience. Consequently, this book contains much elementary expository material.

We give here a brief preview of the book. Each of the four chapters has its own introduction, with a more thorough description of the contents.

Chapter 1 begins with a (slightly new guise of) a familiar combinatorial problem: to classify finite matrices over the non–negative integers which have Euclidean norm no greater than 2. These are classified by the ubiquitous *Coxeter graphs* of type A, D, or E (see [HHSV] for other occurrences of these graphs) and the set of possible norms is $\{2\} \cup \{2\cos\pi/q : q \geq 2\}$.

The central theme of the book — the discussion of which begins in Chapter 2 — is the tower of algebras $M_0 \subset M_1 \subset \cdots \subset M_k \subset \cdots$ determined by a pair $M_0 \subset M_1$ of algebras (with the same identity). The tower can be used to define various invariants of the pair, including the index $[M_1:M_0]$

In Chapters 2 and 3, we study two cases of the tower construction in detail.

In Chapter 2, the algebras are finite direct sums of full matrix algebras over some field. A pair $M_0 \subset M_1$ is described, up to isomorphism, by an inclusion matrix Λ with non–negative integer entries. This matrix may be encoded as a graph, known as the Bratteli diagram of the pair. It turns out that the index $[M_1:M_0]$ equals $\|\Lambda\|^2$; thus it follows from Chapter 1 that $[M_1:M_0] \leq 4$ if and only if the Bratteli diagram is a Coxeter graph of type A, D, or E.

In Chapter 3, the algebras are finite von Neumann algebras with finite dimensional centers. Somewhat surprisingly, the results of Chapter 2 essentially extend to this setting. But now a pair $M_0 \subset M_1$ is (partially) described by an inclusion matrix Λ with entries in $\{2\cos\pi/q : q \geq 2\} \cup \{r : r \geq 2\}$, and pairs with index no greater than 4 are associated to Coxeter graphs of arbitrary type, including types B,F,G,H,I.

Finally, Chapter 4 is a further analysis of pairs $N \subset M$ of finite factors of finite index. There are two main themes. The first is the notion of a commuting square, due to Popa [Pop1], and its use in approximating pairs of hyperfinite II_1 factors simultaneously by finite dimensional von Neumann algebras. The second theme is the derived tower of a pair of II_1 factors, which is the chain of (necessarily finite dimensional) relative commutants $M_0' \cap M_k$ in the tower. All the information in the derived tower can be encoded in a (possibly infinite) graph, the principal graph of the pair. When the index is less than 4, the graph is a Coxeter graph of type A, D, or E.

In Chapter 2, we also describe how a certain quotients of the Hecke algebra of type A appear in the tower construction associated to a pair $M_0 \subset M_1$ of, say, finite dimensional semi-simple algebras over the field $\mathbb{C}$. For each choice of a positive faithful trace tr on M_1, there is a unique trace preserving conditional expectation $E_1 : M_1 \to M_0$, and it turns out that M_2 is naturally generated by M_1 and E_1. Now if the trace tr satisfies the so-called Markov condition, then the situation propagates up the tower, and each algebra M_{k+1} is naturally generated by M_k and a conditional expectation $E_k : M_k \to M_{k-1}$, for all $k \geq 1$. Moreover the E_k's are idempotents which satisfy the "braiding" relations

$$\beta E_iE_jE_i = E_i \qquad \text{if } |i-j| = 1, \text{ and}$$
$$E_iE_j = E_jE_i \qquad \text{if } |i-j| \geq 2,$$

where $\beta = [M_1:M_0]$. The abstract algebra $\mathcal{A}_{\beta,k}$ presented by generators $\{\mathbf{1},E_1,\cdots E_{k-1}\}$ and relations as above is a quotient of the Hecke algebra $H_k(q)$, where $q \in \mathbb{C}$ satisfies $\beta = 2 + q + q^{-1}$.

Although we do not discuss this in the text, we might remark here that the map

$$\mathbb{B}_\infty \longrightarrow H_\infty(q) \longrightarrow \text{alg}\,\{\mathbf{1},E_1,\cdots\} \xrightarrow{\text{tr}} \mathbb{C}$$

where $\mathbb{B}_\infty$ is the inductive limit of the braid groups $\mathbb{B}_k$, is, up to a normalization, the Jones link invariant [Jo3]. Also let us point out that to obtain the Jones invariant in this way, it is necessary to deal only with finite dimensional algebras, not the less familiar infinite dimensional von Neumann factors.

There are several appendices.

Appendix I extends the computations of Chapter 1.

Appendix II.a relates complex semi-simple algebras and finite dimensional C^*-algebras. Appendix II.b explains one appearance of the algebras $\mathcal{A}_{\beta,k}$ in statistical mechanics. Appendix II.c is a further discussion of $\mathcal{A}_{\beta,k}$ for special values of β.

Appendix III is an exposition of Hecke subgroups in $PSL_2(\mathbb{R})$, and thus another famous occurrence of the sequence $(2\cos(\pi/k))_{k\geq 3}$.

It is a pleasure to record our gratitude to numerous friends and colleagues for their generous help, including:

R. Baldi, D. Bichsel, H. Dherete, M. Kervaire, A. Ocneanu, M. Pimsner, S. Popa, G. Skandalis, C. Skau, R. Steinberg, V. Sunder, A. Valette, and H. Wenzl.

We gratefully acknowledge support from the MSRI in Berkeley, the IHES in Bures, the United States NSF, the IMA in Minneapolis, and our home institutions during our work on this project.

CONTENTS

Preface v

Chapter 1. Matrices over the natural numbers: values of the norm, classification, and variations. 1

1.1. Introduction. 1

1.2. Proof of Kronecker's theorem. 4

1.3. Decomposability and pseudo-equivalence. 8

1.4. Graphs with norms no larger than 2. 13

1.5. The set $\mathbf{E}$ of norms of graphs and integral matrices. 24

Chapter 2. Towers of multi-matrix algebras. 28

2.1. Introduction. 28

2.2. Commutant and bicommutant 36

2.3. Inclusion matrix and Bratteli diagram for inclusions of multi-matrix algebras. 42

2.4. The fundamental construction and towers for multi-matrix algebras. 57

2.5. Traces. 66

2.6. Conditional expectations. 69

2.7. Markov traces on pairs of multi-matrix algebras. 80

2.8. The algebras $\mathcal{A}_{\beta,k}$ for generic β. 86

2.9. An approach to the non-generic case. 102

2.10. A digression on Hecke algebras. 112

2.10.a. The complex Hecke algebra defined by $GL_n(q)$ and its Borel subgroup. 112

2.10.b. The Hecke algebras $H_{q,n}$. 116

2.10.c. Complex representations of the symmetric group. 118

2.10.d. Irreducible representations of $H_{q,n}$ for $q \notin \Omega$. 119

2.11. The relationship between $\mathcal{A}_{\beta,n}$ and the Hecke algebras. 123

Chapter 3. Finite von Neumann algebras with finite dimensional centers. 128

3.1. Introduction. 128

3.2. The coupling constant: definition. 133

3.3. The coupling constant: examples. 142

3.3.a. Discrete series. 142

3.3.b. Factors defined by icc groups. 143

3.3.c. $W^*(\Gamma)$-modules associated to subrepresentations of λ_G. 144

3.3.d. The formula $\dim_\Gamma(H) = \text{covol}(\Gamma)\, d_\pi$. 145

3.3.e. A digression on the Peterson inner product. 148

3.4. Index for subfactors of II_1 factors. 149

3.5. Inclusions of finite von Neumann algebras with finite dimensional centers 156

3.6. The fundamental construction. 161

3.7. Markov traces on $\text{End}_N(M)$, a generalization of index. 172

Chapter 4. Commuting squares, subfactors, and the derived tower. 182

4.1. Introduction. 182

4.2. Commuting squares. 188

4.3. Wenzl's index formula. 199

4.4. Examples of irreducible pairs of factors of index less than 4, and a lemma of C. Skau. 203

4.5. More examples of irreducible paris of factors, and the index value $3 + 3^{1/2}$. 207

4.6. The derived tower and the Coxeter invariant. 212

4.7. Examples of derived towers 219

4.7.a. Finite group actions. 219

4.7.b. The A_n Coxeter graphs. 220

4.7.c. A general method. 221

4.7.d. Some examples of derived towers for index 4 subfactors. 222

4.7.e. The tunnel construction. 224

4.7.f. The derived tower for $R \supset R_\beta$ when $\beta > 4$. 225

Appendix I. Classification of Coxeter graphs with spectral radius just beyond the Kronecker range. 232

I.1. The results. 232

I.2. Computations of characteristic polynomials for ordinary graphs. 235

I.3. Proofs of theorems I.1.2 and I.1.3. 243

Appendix II.a. Complex semisimple algebras and finite dimensional C^*-algebras 253

Appendix II.b. The algebras $\mathcal{A}_{\beta,k}$ in statistical mechanics. 259

Appendix II.c. More on the algebras $\mathcal{A}_{\beta,k}$ for non-generic β. 266

Appendix III. Hecke groups and other subgroups of $PSL(2,\mathbb{R})$ generated by parabolic pairs. 274

References. 281

Index. 287

CHAPTER 1

Matrices over the natural numbers: Values of the norms, classification, and variations

1.1. Introduction.

As already mentioned, the initial problem for this chapter is combinatorial: it is the classification of finite matrices over the nonnegative integers $\mathbb{N} = \{0,1,2,\cdots\}$ which have Euclidean operator norms no larger than 2. The reader should be aware from the start that most matrices below are *not* square.

We establish first some notation. For $m \geq 1$ the real vector space $\mathbb{R}^m$ has the standard basis $\{e_1,\cdots,e_m\}$, the standard inner product $(\xi|\eta) = \sum_{1 \leq i \leq m} \xi_i \eta_i$, and the associated norm $\|\xi\| = (\xi|\xi)^{1/2}$. For $m,n \geq 1$ and for a subset $S \subset \mathbb{R}$, we denote by $\mathrm{Mat}_{m,n}(S)$ the set of m–by–n matrices with entries in S; we write $\mathrm{Mat}_m(S)$ for $\mathrm{Mat}_{m,m}(S)$, and $\mathrm{Mat}_{fin}(S)$ for the disjoint union of the $\mathrm{Mat}_{m,n}(S)$'s over positive integers m,n. For $X \in \mathrm{Mat}_{m,n}(\mathbb{R})$, the transpose of X is X^t and $X_{i,j}$ is the entry of X in row i and column j. We think of vectors in $\mathbb{R}^m$ as column vectors, and consequently we identify $X \in \mathrm{Mat}_{m,n}(\mathbb{R})$ with a linear map $\mathbb{R}^n \longrightarrow \mathbb{R}^m$. The Euclidean operator norm of X is defined to be

$$\|X\| = \mathrm{Max}\{\|X\xi\| : \xi \in \mathbb{R}^n \text{ and } \|\xi\| \leq 1\}.$$

For $S \subset \mathbb{R}$ we set

$$\mathcal{N}(S) = \{t \in \mathbb{R}_+ : t = \|Y\| \text{ for some } Y \in \mathrm{Mat}_{fin}(S)\}.$$

Our first result is essentially due to Kronecker [Kro]; it shows that $\mathcal{N}(\mathbb{Z}) \cap [0,2]$ is infinite and easy to describe. Though not necessary for the logical understanding of the classification of integer matrices with small norm, the result helps us to believe that the matrices under study admit a comprehensible classification.

Theorem 1.1.1. *Let* X *be a finite matrix over* $\mathbb{Z}$. *Then either* $\|X\| = 2\cos\frac{\pi}{q}$ *for some natural number* $q \geq 2$ *or* $\|X\| \geq 2$.

Moreover, for any integer $q \geq 3$, *there exists a* p–by–p *matrix* X *over* $\{0,1\}$ *with* $\|X\| = 2\cos\frac{\pi}{q}$, *where* p *is the largest integer with* $p \leq q/2$.

This theorem is proved in section 1.2. Let us now describe the classification of matrices

$X \in \mathrm{Mat}_{\mathrm{fin}}(\mathbb{N})$ with $\|X\| \leq 2$. Obviously, we need only consider $X \in \mathrm{Mat}_{\mathrm{fin}}(\{0,1,2\})$.

The first step is to encode the matrix as a bicolored labelled graph, some edges being marked with a sign ∞. (The choice of ∞ fits with the marking conventions for Coxeter graphs, as will be explained below. We distinguish labelled graphs, in which each vertex has a well-defined name, from marked graph, in which some edges are decorated with various signs.) More generally, a matrix $X \in \mathrm{Mat}_{m,n}(\mathbb{R})$ is encoded in a labelled bicolored graph $\tilde{\Gamma}(X)$ which has m black vertices $b_1,\cdots,b_m$ and n white vertices $w_1,\cdots,w_n$; there is no edge joining either two b's or two w's; there is an edge joining b_i and w_j if and only if $X_{i,j} \neq 0$; finally an edge corresponding to $X_{i,j} \notin \{0,1\}$ is marked with some sign conveying information about the value of $X_{i,j}$. In particular, an edge corresponding to $X_{i,j} = 2$ is marked with ∞ and thus looks like $\circ\overset{\infty}{\text{——}}\bullet$. For example:

$$X = \begin{bmatrix} 1 & 0 & 1 & 2 & 0 & 0 \\ 0 & 1 & 0 & 0 & 1 & 0 \end{bmatrix}$$

$\tilde{\Gamma}(X)$:

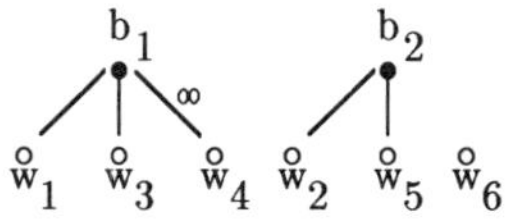

As one may expect for any classification, the next step is to establish appropriate notions of indecomposability and equivalence. We say that X is indecomposable if $\tilde{\Gamma}(X)$ is connected; two matrices X and X′ are pseudo-equivalent if $\tilde{\Gamma}(X)$ and $\tilde{\Gamma}(X')$ are isomorphic as (unlabeled) marked bicolored graphs. These notions will be discussed in greater detail in section 1.3, which also contains a proposition reducing the classification to the indecomposable case. Now comes the solution to the initial problem.

Theorem 1.1.2. *The encoding described above sets up a bijection between:*

(i) *Indecomposable matrices in* $\mathrm{Mat}_{\mathrm{fin}}(\{0,1\})$ *of norms smaller than* 2, *up to pseudo-equivalence, and*

(ii) *Irreducible Coxeter graphs from the following list, together with a bicoloration, up to isomorphism of bicolored graphs. The list is*

$$A_\ell \ (\ell \geq 2), \quad D_\ell \ (\ell \geq 4), \quad E_\ell \ (\ell=6,7,8).$$

Also it sets up a bijection between

(iii) *Indecomposable matrices in* $\mathrm{Mat}_{\mathrm{fin}}(\{0,1,2\})$ *of norm equal to* 2, *up to pseudo-equivalence, and*

(iv) *Irreducible Coxeter graphs from the following list, together with a bicoloration, up to isomorphism of bicolored graphs. The list is*

$$A_\ell^{(1)} \ (\ell \ odd, \ \ell \geq 1), \quad D_\ell^{(1)} (\ell \geq 4), \quad E_\ell^{(1)} (\ell=6,7,8).$$

This theorem is proved in section 1.4, which also contains pictures of the graphs and tables of matrices. The theorem is also essentially proved in numerous sources, including, for example, [Smi] and [CGSS]. One point should be stressed: the combinatorics underlying Theorem 1.1.2 is the same as that which enters into the classification of simple Lie algebras or of reflection groups of the corresponding types, but the theory of Lie algebras is not needed, nor that of reflection groups. On the other hand, these theories immediately suggest both statement and proof of the next result.

We set

$$\mathbb{K} = \{0,1,2\} \cup \{2 \cos \tfrac{\pi}{q}\}_{q=4,5,6\cdots}$$

so that $\mathbb{K} = [0,2] \cap \mathcal{N}(\mathbb{Z})$ by Theorem 1.1.1. To encode $X \in \mathrm{Mat}_{m,n}(\mathbb{K})$, we mark an edge with ∞ if it corresponds to $X_{i,j} = 2$, with q if it corresponds to $X_{i,j} = 2 \cos \frac{\pi}{q}$ for some integer $q \geq 4$, and not at all if $X_{i,j} = 2 \cos \frac{\pi}{3} = 1$. For example:

$$X = \begin{bmatrix} \sqrt{2} & 0 \\ 1 & \sqrt{2} \end{bmatrix} \qquad \tilde{\Gamma}(x) :$$

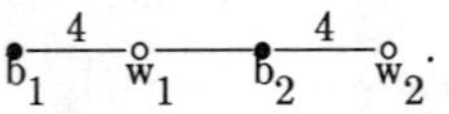

Such a marked graph is known as a Coxeter graph. Theorem 1.1.2 may now be generalized as

Theorem 1.1.3. *The encoding above sets up a bijection between:*

(i) *Indecomposable matrices in* $\mathrm{Mat}_{\mathrm{fin}}(\mathbb{K})$ *of norm smaller than* 2, *up to pseudo–equivalence, and*

(ii) *Irreducible Coxeter graphs from the following list, together with a bicoloration, up to isomorphism of bicolored graphs. The list is*

$$A_\ell\ (\ell \geq 2),\quad B_\ell\ (\ell \geq 2),\quad D_\ell\ (\ell \geq 4),\quad E_\ell\ (\ell{=}6,7,8),$$

$$F_4,\quad G_2,\quad H_\ell\ (\ell{=}3,4),\quad I_2(p),\quad (p{=}5 \text{ or } p \geq 7).$$

Also it sets up a *bijection between:*

(iii) *Indecomposable matrices in* $\mathrm{Mat}_{\mathrm{fin}}(\mathbb{K})$ *of norm equal to* 2, *up to pseudo–equivalence, and*

(iv) *Irreducible Coxeter graphs from the following list, together with a bicoloration, up to isomorphism of bicolored graphs. The list is*

$$A_\ell^{(1)}\ (\ell \text{ odd}, \ell \geq 1),\quad B_\ell^{(1)}\ (\ell \geq 2),\quad C_\ell^{(1)}\ (\ell \geq 3),$$

$$D_\ell^{(1)}\ (\ell \geq 4),\quad E_\ell^{(1)}\ (\ell{=}6,7,8),\quad F_4^{(1)},\quad G_2^{(1)}.$$

This is also explained in Section 1.4. Theorems 1.1.2 and 1.1.3 imply an interesting fact:

$$\mathcal{N}(\mathcal{N}(\mathbb{N})) \cap [0,2] = \mathcal{N}(\mathbb{N}) \cap [0,2].$$

This generalizes to the following theorem, which was shown to us by G. Skandalis.

Theorem 1.1.4. *For any* $S \subset \mathbb{R}$ *one has* $\mathcal{N}(\mathcal{N}(S)) = \mathcal{N}(S)$.

This is proved in Section 1.5. The set $\mathcal{N}(\mathbb{N})$ is also the set $\mathcal{N}(\{0,1\})$ of possible spectral radii of graphs, as has been observed by Hoffman [Hof]. This $\mathcal{N}(\mathbb{N})$ is an interesting set of totally real algebraic integers. Here is a sampling of some known facts regarding $\mathcal{N}(\mathbb{N})$.

Proposition 1.1.5. *The set* $\mathbb{E} = \mathcal{N}(\mathbb{N})$ *has the following properties*:

(i) *If* $s,t \in \mathbb{E}$ *then* $s+t$, st, $(s^2+t^2)^{1/2}$, and $\left[t^2+\frac{s^2}{4}\right]^{1/2}+\frac{s}{2}$ are in $\mathbb{E}$.

(ii) *If* $s \in \mathbb{E}$, $s > 1$, *then* $s+s^{-1}$, $s^2(s^2-1)^{-1/2}$ *are in the limit set of* $\mathbb{E}$.

(iii) *The smallest limit point of* $\mathbb{E}$ *is* $2 = \lim_{q\to\infty} 2\cos(\frac{\pi}{q})$; *there exists* $d > 2$ *with* $\mathbb{E} \cap]2,d[= \phi$.

(iv) *The smallest limit point of limit points of* $\mathbb{E}$ *is* $\lambda_\infty = \sqrt{2+5^{1/2}} \approx 2{,}058171$. *Limit points of* $\mathbb{E}$ *in the interval* $]2,\lambda_\infty[$ *constitute an increasing infinite sequence* $(\lambda_q)_{q\geq 3}$; *furthermore for each* $q \geq 3$, *there are increasing sequences and decreasing sequences in* $\mathbb{E}$ *converging towards* λ_q.

(v) *The closure of* $\mathbb{E}$ *contains* $[\lambda_\infty, \infty[$.

(vi) $\mathbb{E}$ *is not closed.*

Statement (i) is proved in Section 1.5. See also [HW] for other operations defined inside $\mathbb{E}$. Statements (ii) to (iv) are proved in Appendix I. Statement (v) is a recent result of J. Shearer [She]. For (vi), see Appendix I, after I.3.6. What is known of $\mathbb{E}$ makes it look somehow similar to another set recently studied by [Smy]; see also Lehmer's problem, referred to below in remark (3) following Proposition I.3.4, in Appendix I.

1.2. Proof of Kronecker's theorem.

We repeat first the two results of [Kro], with their original proof.

Proposition 1.2.1. *Let* λ *be a nonzero complex number which is a root of a monic polynomial* $P \in \mathbb{Z}[T]$. *If all the roots of* P *are in the disc* $\{z \in \mathbb{C} : |z| \leq 1\}$, *then* λ *is a root of unity.*

Proof. Let ℓ denote the degree of P. The set of all monic polynomials in $\mathbb{Z}[T]$ of degree ℓ having all their roots in the unit disc is finite. Indeed, such a polynomial is of the form

$$Q(T) = T^{\ell} - a_1 T^{\ell-1} + \cdots + (-1)^{\ell} a_{\ell} = \prod_{1 \leq j \leq \ell} (T - \nu_j)$$

with $|\nu_1| \leq 1, \cdots, |\nu_\ell| \leq 1$, and consequently with

$$\begin{aligned} |a_1| &= |\nu_1 + \cdots + \nu_\ell| \leq \ell = \binom{\ell}{1} \\ |a_2| &= \Big| \sum_{1 \leq j,k \leq \ell} \nu_j \nu_k \Big| \leq \binom{\ell}{2} \\ &\cdots\cdots\cdots\cdots\cdots \\ |a_\ell| &= |\nu_1 \nu_2 \cdots \nu_\ell| \leq 1 = \binom{\ell}{\ell} \end{aligned}$$

As the a_j's are in $\mathbb{Z}$, the claim is clear.

Write $P(T) = P_1(T) = \prod_{1 \leq j \leq \ell} (T - \mu_j)$ and, for any $k \geq 1$, set $P_k(T) = \prod_{1 \leq j \leq \ell} (T - \mu_j^k)$. Then $P_k \in \mathbb{Z}[T]$ by Newton's formulas for symmetric polynomials (see, e.g., [BA4], page A.IV.57). It follows from the beginning of the proof that $P_j = P_k$ for some j,k with $1 \leq j < k$, and thus that there exists a permutation σ of $\{1,2,\cdots,\ell\}$ with

$$\mu_1^j = \mu_{\sigma(1)}^k, \cdots, \mu_\ell^j = \mu_{\sigma(\ell)}^k.$$

If σ^i denotes σ iterated i times (thus $\sigma^m(1) = 1$ for some $m \leq \ell$) and if notations are such that $\mu_1 = \lambda$, one obtains after elimination

$$\lambda^{j^m} = \lambda^{jj^{m-1}} = \mu_{\sigma(1)}^{kj^{m-1}} = \mu_{\sigma(\sigma(1))}^{k^2 j^{m-2}} = \cdots = \lambda^{k^m}.$$

In other words, λ is a $(k^m - j^m)^{th}$ root of unity. #

Proposition 1.2.2. *Let* λ *be a nonzero real number which is a root of a monic polynomial* $P \in \mathbb{Z}[T]$. *If all the roots of* P *are real and in* $[-2,2]$, *then* $\lambda = 2\cos(2\pi r)$ *for some rational number* r.

Proof. Let ℓ be the degree of P and set

$$Q(T) = T^{\ell}P(T+\tfrac{1}{T}) \in \mathbb{Z}[T].$$

Denote by $\lambda = 2\cos\theta_1, 2\cos\theta_2, \cdots, 2\cos\theta_\ell$ the roots of P. Then

$$P(T) = \prod_{1\le j\le \ell}(T-2\cos\theta_j)$$

$$Q(T) = \prod_{1\le j\le \ell}(T^2-2T\cos\theta_j+1) = \prod_{1\le j\le \ell}(T-e^{i\theta_j})(T-e^{-i\theta_j}).$$

The previous proposition shows that $e^{i\theta_1}$ is a root of 1, namely, that $\theta_1/2\pi$ is rational.

#

Corollary 1.2.3. *Let* $P \in \mathbb{Z}[T]$ *be a monic polynomial of degree* ℓ, *of which the roots* $\lambda_1, \cdots, \lambda_\ell$ *are real and in* $]-2,2[$; *suppose* $P(T) \neq T^{\ell}$. *Then*

$$\max\{|\lambda_j| : 1 \le j \le \ell\} = 2\cos\frac{\pi}{q}$$

for some natural number $q \ge 3$.

Proof. By the previous proposition, there exists $p_1, \cdots, p_\ell, q_1, \cdots, q_\ell \in \mathbb{Z}$, with $(p_j, q_j) = 1$, such that $\lambda_j = 2\cos(2\pi p_j/q_j)$. But $2\cos(2\pi p_j/q_j)$ and $2\cos(2\pi/q_j)$ are conjugate in the field $\mathbb{Q}(\exp(i2\pi p_j/q_j))$, and as P has coefficients in $\mathbb{Z}$, the number $2\cos(2\pi/q_j)$ is also a root of P. In case $q_j = 2k_j+1$ is odd, $2\cos(2\pi k_j/q_j)$ is also a root of P, and $|2\cos(2\pi k_j/q_j)| = |2\cos(\pi - \pi/q_j)| = |2\cos(\pi/q_j)|$. Thus $\max\{|\lambda_j| : 1 \le j \le \ell\} = 2|\cos(\pi/q)|$ with

$$q = \max\{q', q''\}$$
$$q' = \max\{|q_j| : q_j \text{ is odd}\}$$
$$q'' = \frac{1}{2}\max\{|q_j| : q_j \text{ is even}\}.$$

Finally $q \ge 3$ because $2|\cos\pi| = 2$ and $2|\cos\frac{\pi}{2}| = 0$. #

The last ingredient in the proof of Theorem 1.1.1 is a well known exercise in matrix algebra:

Lemma 1.2.4. *Consider integers* $\ell,m,n \geq 1$ *with* $\ell = m+n$ *and a matrix* $X \in \mathrm{Mat}_{m,n}(\mathbb{R})$. *Set*

$$Y = \begin{bmatrix} 0 & X \\ X^t & 0 \end{bmatrix} \in \mathrm{Mat}_\ell(\mathbb{R}).$$

Then

$$\|X\| = \|X^t\| = \|Y\| = \|X^tX\|^{1/2} = \|XX^t\|^{1/2}.$$

Proof. For any $\xi \in \mathbb{R}^n$ one has

$$\|X\xi\|^2 = (X^tX\xi \mid \xi) \leq \|X^tX\|\|\xi\|^2 \leq \|X^t\|\|X\|\|\xi\|^2. \qquad (*)$$

It follows that $\|X\| \leq \|X^t\|$. As this holds with X and X^t exchanged, $\|X\| = \|X^t\|$. Now (*) implies $\|X\|^2 = \|X^tX\|$, and then $\|X\|^2 = \|X^t\|^2 = \|XX^t\|$, Finally

$$\|Y\|^2 = \|Y^tY\| = \left\| \begin{bmatrix} XX^t & 0 \\ 0 & X^tX \end{bmatrix} \right\| \text{ and}$$

$\|Y\| = \|X\|$. #

Proof of Theorem 1.1.1. Let X be a finite matrix over $\mathbb{Z}$ with $\|X\| < 2$ and let $Y \in \mathrm{Mat}_\ell(\mathbb{Z})$ be as above. As Y is symmetric, its eigenvalues $\lambda_1,\cdots,\lambda_\ell$ are real and $\|X\| = \|Y\| = \max\{|\lambda_1|,\cdots,|\lambda_\ell|\}$. These eigenvalues being the roots of the characteristic polynomial of Y, one has $Y = 0$ or $\|Y\| = 2\cos\frac{\pi}{q}$ for some $q \geq 3$ by corollary 1.2.3. The last claim in the theorem follows from the example below. #

Example 1.2.5. Given an integer $\ell \geq 2$, consider

$$Y = \begin{bmatrix} 0 & 1 & 0 & \cdot & 0 & 0 \\ 1 & 0 & 1 & \cdot & 0 & 0 \\ 0 & 1 & 0 & \cdot & 0 & 0 \\ \cdot & \cdot & \cdot & \cdot & \cdot & \cdot \\ 0 & 0 & 0 & \cdot & 0 & 1 \\ 0 & 0 & 0 & \cdot & 1 & 0 \end{bmatrix} \in \mathrm{Mat}_\ell(\{0,1\}).$$

For $j = 1,\cdots,\ell$, one checks that $Y\xi_j = \lambda_j\xi_j$ with

$$\lambda_j = 2\cos\frac{j\pi}{\ell+1} \in \mathbb{R} \qquad \xi_j = \left[\sin\left[\frac{j\pi}{\ell+1}k\right]\right]_{1\leq k\leq\ell} \in \mathbb{R}^\ell.$$

As Y is symmetric one has

$$\|Y\| = \max\{|\lambda_1|,\cdots,|\lambda_\ell|\} = 2\cos\frac{\pi}{\ell+1}.$$

One may rewrite the rows and columns of Y in the order

$$2,4,\cdots,\ell,1,3,\cdots,\ell-1 \quad \text{if } \ell \text{ is even}$$
$$2,4,\cdots,\ell-1,1,3,\cdots,\ell \quad \text{if } \ell \text{ is odd,}$$

and obtain a matrix of the form $\begin{bmatrix} 0 & X \\ X^t & 0 \end{bmatrix}$ with

$$X = \begin{bmatrix} 1 & 1 & 0 & \cdot & 0 & 0 \\ 0 & 1 & 1 & \cdot & 0 & 0 \\ 0 & 0 & 0 & \cdot & 0 & 0 \\ \cdot & \cdot & \cdot & \cdot & \cdot & \cdot \\ 0 & 0 & 0 & \cdot & 1 & 1 \\ 0 & 0 & 0 & \cdot & 0 & 1 \end{bmatrix} \in \mathrm{Mat}_m(\{0,1\}) \quad \text{if } \ell = 2m;$$

$$X = \begin{bmatrix} 1 & 1 & 0 & \cdot & 0 & 0 & 0 \\ 0 & 1 & 1 & \cdot & 0 & 0 & 0 \\ 0 & 0 & 1 & \cdot & 0 & 0 & 0 \\ \cdot & \cdot & \cdot & \cdot & \cdot & \cdot & \cdot \\ 0 & 0 & 0 & \cdot & 1 & 1 & 0 \\ 0 & 0 & 0 & \cdot & 0 & 1 & 1 \end{bmatrix} \in \mathrm{Mat}_{m,m+1}(\{0,1\}) \quad \text{if } \ell = 2m+1.$$

In both cases, $\|X\| = 2\cos\frac{\pi}{\ell+1}$.

Remarks.

(1) The ideas of Kronecker in [Kro] have motivated much further research; see, for example, [Boy] and [Rob].

(2) The matrix in the example above is "well known". Indeed, it could well be the first matrix in the history of mathematics of which the eigenvalues have ever been computed! This occurred in 1759, when Lagrange was studying a system of ordinary differential equations which "approach" the equation for a vibrating string ([Lag], n°s 19–20). We owe this observation to G. Wanner.

(3) The set of positive algebraic integers which arise as spectral radii of (not necessarily symmetric) integral matrices is considerably larger than the set $\mathcal{N}(\mathbb{Z})$. Indeed, the former is dense in $[1,\infty[$, even if one considers only aperiodic matrices with entries in $\mathbb{N}$ [Lin]; but the latter is not, as shown by Theorem 1.1.1.

(4) It is known that the degree of the algebraic integer $2\cos\frac{\pi}{q}$ is $\frac{1}{2}\varphi(q)$, where φ is Euler's function [Le2]. It is easy to see that the estimate $p \leq q/2$ in Theorem 1.1.1 is not sharp.

1.3. Decomposability and pseudo-equivalence.

We consider first matrices in $\mathrm{Mat}_{m,n}(\mathbb{R})$ for some $m,n \geq 1$. Any permutation α of $\{1,2,\cdots,m\}$ will be identified with the m-by-m matrix of the linear automorphism of $\mathbb{R}^m$ mapping e_i to $e_{\alpha(i)}$; with this convention $\alpha^t = \alpha^{-1}$. Similarly any permutation

$\beta \in \mathfrak{S}_n$ is identified with a matrix in $\mathrm{Mat}_n(\{0,1\})$. Matrices X_1 and X_2 in $\mathrm{Mat}_{m,n}(\mathbb{R})$ are defined to be pseudo- equivalent if there exist $\alpha \in \mathfrak{S}_m$ and $\beta \in \mathfrak{S}_n$ with $X_2 = \alpha X_1 \beta$, namely, if appropriate exchanges of rows and columns convert X_1 to X_2. A matrix is irredundant if none of its rows and columns is zero. An irredundant matrix $X \in \mathrm{Mat}_{m,n}(\mathbb{R})$ is decomposable if there exist integers $m', m'', n', m'' \geq 1$ with $m = m'+m''$ and $n = n'+n''$, as well as $X' \in \mathrm{Mat}_{m',n'}(\mathbb{R})$ and $X'' \in \mathrm{Mat}_{m'',n''}(\mathbb{R})$ such that X is pseudo-equivalent to $\begin{bmatrix} X' & 0 \\ 0 & X'' \end{bmatrix}$. A matrix is indecomposable if it is irredundant and not decomposable. For example, it is not difficult to check that

$$\begin{bmatrix} 1 & 0 & 0 \\ 1 & 1 & 0 \\ 0 & 1 & 1 \end{bmatrix} \quad \text{and} \quad \begin{bmatrix} 0 & 1 & 0 \\ 1 & 1 & 0 \\ 0 & 1 & 1 \end{bmatrix}$$

are indecomposable, but that

$$\begin{bmatrix} 0 & 1 & 0 \\ 1 & 0 & 1 \\ 0 & 1 & 0 \end{bmatrix}$$

is decomposable and pseudo-equivalent to

$$\begin{bmatrix} 1 & 0 & 0 \\ 1 & 0 & 0 \\ 0 & 1 & 1 \end{bmatrix}.$$

These notions should be distinguished from those defined by Frobenius [Fro] for square matrices. Recall that $Z \in \mathrm{Mat}_m(\mathbb{R})$ is reducible if there exists one permutation $\gamma \in \mathfrak{S}_m$ with $\gamma Z \gamma^{-1}$ of the form $\begin{bmatrix} Z' & Z''' \\ 0 & Z'' \end{bmatrix}$, where Z' and Z'' are square blocs of sizes $m', m'' \geq 1$ with $m'+m'' = m$. Matrices such as Z and $\gamma Z \gamma^{-1}$ are called equivalent. For example, the matrix

$$\begin{bmatrix} 0 & 1 & 0 \\ 1 & 0 & 1 \\ 0 & 1 & 0 \end{bmatrix}$$

already considered is irreducible. (Frobenius uses "zerfallend" and "zerlegbar" for "reducible," and no specific word for "equivalent.")

Let $X \in \mathrm{Mat}_{m,n}(\mathbb{R})$. Denote by $X^c \in \mathrm{Mat}_{m,n}(\{0,1\})$ the matrix defined by $X^c_{i,j} = 1$ if $X_{i,j} \neq 0$ and $X^c_{i,j} = 0$ for $X_{i,j} = 0$; the superscript c stands for "combinatorial". We have explained in section 1.1 how X^c can be encoded in a bicolored labelled graph $\tilde{\Gamma}(X^c)$ containing m black vertices $b_1, \cdots, b_m$ and n white vertices $w_1, \cdots, w_n$. Now set $\ell = m+n$ and consider the square matrices

$$Y = \begin{bmatrix} 0 & X \\ X^t & 0 \end{bmatrix} \in \mathrm{Mat}_\ell(\mathbb{R}) \qquad Y^c = \begin{bmatrix} 0 & X^c \\ (X^c)^t & 0 \end{bmatrix} \in \mathrm{Mat}_\ell(\{0,1\}).$$

Denote by $\Gamma(X^c)$ the labelled graph underlying $\tilde{\Gamma}(X^c)$: it has uncolored vertices

$$v_1 = b_1, \cdots, v_m = b_m, \quad v_{m+1} = w_1, \cdots, v_\ell = w_n$$

and Y^c is called its adjacency matrix in graph theory. Here is a list of easily established observations.

(a) The following are equivalent:
 (i) one of the matrices X, X^c, Y, Y^c is irredundant;
 (ii) all these matrices are irredundant;
 (iii) the graph $\Gamma(X^c)$ has no isolated vertex.

(b) The following are equivalent:
 (i) one of (hence both) the matrices X, X^c is indecomposable;
 (ii) the graph $\Gamma(X^c)$ is connected;
 (iii) one of (hence both) the matrices Y, Y^c is irreducible.

(c) The following are equivalent:
 (i) X_1^c and X_2^c are pseudo-equivalent;
 (ii) $\tilde{\Gamma}(X_1^c)$ and $\tilde{\Gamma}(X_2^c)$ are isomorphic as bicolored graphs.

(d) The following are equivalent:
 (i) Y_1^c and Y_2^c are equivalent;
 (ii) $\Gamma(X_1^c)$ and $\Gamma(X_2^c)$ are isomorphic as graphs.

Proposition 1.3.1. *For any irredundant matrix* $X \in \mathrm{Mat}_{m,n}(\mathbb{R})$, *there exist an integer* $a \geq 1$, *partitions*

$$m = \sum_{1 \leq r \leq a} m_r, \quad n = \sum_{1 \leq r \leq a} n_r, \quad m_1, \cdots, m_a, n_1, \cdots, n_a \in \{1,2,\cdots\}$$

and irredundant indecomposable matrices $X_r \in \mathrm{Mat}_{m_r, n_r}(\mathbb{R})$, *for* $r = 1, \cdots, a$, *such that* X *is pseudo-equivalent to the matrix*

$$\begin{array}{c} \begin{bmatrix} X_1 & 0 & \cdot & 0 \\ 0 & X_2 & \cdot & 0 \\ \cdot & \cdot & \cdot & \cdot \\ 0 & 0 & \cdot & X_a \end{bmatrix} \begin{array}{l} \}m_1 \\ \}m_2. \\ \\ \}m_a \end{array} \\ \underbrace{\quad}_{n_1} \underbrace{\quad}_{n_2} \quad \underbrace{\quad}_{n_a} \end{array}$$

In this case one has $\|X\| = \max\{\|X_1\|,\cdots,\|X_a\|\}$.

Uniqueness holds as follows: if X *is also pseudo-equivalent to a matrix with indecomposable blocs* $X'_s \in \mathrm{Mat}_{m'_s,n'_s}(\mathbb{R})$, *for* $s = 1,\cdots,b$, *then* $b = a$ *and, after some renumbering of the triples* (m'_s,n'_s,X'_s), *one has*

$$m'_r = m_r\,, \quad n'_r = n_r, \quad \textit{and}$$

$$X'_r \ \textit{pseudo-equivalent to}\ X_r$$

for $r = 1,\cdots, a$.

Proof. For the existence part, consider the connected components $\tilde{\Gamma}(X^c)_r$ of $\tilde{\Gamma}(X^c)$, where $r = 1,\cdots,a$. Denote by m_r [respectively, n_r] the number of black [respectively, white] points in $\tilde{\Gamma}(X^c)_r$. Permute the rows of X^c and obtain a matrix X' in such a way that the black point b_i belongs to $\tilde{\Gamma}(X')_r$ if

$$m_1+\cdots+m_{r-1} < i \le m_1+\cdots+m_r \qquad (r=1,\cdots,a).$$

Similarly, permute the columns of X' and obtain a matrix $X'' \in \mathrm{Mat}_{m,n}(\{0,1\})$ in such a way that the white point w_j belongs to $\tilde{\Gamma}(X'')_r$ if

$$n_1+\cdots+n_{r-1} < j \le n_1+\cdots+n_r \qquad (r=1,\cdots, a).$$

The matrix X'' has the desired form.

Uniqueness follows from the fact that $\tilde{\Gamma}(X^c)$ has well defined connected components.
#

Remarks.

(1) If X corresponds to the bicolored labelled graph $\tilde{\Gamma}(X)$, then X^t corresponds to a graph with colors exchanged. The example at the end of section 1.2 is a Jordan block, say

$$X = \begin{bmatrix} 1 & 1 & 0 & 0 \\ 0 & 1 & 1 & 0 \\ 0 & 0 & 1 & 1 \\ 0 & 0 & 0 & 1 \end{bmatrix}$$

$\tilde{\Gamma}(X)$

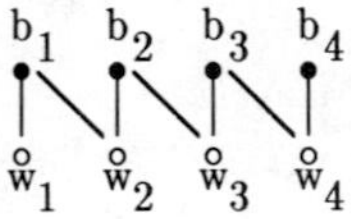

for $m = 4$. As $\tilde{\Gamma}(X)$ is clearly isomorphic (as unlabeled bicolored graph) to the graph with colors exchanged, a Jordan block is pseudo-equivalent to its transpose. But

$$\begin{bmatrix} 0 & 1 & 0 \\ 1 & 1 & 1 \\ 0 & 0 & 1 \end{bmatrix} \quad \text{and} \quad \begin{bmatrix} 0 & 1 & 0 \\ 1 & 1 & 0 \\ 0 & 1 & 1 \end{bmatrix}$$

are not pseudo-equivalent to each other, because the graph corresponding to the former has a white vertex of degree 3.

(2) Consider $X, X' \in \mathrm{Mat}_{m,n}(\mathbb{R})$ and the corresponding $Y, Y' \in \mathrm{Mat}_{\ell}(\mathbb{R})$, with $\ell = m+n$. If X' is pseudo-equivalent to X, say $X' = \alpha X \beta$. then Y' is equivalent to Y, because $Y' = \gamma Y \gamma^{-1}$ with $\gamma = \begin{bmatrix} \alpha & 0 \\ 0 & \beta^{-1} \end{bmatrix}$. But one may have Y' equivalent to Y without X' being pseudo-equivalent to X. The reader can check this with X and X' the two 3-by-3 matrices of the previous remark, or with $X \in \mathrm{Mat}_4(\{0,1\})$ and $X' \in \mathrm{Mat}_{6,2}(\{0,1\})$ corresponding to

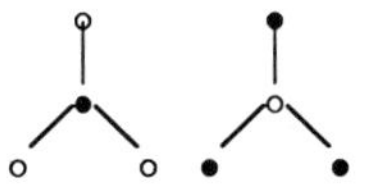

and

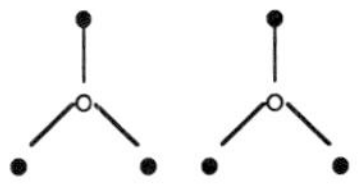

(3) There is of course a proposition analogous to 1.3.1 in terms of irreducible square matrices and equivalence à la Frobenius. See [Gan], § XIII.4.

For future reference, we also note the following fact. A square matrix $X \in \mathrm{Mat}_m(\mathbb{R}_+)$ with nonnegative entries is irreducible if and only if, for each $i,j \in \{1,\cdots,m\}$, there exists an integer p (depending on i and j) such that the $(i,j)^{\underline{th}}$ entry of X^p is strictly positive (see [Gan], § XIII.1). It is aperiodic ("primitive" is also used) if there exists an integer p such that all entries of X^p are strictly positive.

Lemma 1.3.2.

(a) *For* $X \in \mathrm{Mat}_{m,n}(\mathbb{R})$, *the following are equivalent:*

(i) X *is irredundant;*

(ii) $X^tX \in \mathrm{Mat}_m(\mathbb{R})$ *and* $XX^t \in \mathrm{Mat}_n(\mathbb{R})$ *are irredundant.*

(b) *For* $X \in \mathrm{Mat}_{m,n}(\mathbb{R}_+)$, *the following are equivalent:*

(i) X *is indecomposable;*

(ii) XX^t *is irreducible;*

(iii) XX^t *is aperiodic;*

(iv) X^tX *is irreducible;*

(v) X^tX *is aperiodic.*

Proof. Claim (a) is immediate from multiplication rules.

For (b), if $b_1,\cdots,b_m,w_1,\cdots,w_n$ are the vertices of $\tilde{\Gamma}(X)$, then $\tilde{\Gamma}(XX^t)$ has m black vertices $b_1,\cdots,b_m$ and m white vertices $b_1',\cdots,b_m'$. There exists a line between b_i and b_k' if and only if there exist a line in $\tilde{\Gamma}(X)$ between b_i and w_j as well as between w_j and b_k, for some $j \in \{1,\cdots,n\}$. This shows equivalence of (i) and (ii); the same works for (i) and (iv).

Now a matrix such as X^tX, which is positive both in the sense of Perron–Frobenius theory and in the sense of operator theory, is irreducible if and only if it is aperiodic. This follows from Theorem 8 in § XIII.5 of [Gan]. #

About (a), observe that $X = \begin{bmatrix} 1 & 0 \\ 1 & 0 \end{bmatrix}$ is redundant, with XX^t irredundant and X^tX redundant.

Let Γ be a finite graph with ℓ vertices. The adjacency matrix of Γ is (the equivalence class of) a symmetric matrix $Y = Y_\Gamma \in \mathrm{Mat}_\ell(\{0,1\})$, with $Y_{i,j} = 1$ if there is an edge between vertices i and j and with $Y_{i,j} = 0$ if not. The characteristic polynomial of Γ is that of Y_Γ, and the spectral radius or norm of Γ is

$$\|\Gamma\| = \max\{|\lambda| : \lambda \text{ is an eigenvalue of } Y_\Gamma\}.$$

From Proposition 1.3.1, it is clear that the following program will solve the problem of classifying matrices $X \in \mathrm{Mat}_{fin}(\{0,1\})$ satisfying $\|X\| \le 2$:

(1) classify finite connected graphs satisfying $\|\Gamma\| \le 2$;
(2) classify bicolorations on these.

Consider now an indecomposable matrix $X \in \mathrm{Mat}_{fin}(\{0,1,2\})$ and assume that 2 is an entry of X. Then one has either $X = 2$ (as 1-by-1 matrix) or $\|X\| \ge 5^{1/2}$: this is clear from the definition of $\|X\|$, and follows also from Lemma 2 in § XIII.2 of [Gan]. Consequently the program above will also solve our initial problem of classifying $X \in \mathrm{Mat}_{fin}(\{0,1,2\})$ with $\|X\| \le 2$.

1.4. Graphs with norms no larger than 2.

The elementary classification of the present section is based on two fundamental facts of Perron–Frobenius theory, for the proof of which we refer to [Gan].

The first one is about an irreducible square matrix $Y \in \mathrm{Mat}_k(\mathbb{R}_+)$ with nonnegative entries. A Perron–Frobenius vector for Y is an eigenvector $\xi \in \mathbb{R}_+^k$ of Y with nonnegative entries. The fact is that such a vector

(i) always exist;
(ii) is unique up to multiplication by a positive scalar;

(iii) corresponds to an eigenvalue which is simple and which is the spectral radius of Y.

In particular, if Γ is a finite connected graph, and $Y \in \mathrm{Mat}_k(\{0,1\})$ is its adjacency matrix, then Y is irreducible, although in general not aperiodic; a Perron– Frobenius vector ξ for Γ (by which we mean for Y) satisfies

$$Y\xi = \|\Gamma\|\xi.$$

As the vector space where Y acts has its natural basis indexed by the vertices of Γ, it is convenient to represent such a ξ on the graph Γ, with the value of each coordinate of ξ written near the corresponding vertex of Γ. For example, if $Y = \begin{bmatrix} 0 & 1 \\ 1 & 0 \end{bmatrix}$, then $\begin{bmatrix} 1 \\ 1 \end{bmatrix} \in \mathbb{R}^2$ is represented by $\overset{1}{\circ}\!\!-\!\!-\!\!\overset{1}{\circ}$.

We consider the following graphs, each given with a Perron–Frobenius vector. (Note that the subscript ℓ means that the graph has $\ell+1$ vertices (sic).)

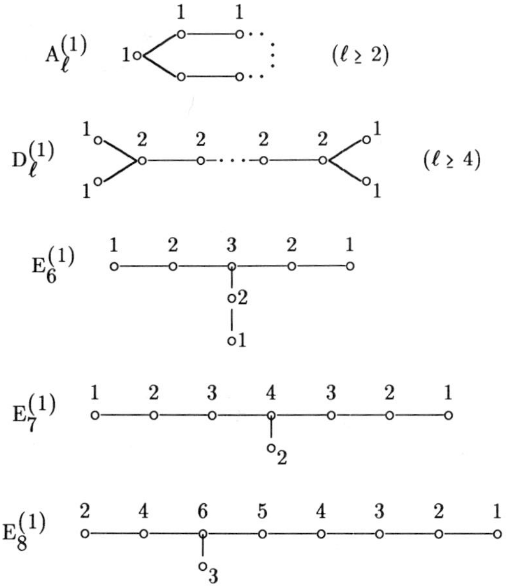

Lemma 1.4.1. *Each graph in the list above has norm* 2.

Proof. Check that the indicated vector is a Perron–Frobenius vector corresponding to the eigenvalue 2. #

The second fact from Perron–Frobenius theory concerns two matrices $Y', Y \in \mathrm{Mat}_k(\mathbb{R}_+)$ with Y irreducible, $Y' \leq Y$ (this means $(Y-Y')_{i,j} \geq 0$ for i,j = 1,···,k) and $Y' \neq Y$. Then the fact is that the spectral radius of Y' is strictly smaller than that of Y. In particular, this implies the following lemma.

<u>Lemma 1.4.2</u>. *Let* Γ *be a finite connected graph and let* Γ' *be a proper subgraph of* Γ; *then* $\|\Gamma'\| < \|\Gamma\|$.

(By a subgraph of Γ, we mean one obtained by erasing some edges and some vertices together with all edges emanating from these vertices; for example, o—o o is a subgraph of o—o—o—o.)

The graphs listed in the next theorem are the ones above together with the following (Now ℓ is the number of vertices!):

A_ℓ  $(\ell \geq 1)$

D_ℓ 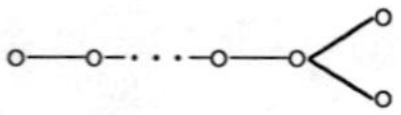 $(\ell \geq 4)$

E_ℓ 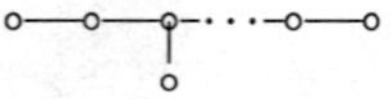 $(\ell = 6,7,8)$

Given integers p,q,r with $2 \leq p \leq q \leq r$, we consider also

$T_{p,q,r}$ 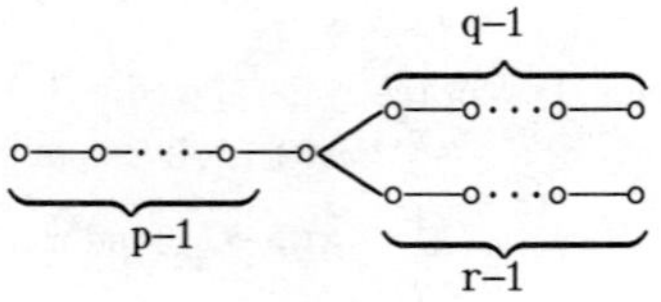

(p+q+r–2 vertices).

One has in particular $D_\ell = T_{2,2,\ell-2}$ and $E_\ell = T_{2,3,\ell-3}$.

<u>Theorem 1.4.3.</u> *Let* Γ *be a nonempty finite connected graph.*

(a) $\|\Gamma\| < 2$ *if and only if* Γ *is one of*

$$A_\ell \ (\ell \geq 1) \quad D_\ell \ (\ell \geq 4) \quad E_\ell \ (\ell = 6,7,8)$$

In this case, $\|\Gamma\| = 2\cos(\pi/h_\Gamma)$ *where* h_Γ *is the so-called Coxeter number of* Γ *given by*

$$\ell+1 \text{ for } A_\ell, \quad 2\ell-2 \text{ for } D_\ell, \quad 12,18,30 \text{ for } E_6,E_7,E_8.$$

(b) $\|\Gamma\| = 2$ *if and only if* Γ *is one of*

$$A_\ell^{(1)} \ (\ell \geq 2), \quad D_\ell^{(1)} \ (\ell \geq 4), \quad E_\ell^{(1)} \ (\ell = 6,7,8).$$

Proof. Assume first that $\|\Gamma\| < 2$. By Lemma 1.4.2, the graph Γ contains no cycle (because $\|A_\ell^{(1)}\| = 2$ for $\ell \geq 2$), no point of degree 4 or more (because $\|D_4^{(1)}\| = 2$), and at most one point of degree 3 (because $\|D_\ell^{(1)}\| = 2$ for $\ell \geq 5$). Consequently Γ is either a segment, namely A_ℓ for some $\ell \geq 1$, or one of $T_{2,2,\ell-2} = D_\ell$ for some $\ell \geq 4$, or some graph $T_{p,q,r}$ with $q \geq 3$. In this last case one cannot have $p \geq 3$ (because $E_6^{(1)} = T_{3,3,3}$ is of norm 2), nor $r \geq 6$ (because $E_8^{(1)} = T_{2,3,6}$ is of norm 2). Thus one is left with the list of (a).

Assume now $\|\Gamma\| \leq 2$. The same argument shows that Γ does not properly contain a cycle, a vertex of degree 4 together with its nearest neighbors, or one of the $D_\ell^{(1)}$ for $\ell \geq 5$. If, moreover, $\|\Gamma\| = 2$, it follows readily that Γ is in the list of (b).

If $\Gamma = A_\ell$ then $\|\Gamma\| = 2\cos(\pi/(\ell+1))$ by the example of section 1.2. For the other values of h , see the literature or Proposition I.2.5 of Appendix I. For various meanings of the number h_Γ, see [Ste1]. #

Define the bicoloration number of a graph Γ to be the number of possible partitions of its vertices in two nonempty sets, with no pair of adjacent vertices in one of the parts. It is both elementary and well known that this number is 0 if and only if Γ is a point or contains an odd cycle. For the graphs of the previous theorem, one has

Table 1.4.4. The bicoloration numbers of the graphs listed in 1.4.3 are as follows:

A_1:	0	$A_\ell^{(1)}$ with $\ell \geq 2$ even:	0
A_ℓ with ℓ even:	1	$A_\ell^{(1)}$ with $\ell \geq 3$ odd:	1
A_ℓ with $\ell \geq 3$ odd:	2	$D_\ell^{(1)}$ with $\ell \geq 4$ even:	2
D_ℓ with $\ell \geq 4$:	2	$D_\ell^{(1)}$ with $\ell \geq 5$ odd:	1
E_ℓ with $\ell = 6,7,8$:	2	$E_\ell^{(1)}$ with $\ell = 6,7,8$:	2

This completes the program set up at the end of Section 1.3, and consequently the proof of Theorem 1.1.2. The matrices listed in the following tables (Tables 1.4.5 and 1.4.6) correspond to the graphs of Theorem 1.4.3; the one additional matrix $[2] \in \mathrm{Mat}_1(\mathbb{R})$ is conveniently made to correspond with the marked graph $A_1^{(1)}$, pictured as $\circ\overset{\infty}{—}\circ$. The tables list Coxeter exponents of the graphs Γ; these are integers m_j such that the adjacency matrix of the graph Γ has spectrum $\{2\cos\left[m_j \frac{\pi}{h_\Gamma}\right]\}$, with $j = 1,\cdots,m+n$.

More about Coxeter exponents can be found in section 2 of [Cox] and in exercise V.6.4 of [BLie].

Although the theory above is independent of that of semisimple Lie algebras, root systems are very close at hand. In fact, consider $X \in Mat_{m,n}(\{0,1\})$ with $\|X\| \leq 2$, and set $Y = \begin{bmatrix} 0 & X \\ X^t & 0 \end{bmatrix} \in Mat_\ell(\{0,1\})$ as above. Then $2-Y$ is a real symmetric matrix with positive eigenvalues; as such it can be written uniquely as $2-Y = Z^2$, with $Z \in Mat_\ell(\mathbb{R})$ a symmetric matrix with non-negative eigenvalues. Let $\alpha_1,\cdots,\alpha_\ell$ denote the columns (or rows) of Z. Then $(Z^2)_{i,j}$ is the scalar product $(\alpha_i|\alpha_j)$; since it is 2 if $i = j$ and $-Y_{i,j} \in \{0,-1\}$ if $i \neq j$, the vectors $\alpha_1,\cdots,\alpha_\ell$ are of equal length $\sqrt{2}$ and their mutual angles are $\pi/2$ or $2\pi/3$.

Assume moreover $\|X\| < 2$, so that Y and Z are invertible, and thus $\alpha_1,\cdots,\alpha_\ell$ linearly independent. Then $\{\alpha_1,\cdots,\alpha_\ell\}$ is what is called a reduced system of roots. It is clearly irreducible if and only if X is indecomposable; in this case it is of one of the types A,D,E because all roots have the same length. See [CGSS] for more along these lines.

For the remainder of this section, we assume that the reader has some familiarity with Coxeter graphs. Recall from section 1.1 that

$$\mathbb{K} = \{0,1,2\} \cup \{2 \cos \tfrac{\pi}{q}\}_{q=4,5,6}$$

and that a matrix $X \in Mat_{m,n}(\mathbb{K})$ is encoded as a Coxeter graph, say Γ, plus a bicoloration. In fact, $2 - \begin{bmatrix} 0 & X \\ X^t & 0 \end{bmatrix}$ is twice the matrix of the canonical bilinear form associated to Γ (the form denoted by B_M in [BLie], chap. V, §4, no 1). Any Coxeter graph Γ corresponds to such a matrix X, and $\|\Gamma\| = \|X\| < 2$ [respectively $\|X\| = 2$] if and only if $2 - \begin{bmatrix} 0 & X \\ X^t & 0 \end{bmatrix}$ is positive definite [resp., positive semidefinite]. *Consequently the classification of Theorem 1.1.3 is nothing but another phrasing of the classification of irreducible Coxeter systems which are of the finite type [resp., the affine type].* We refer to [BLie] for further details. It is again true that an indecomposable matrix $X \in Mat_{m,n}(\mathbb{K})$ corresponding to a Coxeter graph Γ has norm $2\cos(\pi/h_\Gamma)$, where h_Γ is either as in Theorem 1.4.3 or as in the following list:

2ℓ for B_ℓ 12 for F_4 6 for G_2

10 for H_3 30 for H_4 p for $I_2(p)$.

Table 1.4.5.

Type of Y		size	matrix Y	bicolored graph	Coxeter number	exponents	${}^tY$ figures in table
A_ℓ, $\ell = 2m$	$m \geqslant 1$	m-by-m	$\begin{pmatrix} 1 & 1 & & 0 \\ & \ddots & \ddots & \\ & & \ddots & 1 \\ 0 & & & 1 \end{pmatrix}$		$\ell + 1$	$1, 2, \ldots, \ell$	no
A_ℓ, $\ell = 2m + 1$	$m \geqslant 1$	m-by-(m + 1)	$\begin{pmatrix} 1 & 1 & & & 0 \\ & \ddots & \ddots & & \\ 0 & & & 1 & 1 \end{pmatrix}$		$\ell + 1$	$1, 2, \ldots, \ell$	yes
D_ℓ, $\ell = 2m$	$m \geqslant 2$	(m − 1)-by-(m + 1)	$\begin{pmatrix} 1 & 1 & & & 0 \\ & \ddots & \ddots & & \\ 0 & & 1 & 1 & 1 \end{pmatrix}$		$2\ell - 2$	$1, 3, 5, \ldots, 2\ell - 3$, and $\ell - 1$	yes
D_ℓ, $\ell = 2m + 1$	$m \geqslant 2$	m-by-(m + 1)	$\begin{pmatrix} 1 & & & & 0 \\ 1 & \ddots & \ddots & & \\ 0 & & 1 & 1 & 1 \end{pmatrix}$	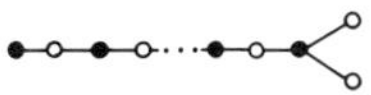	$2\ell - 2$	$1, 3, 5, \ldots, 2\ell - 3$, and $\ell - 1$	yes
E_6		3-by-3	$\begin{pmatrix} 1 & 0 & 0 \\ 1 & 1 & 1 \\ 0 & 0 & 1 \end{pmatrix}$		12	1, 4, 5, 7, 8, 11	yes
E_7		3-by-4	$\begin{pmatrix} 1 & 0 & 0 & 0 \\ 1 & 1 & 1 & 0 \\ 0 & 0 & 1 & 1 \end{pmatrix}$		18	1, 5, 7, 9, 11, 13, 17	yes
E_8		4-by-4	$\begin{pmatrix} 1 & 0 & 0 & 0 \\ 1 & 1 & 1 & 0 \\ 0 & 0 & 1 & 1 \\ 0 & 0 & 0 & 1 \end{pmatrix}$	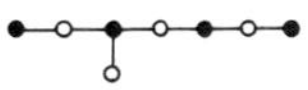	30	1, 7, 11, 13, 17, 19, 23, 29	yes

N.B.: type A_1 does not appear; it would correspond to the empty matrix, identified to the (linear!) map $\mathbb{R}^\circ \to \mathbb{R}^\circ$.

Table 1.4.6.

Type of Y		size	matrix Y	bicolored graph	${}^tY$ figures in table
$A_1^{(1)}$		1-by-1	2		no
$A_\ell^{(1)}$	$\ell = 2m+1 \geqslant 3$	(m + 1)-by-(m + 1)	$\begin{pmatrix} 1 & 1 & & \\ & \ddots & \ddots & \\ & & & 1 \\ 1 & & & 1 \end{pmatrix}$	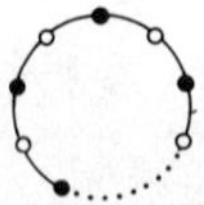	no
$D_\ell^{(1)}$	$\ell = 2m \geqslant 4$	(m − 1)-by-(m + 2)	$\begin{pmatrix} 1 & 1 & 1 & & & \\ & & 1 & \ddots & & \\ & & & \ddots & 1 & \\ & & & & 1 & 1 & 1 \end{pmatrix}$	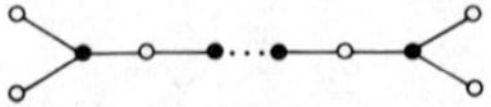	yes
$D_\ell^{(1)}$	$\ell = 2m+1 \geqslant 5$	(m + 1)-by-(m + 1)	$\begin{pmatrix} 1 & & & \\ 1 & \ddots & & \\ 1 & \ddots & & \\ & & 1 & 1 & 1 \end{pmatrix}$	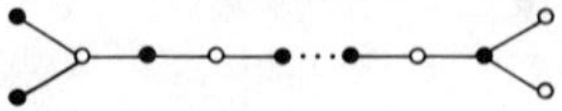	no
$E_6^{(1)}$		4-by-3	$\begin{pmatrix} 1 & 0 & 0 \\ 1 & 1 & 1 \\ 0 & 0 & 1 \\ 0 & 1 & 0 \end{pmatrix}$	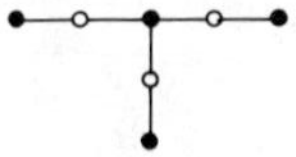	yes
$E_7^{(1)}$		3-by-5	$\begin{pmatrix} 1 & 1 & 0 & 0 & 0 \\ 0 & 1 & 1 & 1 & 0 \\ 0 & 0 & 0 & 1 & 1 \end{pmatrix}$		yes
$E_8^{(1)}$		4-by-5	$\begin{pmatrix} 1 & 0 & 0 & 0 & 0 \\ 1 & 1 & 1 & 0 & 0 \\ 0 & 0 & 1 & 1 & 0 \\ 0 & 0 & 0 & 1 & 1 \end{pmatrix}$		yes

N.B.: $\ell = m + n - 1$ in this Table, but $\ell = m + n$ in Table 1.4.5.

Table 1.4.7. List of Coxeter graphs which satisfy $\|\Gamma\| \leq 2$ and which do not appear in Tables 1.4.5 and 1.4.6.

B_ℓ	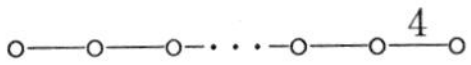	$\ell \geq 2$ ℓ vertices	$h = 2\ell$
F_4	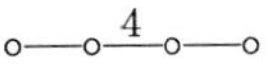	4 vertices	$h = 12$
G_2	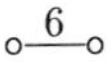	2 vertices	$h = 6$
H_3	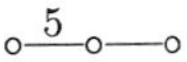	3 vertices	$h = 10$
H_4		4 vertices	$h = 30$
$I_2(p)$	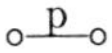 $p = 5$ or $p \geq 7$	2 vertices	$h = p$

For these $\|\Gamma\| = 2\cos(\pi/h)$

$B_\ell^{(1)}$	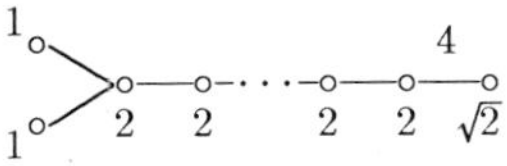	$\ell \geq 3$	$\ell+1$ vertices
$C_\ell^{(1)}$	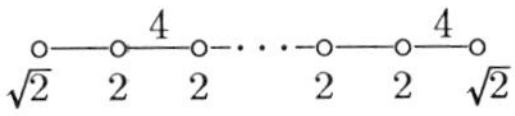	$\ell \geq 2$	$\ell+1$ vertices
$F_4^{(1)} = F_5$	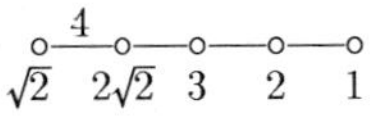		5 vertices
$G_2^{(1)}$	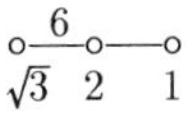		3 vertices

For these, $\|\Gamma\| = 2$
(a Perron–Frobenius vector is indicated)

In Lemma 1.4.1 and in Table 1.4.7, we have indicated a Perron–Frobenius eigenvector for the Coxeter graphs of affine type. For Section 4.5 we will also need to compute the Perron–Frobenius vectors for the connected Coxeter graphs of finite type in the classes A, D, and E. For completeness, we repeat the case of the graphs A_ℓ, already dealt with in Section 1.2, and also give the results for the classes B, F, G, H, and I. Recall that in the standard notation for Coxeter graphs of finite type, the subscript gives the number of vertices. The details of the verifications are straightforward and are left to the reader.

Table 1.4.8. Perron–Frobenius eigenvectors for Coxeter graphs of finite type

Type A_ℓ $(\ell \geq 2)$. Eigenvalue: $2\cos[\pi/(\ell+1)]$

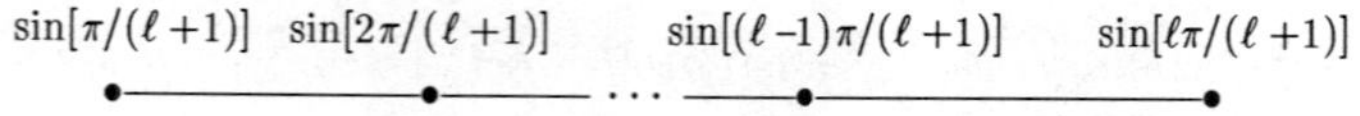

Type B_ℓ $(\ell \geq 2)$. Eigenvalue: $2\cos[\pi/2\ell]$

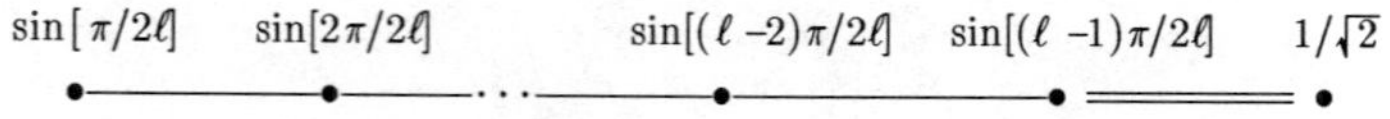

Type D_ℓ $(\ell \geq 4)$. Eigenvalue: $2\cos[\pi/(2\ell-2)]$

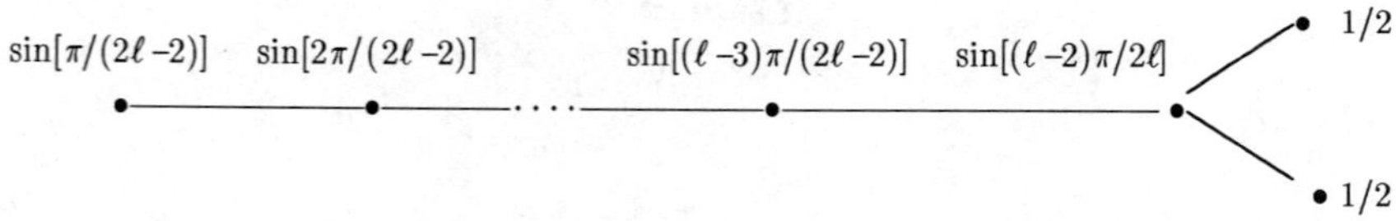

Type E_6. Eigenvalue: $2\cos[\pi/12] = (\sqrt{3}+1)/\sqrt{2}$

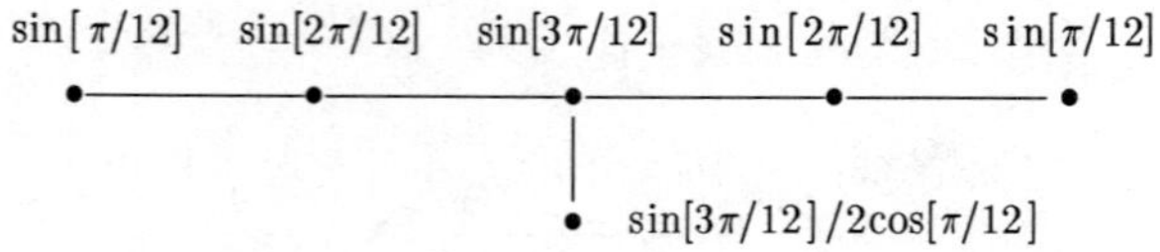

Note: $\sin[\pi/12] = (\sqrt{3}-1)/2\sqrt{2}$ $\sin[2\pi/12] = 1/2$ $\sin[3\pi/12] = 1/\sqrt{2}$
$\sin[3\pi/12]/2\cos[\pi/12] = (\sqrt{3}-1)/2$

Type E_7. Eigenvalue: $2\cos[\pi/18]$

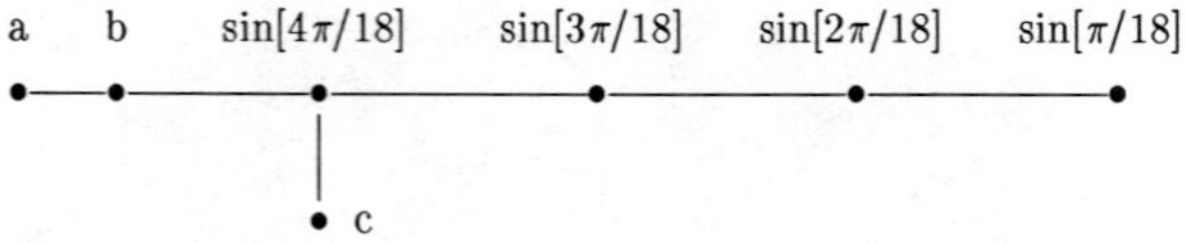

$a = \sin[6\pi/18]/4\cos^2[\pi/18]$, $b = \sin[6\pi/18]/2\cos[\pi/18]$, $c = \sin[4\pi/18]/2\cos[\pi/18]$

Type E_8. Eigenvalue: $2\cos[\pi/30] = \frac{1}{2}\left[\frac{\sqrt{5}+1}{2}\sqrt{3} + \sqrt{\frac{5-\sqrt{5}}{2}}\right]$

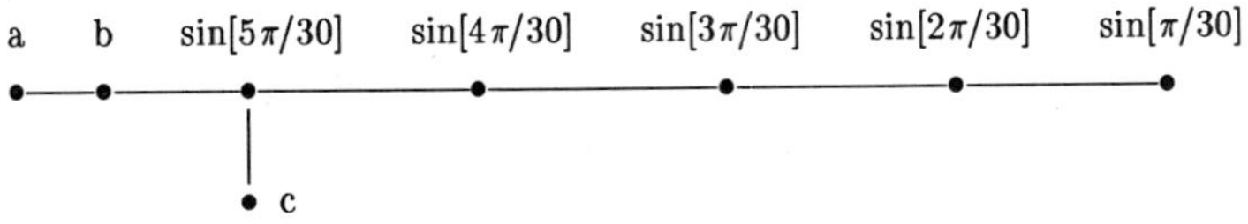

$a = \sin[7\pi/30]/4\cos^2[\pi/30]$, $b = \sin[7\pi/30]/2\cos[\pi/30]$, $c = \sin[5\pi/30]/2\cos[\pi/30]$

Type F_4. Eigenvalue: $2\cos[\pi/12]$

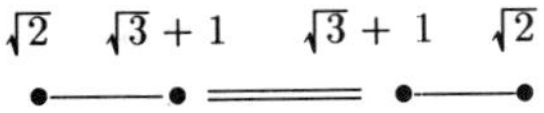

Type G_2. Eigenvalue: $2\cos[\pi/6] = \sqrt{3}$

Type H_3. Eigenvalue: $2\cos[\pi/10]$

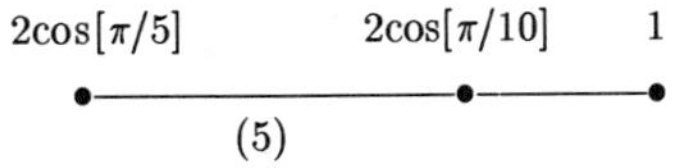

Type H_4. Eigenvalue: $2\cos[\pi/30]$

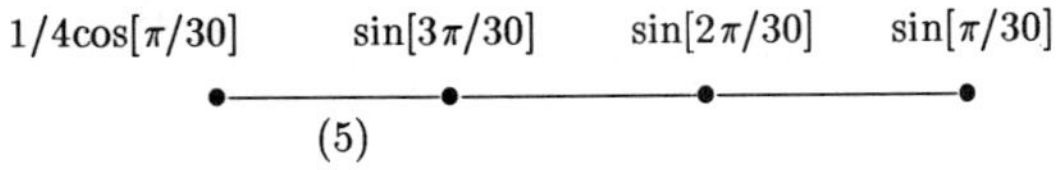

Type $I_2(p)$. Eigenvalue: $2\cos[\pi/p]$

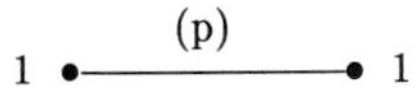

Let Γ be a connected Coxeter graph of finite type A, D, or E with $\ell \geq 2$ vertices. Choose a bicoloration of Γ with m black and n white vertices $(m+n = \ell)$, and let $X \in \mathrm{Mat}_{m,n}(\{0,1\})$ be the corresponding matrix, so that $Y = \begin{bmatrix} 0 & X \\ X^t & 0 \end{bmatrix}$ is the adjacency matrix of Γ. Let χ be the row vector defined by the n white co-ordinates of the Perron–Frobenius vector for Y (as listed in Table 1.4.8); then χ is a Perron–Frobenius row vector for X^tX. For use in Section 4.5, we need to know the square $\|\chi\|^2$ of the Euclidean norm of χ. (In case Γ has two bicolorations, there are two distinct choices for χ, but it follows from the eigenvalue equation for Y that they have the same norm.)

Proposition 1.4.9. *With the notation as above, the values of the square* $\|\chi\|^2$ *of the norm of the Perron–Frobenius eigenvector for* X^tX *are as follows:*

Type A_ℓ:	$(\ell + 1)/4$	*Type* D_ℓ:	$(\ell - 1)/4$
Type E_6:	$(3 - \sqrt{3})/2$	*Type* E_7:	ca. 0,57999
Type E_8:	ca. 0,38502		

Remarks.

(1) Let Γ be a graph with ℓ vertices and let $\lambda_1 \leq \cdots \leq \lambda_\ell$ be the ordered sequence of its eigenvalues. The spectral spread $s(\Gamma)$ of Γ is $\lambda_\ell - \lambda_1$. Of course $\|\Gamma\| \leq 2$ implies $s(\Gamma) \leq 4$; the converse happens to hold with finitely many exceptions which have been classified by Petrović [Pet].

(2) It has been pointed out to us by D. Cetković and C. Godsil that it may also be possible to classify indecomposable matrices $X \in \mathrm{Mat}_{fin}(\{-1,0,1\})$ with $\|X\| \leq 2$. One can write $2 - \begin{bmatrix} 0 & X \\ X^t & 0 \end{bmatrix} = ((\alpha_i | \alpha_j))_{1 \leq i,j \leq \ell}$ as before, so that the problem is equivalent to the classification of irreducible sets of vectors $\{\alpha_1, \cdots, \alpha_\ell\}$ in $\mathbb{R}^\ell$, all of the same length and with mutual angles in $\{\pi/3, \pi/2, 2\pi/3\}$. The possible sets of *lines* spanned by such sets of vectors are classified in [CGSS]. Once this is worked out, the next cases would be $X \in \mathrm{Mat}_{fin}(\mathbb{K} \cup (-\mathbb{K}))$ with $\|X\| \leq 2$ as well as $X \in \mathrm{Mat}_{fin}(\mathbb{Z})$ with $\|X\| > 2$ but $\|X\|$ close to 2.

(3) The subject of spectra of graphs has been extensively investigated. We refer to the excellent early monograph by Biggs [Big], to more recent books by Cvetković, Doob, Gutman, Sachs, and Torgašev [CDS] and [CDGT], as well as to the reviews [CD], [GHM] and [Sch1].

1.5. The set $\mathbb{E}$ of norms of graphs and integral matrices.

The assignment to a subset S of $\mathbb{R}$ of the set $\mathcal{N}(S)$ of norms of matrices on S has interesting properties. Observe that, obviously,

$$S \cap \mathbb{R}_+ \subset \mathcal{N}(S)$$

and also that

$$s,t \in \mathcal{N}(S) \Rightarrow st \in \mathcal{N}(S)$$

because $st = \|X \otimes Y\|$ if $s = \|X\|$ and $t = \|Y\|$. (The inequality $\|X \otimes Y\| \leq \|X\|\|Y\|$ follows from $X \otimes Y = (X \otimes 1)(1 \otimes Y)$; the converse inequality follows from the existence of vectors ξ,η of norm 1 with $\|X\xi\| = \|X\|$ and $\|Y\eta\| = \|Y\|$, hence with $\|(X \otimes Y)(\xi \otimes \eta)\| = \|X\|\|Y\|$.)

Given $S_0 \subset S \subset \mathbb{R}_+$ such that every number in S is a sum of numbers in S_0, one has also

$$\mathcal{N}(S_0) = \mathcal{N}(S).$$

Indeed, the following nice argument of Hoffman shows that $\mathcal{N}(S) \subset \mathcal{N}(S_0)$. Let $X \in \mathrm{Mat}_{\mathrm{fin}}(S)$ and let $Y = \begin{bmatrix} 0 & X \\ X^t & 0 \end{bmatrix} \in \mathrm{Mat}_\ell(S)$. We have to find some $X' \in \mathrm{Mat}_{\mathrm{fin}}(S_0)$ with $\|X'\| = \|Y\|$; we may assume that Y is irreducible. Now there exists a decomposition $Y = Y_1 + \cdots + Y_k$ with Y_j symmetric matrices in $\mathrm{Mat}_\ell(S_0)$ for $j = 1,, \cdots ,k$. Set

$$X' = \begin{bmatrix} Y_1 & Y_2 & \cdots & Y_k \\ Y_2 & Y_3 & \cdots & Y_1 \\ \cdots & \cdots & \cdots & \cdots \\ Y_k & Y_1 & \cdots & Y_{k-1} \end{bmatrix} \in \mathrm{Mat}_{k\ell}(S_0).$$

Let $\xi \in \mathbb{R}_+^\ell$ be a Perron–Frobenius vector for Y and define $\xi' = (\xi,\xi,\cdots,\xi) \in \mathbb{R}_+^{k\ell}$. Then $X'\xi' = \|Y\|\xi'$, so that ξ' is a Perron–Frobenius vector for X'. Consequently $\|X'\| = \|X\|$. In particular one has

$$\mathcal{N}(\{0,1\}) = \mathcal{N}(\mathbb{N})$$

which is Proposition 2.1 of [Hof].

Now we state again the main result of this section, due to G. Skandalis.

<u>Theorem 1.1.4</u>. *For any* $S \subset \mathbb{R}$ *one has* $\mathcal{N}(S) = \mathcal{N}(\mathcal{N}(S))$.

Proof. It is obvious from considering 1–by–1 matrices that $\mathcal{M}(S) \subset \mathcal{M}(\mathcal{M}(S))$. To show the theorem, set $T = \mathcal{M}(S)$ and let $X \in \mathrm{Mat}_{m,n}(T)$; one has to find some $Z \in \mathrm{Mat}_{fin}(S)$ with $\|Z\| = \|X\|$.

For any pair (i,j) with $1 \le i \le m$ and $1 \le j \le n$, we choose an integer $p_{i,j} \ge 1$ and a symmetric matrix $Y_{i,j}$ of size $p_{i,j}$ over S with $\|Y_{i,j}\| = X_{i,j}$. Let p be the product of the $p_{i,j}$'s. Write $Z_{i,j}$ for $1 \otimes \cdots \otimes Y_{i,j} \otimes \cdots \otimes 1 \in \mathrm{Mat}_p(S)$, with the factor $Y_{i,j}$ at the $(i,j)^{\underline{th}}$ place. Consider the matrix $Z \in \mathrm{Mat}_{pm,pn}(S)$ with the $Z_{i,j}$'s as blocks..

Choose for each pair (i,j) some vector $\xi_{i,j} \neq 0$ with $p_{i,j}$ coordinates such that $Y_{i,j}\xi_{i,j} = X_{i,j}\xi_{i,j}$. Set $\xi = \otimes\xi_{i,j} \in \mathbb{R}^p$, so that $Z_{i,j}\xi = X_{i,j}\xi$. Choose also $\eta \in \mathbb{R}^n$ with $\eta \neq 0$ and $\|X\eta\| = \|X\|\|\eta\|$. With $\zeta = \xi \otimes \eta$ one obtains $\|Z\zeta\| = \|\xi\|\|X\|\|\eta\| = \|X\|\|\zeta\|$. it follows that $\|Z\| \ge \|X\|$.

As $\|Z\| \le \|X\|$ by the next lemma, this proves the theorem. #

The integer $p \ge 1$ being given, let B be the algebra $\mathrm{End}(\mathbb{R}^p)$, considered together with the Euclidean operator norm (a "real C^*–algebra"). We identify $\mathrm{Mat}_{m,n}(B)$ with the space of linear maps from $\mathbb{R}^{pn}$ to $\mathbb{R}^{pm}$.

Lemma 1.5.1. *Let* $Z \in \mathrm{Mat}_{m,n}(B)$, *and let* $X \in \mathrm{Mat}_{m,n}(\mathbb{R}_+)$ *be defined by* $X_{i,j} = \|Z_{i,j}\|$ *for* $i = 1,\cdots,m$ *and* $j = 1,\cdots,n$. *Then* $\|Z\| \le \|X\|$.

Proof. For any $Y \in \mathrm{Mat}_n(B)$, set

$$|||Y||| = \sup\{\|Y_{i,j}\| : 1 \le i \le m \text{ and } 1 \le j \le n\}.$$

As

$$Y = \begin{bmatrix} Y_{1,1} & 0 & & 0 \\ 0 & Y_{2,2} & & 0 \\ \cdots & \cdots & \cdots & \cdots \\ 0 & 0 & \cdots & Y_{n,n} \end{bmatrix} + \begin{bmatrix} 0 & Y_{1,2} & & 0 \\ \cdots & \cdots & \cdots & \cdots \\ 0 & 0 & \cdots & Y_{n-1,n} \\ Y_{n,1} & 0 & \cdots & 0 \end{bmatrix}$$

$$+ \begin{bmatrix} 0 & \cdots & 0 & Y_{1,n} \\ Y_{2,1} & \cdots & 0 & 0 \\ \cdots & \cdots & \cdots & \cdots \\ 0 & \cdots & Y_{n,n-1} & 0 \end{bmatrix}$$

one has

$$|||Y||| \le \|Y\| \le n|||Y|||.$$

Now we particularize to $Y = Z^tZ \in Mat_n(B)$. For any integer $k \geq 1$, the entries of Y^k are positive sums of products of entries of Z^t and Z; it follows that

$$|||(Z^tZ)^k||| \leq |||(X^tX)^k|||.$$

Consequently,

$$\|Z^2\| = \|(Z^tZ)^k\|^{1/k} \leq n^{1/k}|||(Z^tZ)^k|||^{1/k}$$
$$\leq n^{1/k}|||(X^tX)^k|||^{1/k} \leq n^{1/k}\|(X^tX)^k\|^{1/k} = n^{1/k}\|X\|^2$$

for all $k \geq 1$. The lemma follows. #

One may deduce from Theorem 1.1.4 numerous properties of the set $\mathcal{N}(S)$. The following is a sample.

Corollary 1.5.2. *Let* $s,t \in \mathcal{N}(S)$. *The following numbers are also in* $\mathcal{N}(S)$:

$$s+t, \quad (s^2+t^2)^{1/2}, \quad \tfrac{1}{2}\{s+(s^2+4kt^2)^{1/2}\} \text{ for every } k \in \mathbb{N}.$$

If $s \neq 0$ *and* $s \neq 1$, *the numbers*

$$s+s^{-1}, \quad s^2(s^2-1)^{-1/2}$$

are in the derived set of $\mathcal{N}(S)$.

Proof. The first claim is a consequence of the following equalities for Perron–Frobenius eigenvectors:

$$\begin{bmatrix} s & t \\ t & s \end{bmatrix}\xi = (s+t)\xi \quad \text{with} \quad \xi = \begin{bmatrix} 1 \\ 1 \end{bmatrix} \in \mathbb{R}^2$$

$$\begin{bmatrix} 0 & s & 0 \\ s & 0 & t \\ 0 & t & 0 \end{bmatrix}\xi = (s^2+t^2)^{1/2}\xi \quad \text{with} \quad \xi = \begin{bmatrix} s \\ (s^2+t^2)^{1/2} \\ t \end{bmatrix} \in \mathbb{R}^3$$

$$\begin{bmatrix} s & t & \cdots & t \\ t & 0 & \cdots & 0 \\ \cdots & \cdots & \cdots & \cdots \\ t & 0 & \cdots & 0 \end{bmatrix}\xi = \tfrac{1}{2}\{s+(s^2+4kt^2)^{1/2}\}\xi \quad \text{with} \quad \xi = \begin{bmatrix} s+(s^2+4kt^2)^{1/2} \\ 2t \\ \vdots \\ 2t \end{bmatrix} \in \mathbb{R}^{k+1}.$$

The second claim is proved in Lemma I.3.7 of Appendix I. #

Remarks about $\mathbf{E} = \mathcal{M}(\{0,1\})$

(1) Given $s,t \in \mathbf{E}$, one may look for an explicit graph with spectral radius $s+t$ (respectively st, and $\frac{1}{2}\{s+(s^2+4kt^2)^{1/2}\}$). Some solutions can be found in [Sch1] (respectively [We], and Theorem 2.13 of [CDS]).

(2) Say $s \in \mathbf{E}$ is irreducible if $s \neq 0,1$ and if $s = s_1 s_2$ with $s_1, s_2 \in \mathbf{E}$ implies $s_1 = 1$ or $s_2 = 1$. Any number in $\mathbf{E}$ can be factored as a product of finitely many irreducibles. Are there only finitely many factorizations? (The answer is obviously yes for $s \leq 4$.) If yes, does the number of factorizations relate simply to the minimal $\ell \in \mathbb{N}$ for which there exists $X \in \mathrm{Mat}_\ell(\mathbb{N})$ with $s = \|X\|$?

(3) Is it true that $\mathcal{M}(\mathbb{N}) = \mathcal{M}(\mathbb{Z})$?

CHAPTER 2
Towers of multi–matrix algebras

2.1. Introduction.

The first purpose of this chapter is to study inclusions of one finite dimensional semi–simple algebra in another. Following [Jo2], we introduce a real–valued invariant, called <u>index</u>, for a pair $\mathbb{1} \in N \subset M$ of arbitrary — not necessarily semi–simple or even finite dimensional — algebras over a field $\mathbb{K}$, as follows:

First, the <u>fundamental construction</u> associates to $N \subset M$ the pair $M \subset L$ where $L = \mathrm{End}_N^r(M)$ is the algebra of endomorphisms of M viewed as a right N–module; M is identified with a subalgebra of L, each $x \in M$ being identified with the left multiplication operator $(y \mapsto xy) \in L$. Second, the <u>tower</u> induced by the pair $N \subset M$ is the nested sequence

$$\mathbb{1} \in M_0 = N \subset M_1 = M \subset \cdots \subset M_k \subset M_{k+1} \subset \cdots$$

of $\mathbb{K}$–algebras, where $M_k \subset M_{k+1}$ is obtained from $M_{k-1} \subset M_k$ by the fundamental construction $(k \geq 1)$. Third, the <u>rank</u> $\mathrm{rk}(M_k | M_0)$ of M_k over M_0 is defined to be the smallest possible number of generators of M_k viewed as a right M_0–module (this rank lies in $\mathbb{N} \cup \{\infty\}$). And finally the <u>index of N in M</u> is the growth rate

$$[M:N] = \lim_{k\to\infty} \sup[\mathrm{rk}(M_k | M_0)]^{1/k}.$$

Two comments on this definition: First, we could exchange the words left and right and obtain a rank and an index "from the other side"; but we shall not study this variation here. Secondly, a more interesting variation comes from using tensor products $M \otimes_N M \otimes_N \cdots$ (a Tor–like idea) instead of endomorphisms $\cdots \mathrm{End}_M^r(\mathrm{End}_N^r(M))$ (an Ext–like idea); we refer to [Jo4] for this. One could check that these variations give the same index for semi–simple pairs, but more general examples may have several indices.

The connection of this subject with Chapter 1 is this: For inclusions $\mathbb{1} \in N \subset M$ of semi–simple algebras over a perfect field, the index turns out to be the square of the norm of a certain matrix of natural numbers Λ_N^M associated with the pair of algebras. Thus the results of Chapter 1 yield restrictions on the possible values of the index.

Recall that a semi–simple algebra over a field $\mathbb{K}$ is the direct sum of its minimal two sided ideals, and each of these is isomorphic to a matrix algebra $\mathrm{Mat}_\mu(\Delta)$ for some $\mu \geq 1$ and some $\mathbb{K}$–division algebra Δ. However, for studying index, we can for the most part restrict our attention to semi–simple algebras for which the simple components are central (have center equal to $\mathbb{K}$). In fact we can even restrict attention to multi–matrix algebras over $\mathbb{K}$, semi–simple algebras whose simple components are isomorphic to matrix algebras over the ground field $\mathbb{K}$. Note that if $\mathbb{K}$ is algebraically closed, then every semi–simple $\mathbb{K}$–algebra is a multi–matrix algebra, since $\mathbb{K}$ has no proper finite dimensional division algebras. We will call an algebra which is isomorphic to $\mathrm{Mat}_\mu(\mathbb{K})$ for some $\mu \geq 1$ a factor. Some authors refer to multi–matrix algebras as "split semi–simple algebras".

The reason why it suffices to study multi–matrix algebras is that index is stable under change of the ground field. Let $\mathbb{K}$ be a perfect field and $\mathbb{E}$ an extension field; for any $\mathbb{K}$–algebra M, let $M^{\mathbb{E}}$ denote $M \otimes_{\mathbb{K}} \mathbb{E}$, an algebra over $\mathbb{E}$. If $\mathbf{1} \in N \subset M$ is a pair of finite dimensional $\mathbb{K}$–algebras, then $[M:N] = [M^{\mathbb{E}}:N^{\mathbb{E}}]$ (Proposition 2.4.4). If M and N are in addition semi–simple over $\mathbb{K}$, then it is possible to choose $\mathbb{E}$ so that $M^{\mathbb{E}}$ and $N^{\mathbb{E}}$ are multi–matrix algebras over $\mathbb{E}$. Taking $\mathbb{E}$ to be an algebraic closure of $\mathbb{K}$ will do, but one can also accomplish this with a finite dimensional field extension.

We now come to the definition of the index matrix Λ_N^M for a pair of semi–simple $\mathbb{K}$–algebras $\mathbf{1} \in N \subset M$. First for a pair of factors $\mathbf{1} \in N \subset M$, the reader can easily check (after looking at Section 2.2) that the index $[M:N]$ is just the ratio of dimensions,

$$[M:N] = \dim_{\mathbb{K}}(M) \div \dim_{\mathbb{K}}(N).$$

In this case Λ_N^M is the 1–by–1 matrix with sole entry $[M:N]^{1/2}$. Next consider a pair of multi–matrix algebras. Let $\{p_i : 1 \leq i \leq m\}$ be the minimal central idempotents in M, so that $M = \bigoplus_{i=1} Mp_i$ and each Mp_i is a factor. Similarly let $\{q_j : 1 \leq j \leq n\}$ be the minimal central idempotents in N. For each pair i,j $(1 \leq i \leq m, 1 \leq j \leq n)$ set

$$M_{i,j} = p_i q_j M p_i q_j \quad \text{and} \quad N_{i,j} = p_i q_j N p_i q_j.$$

Since p_i is central in M, the product $p_i q_j$ is an idempotent in M_i. If $p_i q_j \neq 0$ then $M_{i,j}$ is a factor (Proposition 2.2.3); and, since the map $x \mapsto p_i x p_i$ from Nq_j to $N_{i,j}$ is a non–zero homomorphism, $N_{i,j}$ is also a factor with identity element $p_i q_j$. Define Λ_N^M to be the m–by–n matrix with entries

$$\begin{aligned} \lambda_{i,j} &= 0 && \text{if } p_i q_j = 0, \text{ and} \\ \lambda_{i,j} &= [M_{i,j}:N_{i,j}]^{1/2} && \text{if } p_i q_j \neq 0. \end{aligned}$$

$[M_{i,j}:N_{i,j}]$ is a square integer, so $\lambda_{i,j}$ is a natural number; in fact $M_{i,j} \cong N_{i,j} \otimes \mathrm{Mat}_{\lambda_{i,j}}(\mathbb{K})$, by Proposition 2.2.2. As the set of p_i's is well defined by M and the set of q_j's by N, the inclusion matrix is well defined by the pair $N \subset M$ up to pseudo–equivalence. Obviously one has $\Lambda^{M}_{\theta(N)}$ pseudo–equivalent to Λ^{M}_{N} for any automorphism θ of M. (See also the discussion following Corollary 2.3.2.)

Next consider an arbitrary semi–simple pair $\mathbb{1} \in N \subset M$. The obvious thing would be to define Λ^{M}_{N} just as before, using the simple components of M and N, but this does not suit our purpose. Instead let $\mathbb{E}$ be an algebraic closure of $\mathbb{K}$ and set

$$\Lambda^{M}_{N} = \Lambda^{M^{\mathbb{E}}}_{N^{\mathbb{E}}}.$$

Again Λ^{M}_{N} is well defined up to pseudo–equivalence by the pair $N \subset M$, as the set of factors in $N^{\mathbb{E}}$ and $M^{\mathbb{E}}$ are determined by $N \subset M$. (In case the simple components of M and N are central, we would in fact obtain the same result without extending the ground field; thus an alternative definition of Λ^{M}_{N} is this: let $\mathbb{E}$ be any field extension of $\mathbb{K}$ such that the simple components of $N^{\mathbb{E}}$ and $M^{\mathbb{E}}$ are central. Then $\Lambda^{M}_{N} = \Lambda^{M^{\mathbb{E}}}_{N^{\mathbb{E}}}$.)

For pairs of multi–matrix algebras, $\mathbb{1} \in N \subset M$, the index matrix, together with the dimensions of the minimal ideals of N and M, determine the inclusion up to an inner automorphism of M (Proposition 2.3.3). This is not true for arbitrary semi–simple pairs. We will also refer to the index matrix Λ^{M}_{N} as the <u>inclusion matrix</u>, when $N \subset M$ is a pair of multi–matrix algebras.

For a semi–simple pair, the index is related to the inclusion matrix in the following simple fashion (see Section 2.4):

<u>Theorem 2.1.1</u>. *Let* $N \subset M$ *be a semi–simple pair. Then the index of* N *in* M *is given by*

$$[M{:}N] = \|\Lambda^{M}_{N}\|^{2}.$$

Moreover, for any irredundant matrix $\Lambda \in \mathrm{Mat}_{\mathrm{fin}}(\mathbb{N})$, *there exists a multi–matrix pair* $N \subset M$ *with* $\Lambda = \Lambda^{M}_{N}$.

<u>Corollary 2.1.2</u>. *Let* $N \subset M$ *be a pair of semi–simple algebras. Then either* $[M{:}N] = 4\cos^2(\pi/q)$ *for some integer* $q \geq 3$, *or* $[M{:}N] \geq 4$.

Let $N \subset M$ be a multi–matrix pair and let $M \subset L = \mathrm{End}^{r}_{N}(M)$ be the pair obtained by the fundamental construction. We will show that $M \subset L$ is again a multi–matrix pair and that Λ^{L}_{M} is the transpose of Λ^{M}_{N}. Consequently the inclusion matrices of the tower

$$M_0 = N \subset M_1 = M \subset \cdots \subset M_k \subset M_{k+1} \cdots$$

are immediately determined by that of $N \subset M$. These towers have a rich structure, the further study of which requires the introduction of traces.

A <u>trace</u> on M is a linear map tr: $M \to \mathbb{K}$ such that $\mathrm{tr}(xy) = \mathrm{tr}(yx)$ for $x,y \in M$. It is <u>faithful</u> if the bilinear form $(x,y) \mapsto \mathrm{tr}(xy)$ is non-degenerate. If $\mathbb{K}$ is given as an extension of the real field $\mathbb{R}$, a trace tr is <u>positive</u> if $\mathrm{tr}(e) \geq 0$ for any idempotent e in M. A trace tr on $M = \bigoplus_{i=1}^{m} \mathrm{Mat}_{\mu_i}(\mathbb{K})$ is completely described by the <u>row vector</u> $\overleftarrow{s} = (\mathrm{tr}(f_i))_{1 \leq i \leq m}$ where f_i is a minimal idempotent in the factor $\mathrm{Mat}_{\mu_i}(\mathbb{K})$.

Consider a multi-matrix pair $N \subset M$ and assume that there exists a faithful trace tr on M with faithful restriction to N. (This is always possible if $\mathbb{K}$ is of characteristic zero.) Then there exists a unique $\mathbb{K}$-linear map $E : M \to N$, called a <u>conditional expectation</u> from M onto N, such that

$$\begin{aligned} &E(y) = y && \text{for all } y \in N \\ &E(y_1 x y_2) = y_1 E(x) y_2 && \text{for all } x \in M \text{ and } y_1, y_2 \in N \\ &\mathrm{tr}(E(x)) = \mathrm{tr}(x) && \text{for all } x \in M. \end{aligned}$$

One has of course $E \in L = \mathrm{End}_N^r(M)$, and we will show that L is generated as a vector space by the elements $x_1 E x_2$ for $x_1, x_2 \in M$; in short: $L = \langle M,E \rangle$.

Although any multi-matrix pair $N \subset M$ generates a tower by iterating the fundamental construction, traces and conditional expectations in general do not propagate up the tower. That is why we single out a special class of traces as follows. Given $\beta \in \mathbb{K}^*$ and $N \subset M$ as above, define a <u>Markov trace of modulus</u> β on $N \subset M$ to be a faithful trace tr on M with faithful restriction to N for which there exists a (necessarily unique) trace Tr on $L = \mathrm{End}_N^r(M)$ such that

$$\left.\begin{aligned} \mathrm{Tr}(x) &= \mathrm{tr}(x) \\ \beta\ \mathrm{Tr}(xE) &= \mathrm{tr}(x) \end{aligned}\right\} \text{ for all } x \in M$$

To know whether or not such traces exists on a pair $N \subset M$, one has again to consider the inclusion matrix Λ, and its natural action (here from the right) on vectors with coordinates in $\mathbb{K}$ (after reduction of Λ modulo the characteristic of $\mathbb{K}$).

Theorem 2.1.3. *Let* $N \subset M$ *be a multi-matrix pair with inclusion matrix* Λ *and let* $\beta \in \mathbb{K}^*$

(i) There exists a Markov trace tr *of modulus* β *on* $N \subset M$ *if and only if there exists a row vector* $\overleftarrow{s} \in \mathbb{K}^m$ *with*

$$\begin{cases} \overleftarrow{s}\Lambda\Lambda^t = \beta\overleftarrow{s}, \\ \text{all coordinates of } \overleftarrow{s} \text{ distinct from } 0, \text{ and} \\ \text{all coordinates of } \overleftarrow{s}\Lambda \text{ distinct from } 0. \end{cases}$$

If this holds, tr *is described by a scalar multiple of* $\overleftarrow{s}$.

(ii) Let tr *be a Markov trace of modulus* β *on* $N \subset M$, *and let* Tr *be the extension of* tr *to* $L = \mathrm{End}_N^r(M) = \langle M,E \rangle$ *such that* $\beta\mathrm{Tr}(xE) = \mathrm{tr}(x)$ *for all* $x \in M$. *Then* Tr *is again a Markov trace of modulus* β *on* $M \subset L$.

(iii) If $\mathbb{K}$ *is of characteristic* 0, *the modulus of any Markov trace on a pair* $N \subset M$ *is a totally positive algebraic integer.*

A pair $N \subset M$ is called connected if the intersection $Z_N \cap Z_M$ of the centers of M and N consists only of scalar multiples of the identity. Using the Perron–Frobenius theory of matrices with positive entries, one has:

Theorem 2.1.4. *Assume that* $\mathbb{K}$ *is given as an extension of the real field* $\mathbb{R}$. *Let* $N \subset M$ *be a connected multi-matrix pair with inclusion matrix* Λ *and let* $\beta \in \mathbb{K}^*$. *There exists a positive Markov trace of modulus* β *on* $N \subset M$ *if and only if*

$$\beta = [M:N] = \|\Lambda\|^2.$$

Any two positive Markov traces on $N \subset M$ *are proportional.*

This implies:

Corollary 2.1.5. *The set* $\mathbf{E} = \mathcal{N}(\mathbb{N})$ *of Chapter 1 (see Proposition 1.1.5) is also the set of square roots of possible moduli of positive Markov traces.*

We now return to the tower

$$M_0 = N \subset M_1 = M \subset \cdots \subset M_k \subset M_{k+1} \subset \cdots$$

generated by a multi-matrix pair $N \subset M$. We assume that there exists a Markov trace tr of some modulus $\beta \in \mathbb{K}^*$ on $N \subset M$. As Markov traces propagate according to claim (iii) of Theorem 2.1.3, one has for each $k \geq 1$ a Markov trace tr_k on $M_{k-1} \subset M_k$ and a conditional expectation $E_k : M_k \to M_{k-1}$, which can also be viewed as an element of M_{k+1}. We denote by $A_{tr,k}(M_0 \subset M_1)$ the subalgebra of M_k generated by the unit and $E_1, \cdots, E_{k-1}$.

Theorem 2.1.6. *Let* $M_0 \subset M_1$ *be a multi–matrix pair on which there exists a Markov trace* tr *of modulus* β. *For each* $k \geq 1$, *let* M_k *and* E_k *be as above. Then*

(i) M_k *is generated by* M_1 *and* $E_1, \cdots, E_{k-1}$.

(ii) *The idempotents* $E_1, \cdots, E_{k-1}$ *satisfy*

$$\beta E_i E_j E_i = E_i \quad \textit{if } |i-j| = 1, \textit{ and}$$
$$E_i E_j = E_j E_i \quad \textit{if } |i-j| \geq 2.$$

The proofs of Theorems 2.1.3 to 2.1.6 appear in Section 2.7. In the "generic case" (see below), it is remarkable that (ii) is a complete set of relations for the E_i's. This motivates the introduction of the Temperley–Lieb algebras $\mathcal{A}_{\beta,k}$, which first appeared in statistical physics (see [TL] and appendix II.b.). For any $\beta \in \mathbb{K}^*$ and for any integer $k \geq 1$, the $\mathbb{K}$–algebra (with unit) $\mathcal{A}_{\beta,k}$ is defined by the presentation with generators $\epsilon_1, \cdots, \epsilon_{k-1}$ and with relations:

$$\epsilon_i^2 = \epsilon_i$$
$$\beta \epsilon_i \epsilon_j \epsilon_i = \epsilon_i \quad \text{if } |i-j| = 1$$
$$\epsilon_i \epsilon_j = \epsilon_j \epsilon_i \quad \text{if } |i-j| \geq 2.$$

Choosing a number $q \neq 0,-1$ (in $\mathbb{K}$ or in a quadratic extension of $\mathbb{K}$) with $\beta = 2 + q + q^{-1}$, we say that β is generic if q is not a root of 1 (or if $q = 1$ when $\mathbb{K}$ is of characteristic zero).

First, we describe the structure of $\mathcal{A}_{\beta,k}$.

Theorem 2.1.7. (*i*) *For any* $\beta \in \mathbb{K}^*$ *and for any* $k \geq 1$, *the algebra* $\mathcal{A}_{\beta,k}$ *is of dimension* $\frac{1}{k+1}\begin{bmatrix} 2k \\ k \end{bmatrix}$, *and the natural morphism* $\mathcal{A}_{\beta,k} \to \mathcal{A}_{\beta,k+1}$ *is an injection.*

(*ii*) *If* β *is generic then* $\mathcal{A}_{\beta,k}$ *is a multi-matrix algebra isomorphic to* $\bigoplus_{0 \leq j \leq [k/2]} \mathrm{Mat}_{\left\{ k \atop j \right\}}(\mathbb{K})$, *where* $\left\{ k \atop 0 \right\} = 1$ *and* $\left\{ k \atop j \right\} = \begin{bmatrix} k \\ j \end{bmatrix} - \begin{bmatrix} k \\ j-1 \end{bmatrix}$ *for* $j \geq 1$. *Moreover there exists a faithful normalized trace* tr_k *on* $\mathcal{A}_{\beta,k}$ *such that* $\beta tr_k(w\epsilon_j) = tr_k(w)$ *for* $w \in \mathcal{A}_{\beta,j}$ *and* $j \leq k-1$.

Next, we describe the algebra $A_{tr,k}(M_0 \subset M_1)$ previously introduced.

Theorem 2.1.8. *Let* $M_0 \subset M_1$ *and let* tr *and* β *be as in Theorem 2.1.6.*

(i) If β *is generic, the map* $\mathcal{A}_{\beta,k} \to M_k$ *defined by* $\epsilon_j \mapsto E_j$ *for* $j = 1,\cdots,k-1$ *is an isomorphism onto* $A_{tr,k}(M_0 \subset M_1)$. *Moreover this isomorphism is compatible with the trace* tr_k *on* $\mathcal{A}_{\beta,k}$ *and the Markov trace* tr_k *on* M_k.

(ii) Assume $\mathbb{K}$ *is given as an extension of* $\mathbb{R}$, *and that the Markov trace* tr *is positive (so that* $\beta = [M_1:M_0]$*). Then* $A_{tr,k}(M_0 \subset M_1)$ *is isomorphic to a certain quotient* $B_{\beta,k}$ *of* $\mathcal{A}_{\beta,k}$ *which is explicitly described in Section 2.9.*

The braiding relations, which appear above first in Theorem 2.1.6, suggest a strong connection with Artin's braid groups and with Iwahori's Hecke algebras. This observation has constituted a breakthrough in the study of knots in $\mathbb{R}^3$; see [Jo3], [Jo5], [Fre], and [HKW].

For any $q \in \mathbb{K}^*$ and for any integer $k \geq 1$, define the Hecke algebra $H_{q,k}$ to be the $\mathbb{K}$-algebra with unit presented by generators $g_1,\cdots,g_{k-1}$ and by relations

$$\begin{array}{ll} g_i^2 = (q-1)g_i + q & i = 1,\cdots,k-1 \\ g_i g_{i+1} g_i = g_{i+1} g_i g_{i+1} & i = 1,\cdots,k-2 \\ g_i g_j = g_j g_i \ \text{ if } |i-j| \geq 2 & i,j = 1,\cdots,k-1 \end{array}$$

Assuming $q \neq -1$, one may also set $e_i = \frac{g_i+1}{q+1}$ and $\beta = 2 + q + q^{-1}$, and check that $H_{q,k}$ has a presentation with generators $e_1,\cdots,e_{k-1}$ and with relations

$$\begin{array}{ll} e_i^2 = e_i & i = 1,\cdots,k-1 \\ e_i e_{i+1} e_i - \beta^{-1} e_i = e_{i+1} e_i e_{i+1} - \beta^{-1} e_{i+1} & i = 1,\cdots,k-2 \\ e_i e_j = e_j e_i \ \text{ if } |i-j| \geq 2 & i,j = 1,\cdots,k-1. \end{array}$$

Then $H_{q,k}$ is an algebra of dimension $k!$ over $\mathbb{K}$, and the natural morphism $H_{q,k} \to H_{q,k+1}$ is an injection.

If q is not a root of 1 (with $q = 1$ allowed if the characteristic of $\mathbb{K}$ is zero), then $H_{q,k}$ is a multi-matrix algebra with factors in bijection with the set $\mathcal{P}_k$ of partitions of k. We may thus write

$$H_{q,k} = \bigoplus_{\lambda \in \mathcal{P}_k} I_\lambda$$

where I_λ is a factor for each $\lambda \in \mathcal{P}_k$; we refer to Section 2.10 and 2.11 for a more precise description. That there is a relationship between Hecke algebras and the algebras appearing in towers is clear from the presentation of $\mathcal{A}_{\beta,k}$ with ϵ_i's and from that of $H_{q,k}$ with e_i's. More precisely:

Theorem 2.1.9. *Let* $q \in \mathbb{K}^*$ *be a number which is not a root of* 1 *(with* $q = 1$ *allowed when* $\mathrm{char}(\mathbb{K}) = 0$*), and set* $\beta = 2 + q + q^{-1}$. *Consider an integer* $k \geq 1$. *Then* $\mathcal{A}_{\beta,k}$ *is isomorphic to the quotient of the Hecke algebra* $H_{q,k}$ *by the relation*

$$\text{(S)} \qquad (1+g_1)(1+g_2+g_2g_1) = 0$$

and one has

$$\mathcal{A}_{\beta,k} \cong \bigoplus_{\lambda \in \mathcal{P}_k^2} I_\lambda$$

where $\mathcal{P}_k^2$ *is the subset of* $\mathcal{P}_k$ *consisting of those partitions with Young diagrams having at most 2 rows.*

The relation (S) was pointed out to us by Steinberg. In terms of the generators $e_1, \cdots, e_{k-1}$ of $H_{q,k}$ introduced above, it may also be written as

$$\text{(S)} \qquad \beta\, e_1e_2e_1 = e_1$$

Our exposition is organized as follows.

Section 2.2 collects preliminary material on commutants. Section 2.3 shows how to define the index matrix or inclusion matrix of a multi-matrix pair $N \subset M$, and how to encode the relevant information in a Bratteli diagram. Concerning chains of multi-matrix algebras (a natural generalization of pairs), we have added an exposition of the path model due to Ocneanu and Sunder. Theorem 2.1.1 and Corollary 2.1.2 about the index of a pair $N \subset M$ are proved in Section 2.4.

Section 2.5 concerns traces in general, Section 2.6 conditional expectations, and Section 2.7 Markov traces in particular. Theorem 2.1.3 to 2.1.6 are proved in Section 2.7.

Section 2.8 is about the algebras $\mathcal{A}_{\beta,k}$, with emphasis on generic β, and Section 2.9 discusses non-generic β. Section 2.10 is a leisurely digression introducing Hecke algebras, and the final section shows how $\mathcal{A}_{\beta,k}$ is a quotient of the appropriate Hecke algebra.

If $\mathbb{K}$ is our favorite field of complex numbers, multi- matrix (or semi-simple) algebras "are the same" as finite- dimensional C^*-algebras. This is of course well-known, and made precise in Appendix II.a. If follows that the present Chapter 2 may be looked at as a study of certain C^*-algebras which are AF, namely approximately finite; see [Eff]. In Appendix II.b we report briefly on the appearance of the algebras $\mathcal{A}_{\beta,k}$ in statistical mechanics. Appendix II.c contains additional material on the algebras $\mathcal{A}_{\beta,k}$ for non-generic β.

2.2. Commutant and bicommutant.

Let F be an algebra over a field $\mathbb{K}$. (By an algebra we will always mean an algebra with an identity element $\mathbf{1}$.) The commutant $C_F(S)$ of a subset S of F is the algebra of all elements of F which commute with each element of S. One obviously has $S \subset C_F C_F(S)$ for any subset S, and the reader can check that $C_F C_F(C_F(S)) = C_F(S)$ for any S. If A is a subalgebra of F, then $C_F C_F(A)$ is an algebra containing A, a sort of closure of A in F.

Let W be any $\mathbb{K}$-vector space (not necessarily finite dimensional). We remark that for each $d \in \mathbb{N}$ there are canonical algebra isomorphisms

$$\mathrm{End}_{\mathbb{K}}(W^d) \cong \mathrm{Mat}_d(\mathrm{End}_{\mathbb{K}}(W)) \cong \mathrm{End}_{\mathbb{K}}(W) \otimes_{\mathbb{K}} \mathrm{Mat}_d(\mathbb{K}).$$

(For any algebra A, $\mathrm{Mat}_d(A)$ denotes the algebra of d-by-d matrices over A.) For $1 \leq i \leq d$ let $\pi_i : W^d \to W$ be the projection on the ith component, and for $1 \leq j \leq d$ let $\epsilon_j : W \to W^d$ be the injection such that $\pi_i \circ \epsilon_j = \delta_i^j \mathbf{1}$ on W for all i. Given $x \in \mathrm{End}_{\mathbb{K}}(W^d)$ define a matrix $[x_{i,j}] \in \mathrm{Mat}_d(\mathrm{End}_{\mathbb{K}}(W))$ by $x_{i,j} = \pi_i \circ x \circ \epsilon_j$. One can check that the map $x \mapsto [x_{i,j}]$ is an algebra homomorphism, and that it has an inverse, which takes a matrix $[x_{i,j}]$ to $\sum_{i,j} \epsilon_i \circ x_{i,j} \circ \pi_j \in \mathrm{End}_{\mathbb{K}}(W^d)$. This establishes the first isomorphism. Let $\{e_{i,j}: 1 \leq i,j \leq d\}$ denote the standard system of matrix units in $\mathrm{Mat}_d(\mathbb{K})$; the matrix $e_{i,j}$ has a 1 in the (i,j) position and zeros elsewhere. Then the map

$$[x_{i,j}] \mapsto \sum_{i,j} x_{i,j} \otimes e_{i,j}$$

is an isomorphism of $\mathrm{Mat}_d(\mathrm{End}_{\mathbb{K}}(W))$ onto $\mathrm{End}_{\mathbb{K}}(W) \otimes_{\mathbb{K}} \mathrm{Mat}_d(\mathbb{K})$. We will identify

$\mathrm{End}_{\mathbb{K}}(W^d)$, $\mathrm{Mat}_d(\mathrm{End}_{\mathbb{K}}(W))$, and $\mathrm{End}_{\mathbb{K}}(W) \otimes \mathrm{Mat}_d(\mathbb{K})$ via these isomorphisms. For any subalgebra A of $\mathrm{End}_{\mathbb{K}}(W)$, the algebra $A \otimes_{\mathbb{K}} \mathrm{Mat}_d(\mathbb{K})$ is thus identified with $\mathrm{Mat}_d(A)$, and $A \otimes \mathbf{1}$ with the set of diagonal matrices

$$\mathrm{diag}(a) = \begin{bmatrix} a & & \\ & \ddots & \\ & & a \end{bmatrix}, \quad a \in A,$$

or with the set of diagonal endomorphisms

$$\mathrm{diag}(a)(\xi_1,\cdots,\xi_d) = (a\xi_1,\cdots,a\xi_d).$$

The following unassuming lemma is the basis of both the Jacobson density theorem ([BA8], p.39) and of von Neumann's bicommutant theorem ([Tak], p.74), two cornerstones of non-commutative algebra.

Lemma 2.2.1. *Let* W *be a vector space over* $\mathbb{K}$ *and let* A *be a subalgebra of* $\mathrm{End}_{\mathbb{K}}(W)$ *with* $\mathbf{1} \in A$. *Write* A' *for* $C_{\mathrm{End}_{\mathbb{K}}(W)}(A)$.

(a) *The commutant of* $A \otimes \mathbf{1}$ *in* $\mathrm{End}_{\mathbb{K}}(W^d)$ *is* $A' \otimes_{\mathbb{K}} \mathrm{Mat}_d(\mathbb{K})$.

(b) *The commutant of* $A \otimes_{\mathbb{K}} \mathrm{Mat}_d(\mathbb{K})$ *in* $\mathrm{End}_{\mathbb{K}}(W^d)$ *is* $A' \otimes \mathbf{1}$.

Proof. (a) Since $(a\otimes\mathbf{1})(\sum x_{i,j} \otimes e_{i,j}) = \sum ax_{i,j} \otimes e_{i,j}$, and $(\sum x_{i,j} \otimes e_{i,j})(a\otimes\mathbf{1}) = \sum x_{i,j}a \otimes e_{i,j}$, it follows that $\sum x_{i,j} \otimes e_{i,j}$ lies in the commutant of $A \otimes \mathbf{1}$ if and only if $x_{i,j} \in A'$ for all (i,j).

(b) If $x = \sum_{i,j} x_{i,j} \otimes e_{i,j}$ commutes with $A \otimes_{\mathbb{K}} \mathrm{Mat}_d(\mathbb{K})$, then in particular it commutes with the matrix units $\mathbf{1} \otimes e_{\mu,\nu}$ for all (μ,ν),

$$\sum_j x_{\nu,j} \otimes e_{\mu,j} = (1 \otimes e_{\mu,\nu})x = x(1 \otimes e_{\mu,\nu}) = \sum_i x_{i,\mu} \otimes e_{i,\nu}.$$

It follows that $x_{\nu,j} = 0$ for $j \neq \nu$ and $x_{i,\mu} = 0$ for $i \neq \mu$, and $x_{\nu,\nu} = x_{\mu,\mu}$. Thus $x = \sum_i y \otimes e_{i,i} = y \otimes \mathbf{1}$ for some $y \in \mathrm{End}_{\mathbb{K}}(W)$. Since in addition x commutes with $A \otimes \mathbf{1}$, it follows that $y \in A'$. On the other hand any element of the form $y \otimes \mathbf{1}$, $y \in A'$, commutes with $A \otimes_{\mathbb{K}} \mathrm{Mat}_d(\mathbb{K})$. #

Consider a <u>factor</u> $F \cong End_{\mathbb{K}}(V)$, where V is a finite dimensional $\mathbb{K}$–vector space. The basic facts about the representation theory of F are:

(i) Any finite dimensional F–module is completely reducible, and
(ii) any irreducible F–module is equivalent to the standard module V.

(By a module over a $\mathbb{K}$–algebra A, we will always mean a <u>unital</u> module; that is the identity of A acts as the identity on the module. Thus an A–module is also a $\mathbb{K}$–vector space.) If W is any finite dimensional F–module, it follows from (i) and (ii) that $\dim_{\mathbb{K}}(V)$ divides $\dim_{\mathbb{K}}(W)$, say $\dim_{\mathbb{K}}(W) \div \dim_{\mathbb{K}}(V) = d$, and W is equivalent to the F–module V^d with the diagonal action

$$x(\xi_1,\cdots,\xi_d) = (x\xi_1,\cdots,x\xi_d).$$

We call d the <u>multiplicity</u> of the F–module W.

<u>Lemma 2.2.2.</u> *Let* F *be a factor and let* M *be a subfactor of* F *with* $\mathbf{1} \in M \subset F$. *Then*

(a) $C_F(M)$ *is a factor;*
(b) $C_F C_F(M) = M$;
(c) $M \otimes_{\mathbb{K}} C_F(M)$ *is isomorphic to* F.

<u>Proof</u>. Identify F with $End_{\mathbb{K}}(V)$ for some finite dimensional $\mathbb{K}$–vector space V. Similarly M is isomorphic to $End_{\mathbb{K}}(W)$ for some W, and any isomorphism

$$\alpha : End_{\mathbb{K}}(W) \to M \subset End_{\mathbb{K}}(V)$$

is a representation of $End_{\mathbb{K}}(W)$ on V. It follows that α is equivalent to the diagonal representation of $End_{\mathbb{K}}(W)$ on W^d for some d; that is we can identify F with $End_{\mathbb{K}}(W^d)$ and M with $End_{\mathbb{K}}(W) \otimes \mathbf{1} \subset End_{\mathbb{K}}(W^d)$. Then by the previous lemma $C_F(M) = \mathbf{1} \otimes Mat_d(\mathbb{K})$, which is a factor. Furthermore $C_F C_F(M) = End_{\mathbb{K}}(W) \otimes \mathbf{1} = M$. Finally $M \otimes_{\mathbb{K}} C_F(M) \cong End_{\mathbb{K}}(W) \otimes Mat_d(\mathbb{K}) = F$. #

Let N be a subfactor of a factor M with $\mathbf{1} \in N \subset M$. Recall that the index of N in M is

$$[M : N] = (\dim_{\mathbb{K}} M)(\dim_{\mathbb{K}} N)^{-1}.$$

This index is the square of an integer by Lemma 2.2.2. In case M itself is a subfactor of a factor F, the same lemma implies

$$[C_F(N) : C_F(M)] = \frac{\dim F}{\dim N} \div \frac{\dim F}{\dim M} = [M : N]$$

which is a preview of Proposition 2.3.5 below.

Proposition 2.2.3. *Let* M *be a multi-matrix subalgebra of a factor* F *with* $\mathbf{1} \in M \subset F$. *Then*

(a) $C_F(M)$ *is a multi-matrix algebra, and has the same minimal central idempotents as* M;

(b) $C_F(C_F(M)) = M$.

Proof. Let $p_1, \cdots, p_m$ be the minimal central idempotents of M, and identify M with $\bigoplus_{i=1}^{m} p_i M$; each $p_i M$ is a factor. As the center $M \cap C_F(M)$ of M is contained in the center $C_F(C_F(M)) \cap C_F(M)$ of $C_F(M)$ and as $\sum_{i=1}^{m} p_i = 1$, one has $C_F(M) = \bigoplus_{i=1}^{m} p_i C_F(M) p_i$. It is straightforward to check (using $p_i \in M \cap C_F(M)$) that $p_i C_F(M) p_i = C_{p_i F p_i}(p_i M)$, and this is a factor by Lemma 2.2.2. Thus $C_F(M)$ is a multi-matrix algebra, and its center coincides with that of M, so that (a) holds.

Similary one has

$$p_i C_F(C_F(M)) p_i = C_{p_i F p_i}(C_{p_i F p_i}(p_i M))$$

and the lemma implies

$$C_F(C_F(M)) = \bigoplus_{i=1}^{m} p_i C_F(C_F(M)) p_i = \bigoplus_{i=1}^{m} p_i M = M$$

so that (b) holds. #

Remarks.

(1) If M is a subalgebra of a factor F and if M is not semi-simple, one may have $M \subsetneq C_F(C_F(M))$. An example is given by the algebra M of matrices of the form $\begin{bmatrix} a & 0 \\ c & d \end{bmatrix}$ in the factor $F = \mathrm{Mat}_2(\mathbb{K})$, for which $C_F(C_F(M)) = F$.

(2) The biccommutant theorem is valid for all finite dimensional semi–simple algebras; see [Weyl], Theorem 3.5 B. Brauer [Br] and Weyl (op.cit.) stressed the importance of the theorem for invariant theory. The ultimate form of the theorem is Jacobson's density theorem ([Jac1], [Jac2]). It was noted by Bourbaki ([BA8], p. 39) that Jacobson's theorem follows from von Neumann's proof of his bicommutant theorem for operator algebras ([vN] or [Tak]).

(3) Let V be a vector space such that $F = \mathrm{End}_{\mathbb{K}}(V)$. The usual notation in functional analysis for $C_F(M)$ is M'. In some books of algebra, V viewed as a $C_F(M)$–module is called the counter–module of the M–module V.

Lemma 2.2.4. *Let* M *be a subfactor of a factor* F *with* $\mathbf{1} \in M \subset F$ *and let* $q \in M \cup C_F(M)$ *be a nonzero idempotent. Then*

(a) qMq *is a factor;*

(b) $C_{qFq}(qMq) = qC_F(M)q.$

Proof. Assume first that $q \in M.$

(a) If M is identified with $\mathrm{End}_{\mathbb{K}}(V)$, then qMq is isomorphic to $\mathrm{End}_{\mathbb{K}}(qV)$.

(b) Consider first $x \in qMq \subset M$ and $y \in qC_F(M)q$. Choose $z \in C_F(M)$ with $y = qzq$ and compute

$$xy = q(qxq)zq = qzqxq = yx.$$

It follows that $qC_F(M)q \subset C_{qFq}(qMq)$. Consider then $s = qsq \in C_{qFq}(qC_F(M)q)$ and $t \in C_F(M)$. As $tq = qt$ one has

$$st = sqqt = sqtq = qtqs = tqqs = ts.$$

It follows that $C_{qFq}(qC_F(M)q) \subset qC_F(C_F(M))q$, the last term being qMq by Lemma 2.2.2. Taking now commutants and applying Lemma 2.2.2 in qFq, one has

$$C_{qFq}(qMq) \subset C_{qFq}(C_{qFq}(qC_F(M)q)) = qC_F(M)q.$$

Assume now that $q \in C_F(M)$.

(a) The linear map φ from M to qMq which sends x to qxq is a morphism of algebras and its image contains $q = q\mathbf{1}q$. As M is simple, φ is an isomorphism.

(b) By the first part of the proof one has

$$C_{qFq}(qC_F(M)q) = qC_F(C_F(M))q = qMq,$$

using $C_F(C_F(M)) = M$. Now take commutants and apply Lemma 2.2.2 in qFq to obtain the conclusion. #

Proposition 2.2.5. *Let* M *be a multi-matrix subalgebra of a factor* F *with* $\mathbf{1} \in M \subset F$ *and let* $q \in M \cup C_F(M)$ *be a nonzero idempotent. Then*

(a) qMq *is a multi-matrix algebra.*
(b) $C_{qFq}(qMq) = qC_F(M)q$.

Proof. The notation being as in the proof of Proposition 2.2.3, set $q_i = p_i q$ and observe that $q = \sum_{i=1}^{m} q_i$. One obviously has

$$C_{qFq}(qMq) = \bigoplus_{i=1}^{m} q_i C_{qFq}(qMq) q_i = \bigoplus_{i=1}^{m} C_{q_iFq_i}(q_iMq_i), \text{ and}$$

$$qC_F(M)q = \bigoplus_{i=1}^{m} q_i C_F(M) q_i = \bigoplus_{i=1}^{m} q_i C_{p_iFp_i}(p_iM) q_i .$$

Claim (b) follows because

$$C_{q_iFq_i}(q_iMq_i) = q_i C_{p_iFp_i}(p_iMp_i) q_i$$

by Lemma 2.2.4. We leave details of (a) to the reader. #

Remarks.

(1) The algebra qMq is called the reduction of M by q.

(2) It is easy to find examples which show that one cannot suppress in Proposition 2.2.5 the hypothesis that q is either in M or in its commutant.

Next is a version of the Skolem–Noether theorem:

Proposition 2.2.6. *Suppose that* M *is a factor,* N *and* $\overline{N}$ *are subfactors of* M *containing the identity element of* M, *and* $\varphi : N \to \overline{N}$ *is an isomorphism. Then there is an inner automorphism* θ *of* M *such that* $\theta|_N = \varphi$. *In particular, any automorphism of* M *is inner.*

Proof. We identify M with $\mathrm{End}_{\mathbb{K}}(V)$ for some vector space V over $\mathbb{K}$. Let W be an irreducible N module; then any N module V is equivalent to the direct sum of d

copies of W, where $d = \dim_{\mathbb{K}} V \div \dim_{\mathbb{K}} W$. In particular the two N module structures on V defined by the actions $(n,v) \mapsto nv$ and $(n,v) \mapsto \varphi(n)v$ are equivalent. Hence there is an invertible $u \in \mathrm{End}_{\mathbb{K}}(V) = M$ such that $u(nv) = \varphi(n)u(v)$ for all $n \in N$ and $v \in V$. Thus $\varphi(n) = unu^{-1}$. #

We end this section with the finite dimensional case of the coupling constant theorem of Murray and von Neumann (see Theorem X in [MvN I], and also Theorem X in [MvN IV]). This is a digression motivated by the importance of the theorem for II_1–factors (see Chapter 3).

Consider a factor $M = \mathrm{End}_{\mathbb{K}}(V)$, where V is of dimension μ over $\mathbb{K}$, and a representation π of M in a vector space W. Assume that π is of multiplicity d, so that $\dim_{\mathbb{K}}(W) = d\mu$, and view π as an inclusion $M \subset F$ with $F = \mathrm{End}_{\mathbb{K}}(W)$. Any non zero vector $\xi \in W$ determines a cyclic M–submodule $M\xi$ of W, as well as a cyclic $C_F(M)$–submodule $C_F(M)\xi$ of W.

Proposition 2.2.7. *With the notations above one has*

$$\dim_{\mathbb{K}}(C_F(M)\xi)/\dim_{\mathbb{K}}(M\xi) = d/\mu = [F : M]/\dim_{\mathbb{K}}(W)$$

for all $\xi \in W$ *with* $\xi \neq 0$.

Proof. Identify W with V^d. Let $\xi = (\xi_1,\cdots,\xi_d) \in W$ and let ℓ be the dimension of the subspace U of V generated by $\xi_1,\cdots,\xi_d$; one has $1 \leq \ell \leq \min(\mu,d)$.

One may choose ℓ linear independent ξ_j's and write the $d-\ell$ remaining ones as linear combinations of the chosen ones, so that clearly $\dim(M\xi) = \ell\mu$. On the other hand $C_F(M)\xi$ consists of vectors of the form

$$\left(\sum_{j=1}^{d} a_{1,j}\xi_j, \cdots, \sum_{j=1}^{d} a_{d,j}\xi_j\right)$$

with $(a_{i,j})_{1\leq i,j\leq d} \in \mathrm{Mat}_d(\mathbb{K})$, so that $\dim(C_F(M)\xi) = \ell d$. The conclusion follows. #

2.3. Inclusion matrix and Bratteli diagram for inclusions of multi–matrix algebras.

Let M be a multi–matrix algebra over $\mathbb{K}$, with minimal central idempotents $p_1,\cdots,p_m$. Each minimal two–sided ideal p_iM is a factor, isomorphic to $\mathrm{Mat}_{\mu_i}(\mathbb{K})$ for some integer μ_i. We identify M with $\bigoplus_{i=1}^{m} p_i M$, and the center Z_M of M with $\bigoplus_{i=1}^{m} \mathbb{K} p_i$, and denote by μ or $\vec{\mu}$ the m–tuple $(\mu_1,\cdots,\mu_m)^t$ of dimensions. (The superscript t means "transpose", because we think of μ as a column vector.) The isomorphism class of M is completely described by the class of $\vec{\mu}$ modulo permutation of its coordinates. Observe in particular that the $\mathbb{K}$-dimension of M, which is $\sum_{i=1}^{m} \mu_i^2$, is the square of the Euclidean norm of $\vec{\mu}$.

Let N be a multi–matrix subalgebra of M with $\mathbf{1} \in N \subset M$. Denote by $q_1,\cdots,q_n$ the minimal central idempotents of N; we identify N with $\bigoplus_{j=1}^{n} q_j N$ and Z_N with $\bigoplus_{j=1}^{n} \mathbb{K} q_j$. If $q_j N \cong \mathrm{Mat}_{\nu_j}(\mathbb{K})$, then the isomorphism class of N is described by the vector $\nu = (\nu_1,\cdots,\nu_n)^t$.

Set, moreover,

$$M_{i,j} = p_i q_j M p_i q_j \quad \text{and} \quad N_{i,j} = p_i q_j N p_i q_j \quad (1 \le i \le m,\ 1 \le j \le n).$$

As p_i is central in M, the product $p_i q_j$ is an idempotent in $p_i M$. Lemma 2.2.4 shows that $M_{i,j}$ is either 0 or a factor, depending on whether $p_i q_j$ is 0 or not. If $p_i q_j \ne 0$, the map $x \longrightarrow p_i x p_i$ from Nq_j to $N_{i,j}$ is a morphism of algebras containing $p_i q_j$ in its image; as $q_j N$ is simple, the map is an isomorphism. Recall that the <u>index matrix</u> Λ_N^M for the inclusion $N \subset M$ is the m–by–n matrix with entries

$$\begin{aligned} \lambda_{i,j} &= 0 && \text{if } p_i q_j = 0, \text{ and} \\ \lambda_{i,j} &= ([M_{i,j} : N_{i,j}])^{1/2} && \text{if } p_i q_j \ne 0 \end{aligned}$$

Note that $M_{i,j} \cong N_{i,j} \otimes \mathrm{Mat}_{\lambda_{i,j}}(\mathbb{K})$, using Lemma 2.2.2, and, under this isomorphism, $N_{i,j}$ is identified with $N_{i,j} \otimes 1$. We will see that, for multi–matrix algebras, the matrix Λ_N^M together with the vector of dimensions $\vec{\nu}$ determines the inclusion $N \subset M$ up to isomorphism (Proposition 2.3.3); we therefore also refer to Λ_N^M as the <u>inclusion matrix</u> for the pair $N \subset M$.

We next observe that Λ_N^M can be described in terms of the representation theory or, equivalently, the K–theory of the rings N and M. Identify M with $\bigoplus_{i=1}^{m} \mathrm{End}_{\mathbb{K}}(V_i)$, where V_i is a vector space of dimension μ_i, acting on $V = \bigoplus_{i=1}^{m} V_i$. An isomorphism

$\pi : \bigoplus_{j=1}^{n} \mathrm{Mat}_{\nu_j}(\mathbb{K}) \longrightarrow N \subset M$ with $\pi(\mathrm{Mat}_{\nu_j}(\mathbb{K})) = Nq_j$ for each j is then a representation of $\bigoplus_{j=1}^{n} \mathrm{Mat}_{\nu_j}(\mathbb{K})$ on V. Each V_i is a non-trivial N-submodule and we let π_i denote the corresponding subrepresentation. If $q_j p_i \neq 0$ then $q_j p_i V_i = q_j V_i$ is again a non-trivial N-submodule (although not an M-submodule), and the corresponding subrepresentation $\pi_{i,j}$ is zero on $\mathrm{Mat}_{\nu_\ell}(\mathbb{K})$ $(\ell \neq j)$, while the restriction of $\pi_{i,j}$ to $\mathrm{Mat}_{\nu_j}(\mathbb{K})$ is a unital isomorphism onto $N_{i,j} \subset M_{i,j} = \mathrm{End}_{\mathbb{K}}(q_j V_i)$. The multiplicity of this representation is $\lambda_{i,j}$, so there is a basis of $q_j V_i$ such that the matrix of $\pi_{i,j}(x_1,\cdots,x_n)$ is $\mathrm{diag}(x_j,\cdots,x_j)$, with x_j being repeated $\lambda_{i,j}$ times; cf. the proof of Lemma 2.2.2. Hence there is a basis of V_i such that the matrix of $\pi_i(x_1,\cdots,x_n)$ is

$$\mathrm{diag}(x_1,\cdots,x_1;x_2,\cdots,x_2;\cdots;x_n,\cdots,x_n),$$

with x_j being repeated $\lambda_{i,j}$ times.

Next we introduce some terminology in order to give the K-theoretic interpretation of the inclusion matrix. The K_0 group of a ring R is an abelian group constructed from the representation theory of R; the set of equivalence classes of finitely generated projective R-modules forms an abelian semigroup, the operation being direct sum, and $K_0(R)$ is the quotient group of this semigroup. For a multi-matrix algebra $M \cong \oplus\, p_i M \cong \oplus\, \mathrm{End}_{\mathbb{K}}(V_i)$, $K_0(M)$ is the free abelian group generated by the classes of the standard representations $\alpha_i(x) = xp_i|_{V_i}$. This can also be expressed in terms of idempotents. For any idempotent $e \in M$, let $[e]_M$ denote the class of the left ideal Me regarded as an M-module. If e_i is any minimal idempotent in $p_i M$, then $Me_i = Mp_i e_i$ is a minimal left ideal of M, and as a left M-module it is equivalent to the standard module $\{\alpha_i, V_i\}$; thus $\{[e_i]_M\}$ is a basis of $K_0(M)$. Similarly, choose minimal idempotents f_j of $q_j N$ and denote by β_j the corresponding irreducible representation of N, for $j = 1,\cdots,n$. There is a canonical map $K_0(N) \longrightarrow K_0(M)$ induced by the inclusion $N \subset M$, which is the unique morphism of abelian groups taking $[e]_N$ to $[e]_M$ for any idempotent e in N.

Proposition 2.3.1. *Let $N \subset M$ be a pair of multi-matrix algebras and let the notation be as above.*

(a) Λ_N^M *is the matrix of the map* $K_0(N) \longrightarrow K_0(M)$ *induced by the inclusion* $N \longrightarrow M$, *with respect to the basis* $([e_i]_M)_{1\leq i\leq m}$ *of* $K_0(M) \cong \mathbb{Z}^m$ *and* $([f_j]_N)_{1\leq j\leq n}$ *of*

$K_0(N) \cong \mathbb{Z}^n$.

(b) $\Lambda_N^M \vec{\nu} = \vec{\mu}$.

(c) Λ_N^M *is irredundant.*

(d) *The representation* $\mathrm{Ind}_N^M(\beta_j)$ *obtained by induction is equivalent to* $\sum_i \lambda_{i,j}\alpha_i$ *for* $j = 1,\cdots,n$.

(e) *The representation* $\alpha_i|_N$ *obtained by restriction is equivalent to* $\sum_j \lambda_{i,j}\beta_j$ *for* $i = 1,\cdots,m$.

(f) Λ_N^M *is indecomposable if and only if* $Z_M \cap Z_N = \mathbb{K}$.

Proof. (a) We have to compute the image $[f_j]_M \in K_0(M)$ of the basic element $[f_j]_N \in K_0(N)$, where $j \in \{1,\cdots,n\}$. As $f_j = \sum_{i=1}^m p_i f_j$ one has $[f_j]_M = \sum_{i=1}^m [p_i f_j]_M$. But $f_j p_i = f_j q_j p_i$ is the sum of $\lambda_{i,j}$ minimal idempotents in $M_{i,j}$, as one sees from the proof of 2.2.2 or the analysis of representations above. Furthermore, a minimal projection in $M_{i,j}$ remains minimal in Mp_i, and is thus equivalent to e_i. Thus $[f_j]_M = \sum_{i=1}^m \lambda_{i,j}\,[e_i]_M$, as was to be shown.

(b) This follows because $\vec{\mu} = [\mathbf{1}]_M$ and $\vec{\nu} = [\mathbf{1}]_N$.

(c) Given $i \in \{1,\cdots,m\}$, there exists j with $p_i q_j \neq 0$ because $\sum_{i=1}^n q_j = \mathbf{1}$, hence the $i^{\underline{th}}$ row of Λ_N^M is not zero. Similarly, no column of Λ_N^M can be zero because $\sum_{i=1}^m p_i = \mathbf{1}$.

(d) Observe that $\mathrm{Ind}_N^M(\beta_j)$ acts on $M \otimes_N Nf_j$ which is canonically isomorphic to Mf_j; thus $\mathrm{Ind}_N^M(\beta_j)$ is left multiplication by elements of M on Mf_j. The proof of (a) shows that $\mathrm{Ind}_N^M(\beta_j) = \sum_{i=1}^n \lambda_{i,j}\alpha_i$.

(e) This follows directly from the analysis of representations above, i.e. the diagonalization of the representation π.

(f) If $\dim_{\mathbb{K}}(Z_M \cap Z_N) > 1$, there exists an idempotent $r \in Z_M \cap Z_N$ with $r \neq 0$ and $r \neq 1$. With an appropriate indexing of the p_i's and q_j's, one has

$$p_1,\cdots,p_{m'};q_1,\cdots,q_{n'} \in rMr$$
$$p_{m'+1},\cdots,p_m;q_{n'+1},\cdots,q_n \in (1-r)M(1-r)$$

for some m',n' with $0 < m' < m$ and $0 < n' < n$. It follows that $M_{i,j} = 0$ and $\lambda_{i,j} = 0$ unless

<u>either</u> $1 \leq i \leq m'$ and $1 \leq j \leq n'$
<u>or</u> $m'+1 \leq i \leq m$ and $n'+1 \leq j \leq n$

so that Λ_N^M is decomposable. One checks conversely that, if Λ_N^M is decomposable, then $Z_M \cap Z_N$ contains a nontrivial idempotent which is a sum of some p_i's and also a sum of some q_j's. #

<u>Corollary 2.3.2</u>. *Let* L,M,N *be multi-matrix algebras with* $1 \in N \subset M \subset L$. *Then* $\Lambda_N^L = \Lambda_M^L \Lambda_N^M$.

<u>Proof</u>. By functoriality of K_0 (or by counting multiplicities). #

Here is one detail one has to pay attention to. Let $N \subset M$ be a pair of multi-matrix algebras and let Λ_N^M be the corresponding index matrix; assume that M [respectively N] is a direct sum of m [respectively n] factors. As a matrix with rows [respectively columns] indexed by the minimal central idempotents of M [respectively N], the matrix Λ_N^M is well defined by $N \subset M$. But as a matrix with rows [respectively columns] indexed by $\{1,\cdots,m\}$ [respectively $\{1,\cdots,n\}$], Λ_N^M depends on orderings $p_1,\cdots,p_m$ and $q_1,\cdots,q_n$ of these idempotents, and thus Λ_N^M is only defined up to pseudoequivalence by $N \subset M$. Taking for a moment the first point of view, the following makes sense: let $N \subset M$ and $\overline{N} \subset \overline{M}$ be pairs of multi-matrix algebras; if there exists an isomorphism $\theta : M \longrightarrow \overline{M}$ with $\theta(N) = \overline{N}$, then $\Lambda_N^M = \Lambda_{\overline{N}}^{\overline{M}}$, with strict equality. This has a converse that we formulate as follows.

<u>Proposition 2.3.3</u>. *Let* N, $\overline{N}$ *be two multi-matrix subalgebras of a multi-matrix algebra* M, *given together with an isomorphism* $\varphi : N \longrightarrow \overline{N}$. If $\Lambda_N^M = \Lambda_{\overline{N}}^M$, *then* φ *extends to an automorphism* θ *of* M.

Proof. We show this for M a factor, and the general case will follow; we may thus identify M with $\mathrm{End}_{\mathbb{K}}(V)$ for some vector space V over $\mathbb{K}$. Set

$$\Lambda_N^M = \Lambda_{\overline{N}}^M = (\lambda_1,\cdots,\lambda_n) \in M_{1,n}(\mathbb{N}).$$

Let

$$N = \bigoplus_{j=1}^{n} q_j N, \qquad \overline{N} = \bigoplus_{j=1}^{n} \overline{q}_j \overline{N}$$

be decompositions into factors, where the indices are chosen so that $\varphi(q_j) = \overline{q}_j$ for $j = 1,\cdots,n$. Each idempotent q_j is a sum of (say) ν_j minimal idempotents in $q_j N$, and thus of $\lambda_j \nu_j$ minimal idempotents in $q_j M$ (these are still minimal in M). The same holds for $\overline{q}_j$. Consequently there exists an inner automorphism θ' of M such that $\theta'(q_j) = \overline{q}_j$ for $j = 1,\cdots,n$. It follows that one may assume from the start that $q_j = \overline{q}_j$ for $j = 1,\cdots,n$.

For $j \in \{1,\cdots,n\}$, set $V_j = q_j(V)$. By the Skolem– Noether theorem, Proposition 2.2.6, there exists $g_j \in GL(V_j)$ such that $\varphi(y) = g_j y g_j^{-1}$ for all $y \in q_j N$. As $\sum_{1 \le j \le n} q_j = 1$, one has $V = \bigoplus_{1 \le j \le n} V_j$ and one may define $g = \bigoplus_{1 \le j \le n} g_j \in GL(V)$. Then

$$\theta = \mathrm{Ad}(g) : x \mapsto gxg^{-1}$$

has the desired property. #

Corollary 2.3.4. *Any multi–matrix pair* $N \subset M$ *is isomorphic to* $N^{opp} \subset M^{opp}$.

Proof. This follows from 2.3.3, as the two pairs have clearly the same inclusion matrix. Here is another way to look at it: Let β be the antiautomorphism of $N = \bigoplus_{j=1}^{n} \mathrm{Mat}_{\nu_j}(\mathbb{K})$ which coincides with the transposition on each factor, and view β as an isomorphism $N \longrightarrow N^{opp}$; then β may be extended to an isomorphism $\alpha : M \longrightarrow M^{opp}$. #

Remarks.

(1) Our Λ_N^M is as in [Jo 2], but is the transpose of A_N^M in [Jo 1]. So our proposition 2.3.1.d corresponds to 3.2.1 in [Jo 1], which reads $\vec{m} = \vec{n} A_N^M$.

(2) Proposition 2.3.3 shows that a multi–matrix pair $N \subset M$ is characterized (up to isomorphism) by the data $\vec{\nu}$ (which describes N up to isomorphism) and Λ_N^M (which describes the inclusion). The following simple examples show how partial descriptions fail to be complete.

(i) Consider the two subalgebras (both of dimension 62):

$$N = \mathbb{K} \oplus \mathrm{Mat}_5(\mathbb{K}) \oplus \mathrm{Mat}_6(\mathbb{K})$$

$$\overline{N} = \mathrm{Mat}_2(\mathbb{K}) \oplus \mathrm{Mat}_3(\mathbb{K}) \oplus \mathrm{Mat}_7(\mathbb{K})$$

of the factor $M = \mathrm{Mat}_{12}(\mathbb{K})$, both inclusions being described by

$$(x,y,z) \mapsto \begin{bmatrix} x & 0 & 0 \\ 0 & y & 0 \\ 0 & 0 & z \end{bmatrix}.$$

Then $\Lambda_N^M = \Lambda_{\overline{N}}^M = (1\ 1\ 1)$ though N and $\overline{N}$ are not isomorphic.

(ii) Consider $N = \mathbb{K} \oplus \mathrm{Mat}_2(\mathbb{K})$ included in $M = \mathrm{Mat}_4(\mathbb{K})$ by $(x,y) \mapsto \begin{bmatrix} x & 0 & 0 \\ 0 & x & 0 \\ 0 & 0 & y \end{bmatrix}$ and in $\overline{M} = \mathrm{Mat}_5(\mathbb{K})$ by $(x,y) \mapsto \begin{bmatrix} x & 0 & 0 \\ 0 & y & 0 \\ 0 & 0 & y \end{bmatrix}$. Then Λ_N^M and $\Lambda_{\overline{N}}^M$ are pseudo-equivalent to $(2\ \ 1)$ but M and $\overline{M}$ are not isomorphic.

(iii) Consider finally $N = \mathbb{K} \oplus \mathrm{Mat}_2(\mathbb{K})$ included in $M = \mathrm{Mat}_5(\mathbb{K})$ by $(x,y) \mapsto \begin{bmatrix} x & 0 & 0 & 0 \\ 0 & x & 0 & 0 \\ 0 & 0 & x & 0 \\ 0 & 0 & 0 & y \end{bmatrix}$ and by $(x,y) \mapsto \begin{bmatrix} x & 0 & 0 \\ 0 & y & 0 \\ 0 & 0 & y \end{bmatrix}$. Then the first inclusion matrix $(3\ \ 1)$ is not pseudo-equivalent to the second inclusion matrix $(1\ \ 2)$.

The next proposition is a special case of a statement which appears in [BA 8], §5, exercise 17.

Proposition 2.3.5. *Consider two multi-matrix subalgebras* M,N *of a factor* F *with* $1 \in N \subset M \subset F$. *The inclusion matrix for* $C_F(M) \subset C_F(N)$ *is the transpose of the inclusion matrix for* $N \subset M$.

Proof. The proposition is obvious if M and N are factors (see the Remark following 2.2.2). In general, write

$$M = \bigoplus_{i=1}^{m} p_i M \qquad N = \bigoplus_{j=1}^{n} q_j N \qquad \Lambda_N^M = (\lambda_{i,j})$$

and denote by $\tilde{\lambda}_{j,i}$ the entries of the inclusion matrix for

$$C_F(N) = \bigoplus_{j=1}^{n} q_j C_F(N) \supset C_F(M) = \bigoplus_{i=1}^{n} p_i C_F(M).$$

One has by definition

$$\tilde{\lambda}_{j,i} = [q_j p_i C_F(N) q_j p_i : q_j p_i C_F(M) q_j p_i]^{1/2}$$

and by Proposition 2.2.5.b,

$$\tilde{\lambda}_{j,i} = [C_{q_j p_i F q_j p_i}(N_{i,j}) : C_{q_j p_i F q_j p_i}(M_{i,j})]^{1/2}.$$

As $N_{i,j}$ and $M_{i,j}$ are factors in $q_j p_i F q_j p_i$ one has

$$\tilde{\lambda}_{j,i} = [M_{i,j} : N_{i,j}]^{1/2}$$

by the particular case observed in the remark following 2.2.2. #

<u>The Bratteli diagram.</u>

It is useful to describe a pair of multi-matrix algebras $N \subset M$ by its <u>Bratteli diagram</u> $B(N \subset M)$, which is a bicolored weighted multigraph defined as follows. ("Multigraph" means that two points may be joined by more than one line, "weighted" means that each point is given together with a positive integer, and "bicolored" means that points are given one of two colors, in such a way that any edge in the multigraph connects points of different colors.) If $M = \bigoplus_{i=1}^{m} \mathrm{Mat}_{\mu_i}(\mathbb{K})$ and $N = \bigoplus_{j=1}^{n} \mathrm{Mat}_{\nu_j}(\mathbb{K})$ are as above, then $B(N \subset M)$ has m black vertices $b_1, \cdots, b_m$ with respective weights $\mu_1, \cdots, \mu_m$ and n white vertices $w_1, \cdots, w_n$ with respective weights $\nu_1, \cdots, \nu_n$; moreover, the $i^{\underline{th}}$ black vertex and the $j^{\underline{th}}$ white vertex are joined by $\lambda_{i,j}$ lines. (These diagrams were first introduced in order to study inductive limit systems of finite dimensional C^*-algebras; see [Bra] and [Eff].)

<u>Example 2.3.6</u>. If $N = \mathrm{Mat}_\nu(\mathbb{C}) \otimes 1 \subset M = \mathrm{Mat}_\nu(\mathbb{C}) \otimes \mathrm{Mat}_3(\mathbb{C})$, then $B(N \subset M)$ is 

and $\Lambda_N^M = [3]$.

<u>Example 2.3.7</u>. Let G be a finite group and let H be a subgroup of G. As complex group algebras are semi-simple by Maschke's theorem (example II.2), $\mathbb{C}[H] \subset \mathbb{C}[G]$ is a multi-matrix algebra pair.

In particular, let $\mathfrak{S}_3$ be the group of permutations of $\{1,2,3\}$ and let $\mathfrak{S}_2$ be that of $\{1,2\}$. Minimal central idempotents of $\mathbb{C}[\mathfrak{S}_3]$ correspond to Young frames, and also to irreducible representations of $\mathfrak{S}_3$. We denote them by

$p_{[3]}$ corresponding to the identity representation $\pi_{[3]}$ of $\mathfrak{S}_3$

$p_{[1,1,1]}$ corresponding to the signature representation $\pi_{[1,1,1]}$ of $\mathfrak{S}_3$

$p_{[2,1]}$ coresonding to the 2–dimensional irreducible representation $\pi_{[2,1]}$ of $\mathfrak{S}_3$.

Similarly for

$q_{[2]}$ corresponding to the identity representation $\pi_{[2]}$ of $\mathfrak{S}_2$

$q_{[1,1]}$ corresponding to the signature representation $\pi_{[1,1]}$ of $\mathfrak{S}_2$.

It is easy to check that the representations $\pi_{[3]}$, $\pi_{[2,1]}$, $\pi_{[1,1,1]}$ of $\mathfrak{S}_3$ restrict to $\mathfrak{S}_2$ respectively as $\pi_{[2]}$, $\pi_{[2]} \oplus \pi_{[1,1]}$, $\pi_{[1,1]}$. It follows that the inclusion matrix and the Bratteli diagram for $\mathbb{C}[\mathfrak{S}_2] \subset \mathbb{C}[\mathfrak{S}_3]$ are

$$\begin{bmatrix} 1 & 0 \\ 1 & 1 \\ 0 & 1 \end{bmatrix} \quad \text{and}$$

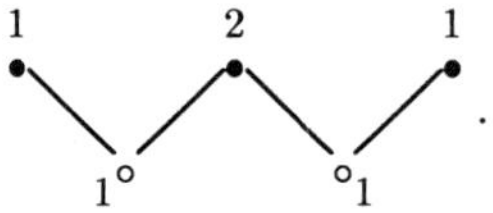

Example 2.3.8. Consider similarly $\mathfrak{S}_3$ as a subgroup of the group $\mathfrak{S}_4$ of permutations of $\{1,2,3,4\}$. The group $\mathfrak{S}_4$ has irreducible representations

$$\pi_{[4]} \quad \pi_{[3,1]} \quad \pi_{[2,2]} \quad \pi_{[2,1,1]} \quad \pi_{[1,1,1,1]}$$

of respective dimensions 1,3,2,3,1, whose restrictions to $\mathfrak{S}_3$ are respectively

$$\pi_{[3]} \quad \pi_{[3]} \oplus \pi_{[2,1]} \quad \pi_{[2,1]} \quad \pi_{[2,1]} \oplus \pi_{[1,1,1]} \quad \pi_{[1,1,1]};$$

see, for example, [Ser1], Example 5.8. It follows that the inclusion matrix and the Bratteli diagram for $\mathbb{C}[\mathfrak{S}_3] \subset \mathbb{C}[\mathfrak{S}_4]$ are as follows. (The reader will check that $\Lambda \vec{\nu} = \vec{\mu}$.)

$$\begin{bmatrix} 1 & 0 & 0 \\ 1 & 1 & 0 \\ 0 & 1 & 0 \\ 0 & 1 & 1 \\ 0 & 0 & 1 \end{bmatrix}$$

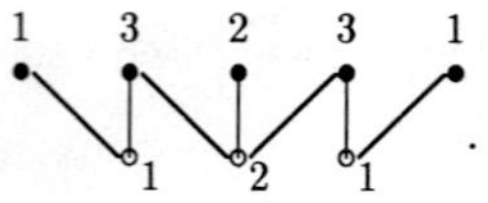

As in the examples, we always draw Bratteli diagrams on two levels, with the upper level representing the larger algebra; the coloring of the vertices is actually superfluous, since the two types of vertices are labelled by their level. The equation $\Lambda\vec{\nu} = \vec{\mu}$ has the following interpretation: For a given black vertex v, consider the set of edges entering v, and for each edge take the weight of the white vertex incident to that edge. The sum of these weights, over all such edges, is the weight of v.

Proposition 2.3.9. (a) *Let* $N \subset M$ *and* $\overline{N} \subset \overline{M}$ *be two multi–matrix algebra pairs with the same Bratteli diagram. Then there exists an isomorphism* $\theta : M \longrightarrow \overline{M}$ *with* $\theta(N) = \overline{N}$.

(b) *A bicolored weighted multigraph* B *(with positive integer weights) is the Bratteli diagram of a multi–matrix algebra pair if and only if the weights and the multiplicities* $\lambda_{i,j}$ *satisfy* $\mu_i = \sum_j \lambda_{i,j}\nu_j$.

Proof. As (a) is nothing but a restatement of Proposition 2.3.3, we are left with the proof of (b).

Let $\mu_1,\cdots,\mu_m$ be the weights of the black points in B and let $\nu_1,\cdots,\nu_n$ be those of the white points and suppose $\mu_i = \sum_j \lambda_{i,j}\nu_j$. Set

$$M = \bigoplus_{j=1}^{m} \mathrm{Mat}_{\mu_i}(\mathbb{K}) \qquad N = \bigoplus_{j=1}^{n} \mathrm{Mat}_{\nu_j}(\mathbb{K}).$$

Let $\lambda_{i,j}$ be the number of lines joining the $i^{\underline{th}}$ black point with the $j^{\underline{th}}$ white point in B. Define a map $N \longrightarrow M$ by associating to $(y_1,\cdots,y_n) \in N$ the element $(x_1,\cdots,x_m) \in M$ with x_i the block–diagonal matrix

$$x_i = \mathrm{diag}(y_1,\cdots,y_1;\cdots;y_n,\cdots,y_n)$$

where y_j is repeated $\lambda_{i,j}$ times. This map identifies N with a subalgebra of M and $B(N \subset M)$ is the B originally given. #

Chains of multi–matrix algebras. Now consider an increasing chain (finite or infinite) of multi–matrix algebras over $\mathbb{K}$,

$$\mathbf{1} \in M_0 \subset M_1 \subset M_2 \subset \cdots.$$

Let $p_1^k, \cdots, p_{m(k)}^k$ denote the minimal central idempotents in M_k, let $\Lambda^{(k)} = (\lambda_{i,j}^k)$ be the inclusion matrix for $M_k \subset M_{k+1}$, and let μ^k be the vector of dimensions of M_k, so that $p_i^k M_k \cong \mathrm{Mat}_{\mu_i^k}(\mathbb{K})$. (Thus $\mu^k = \Lambda^{(k-1)}\Lambda^{(k-2)} \cdots \Lambda^{(0)}\mu^0$.) We associate with the chain of algebras a (finite or infinite) Bratteli diagram B, which is the union of the diagrams $B(M_k \subset M_{k+1})$, the upper (black) vertex of $B(M_k \subset M_{k+1})$ corresponding to p_i^{k+1} being identified with the lower (white) vertex of $B(M_{k+1} \subset M_{k+2})$ corresponding to the same idempotent. For example the diagram for $\mathbb{C}\mathfrak{S}_1 \subset \mathbb{C}\mathfrak{S}_2 \subset \mathbb{C}\mathfrak{S}_3 \subset \mathbb{C}\mathfrak{S}_4$ is

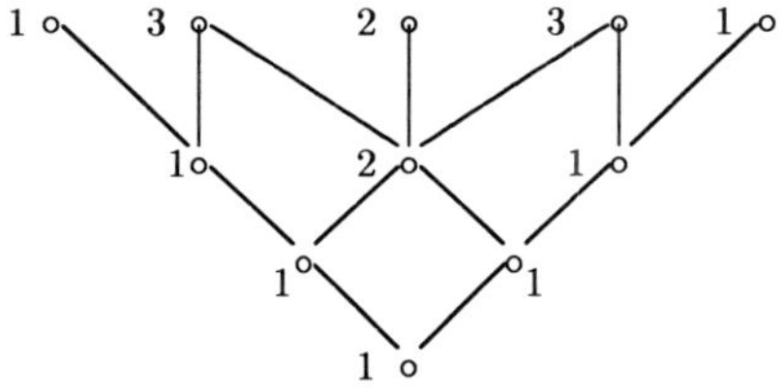

(See examples 2.3.7 and 2.3.8.) We say that the vertices v_i^k corresponding to the minimal central idempotents p_i^k in M_k belong to the kth floor of the diagram. The vertices of the kth and $k+1^{st}$ floors together with the edges joining them — i.e., the image of $B(M_k \subset M_{k+1})$ in B — constitute the kth story of B. The Bratelli diagram B is thus a weighted multigraph with the following features:

(1) There is a function φ from the set of vertices of B to $\mathbb{N} = \{0,1,2,\cdots\}$, which assigns to each vertex the floor which it occupies.

(a) There are only finitely many vertices on each floor; that is $\varphi^{-1}(k)$ is finite for all k. If $\varphi^{-1}(k) \neq \emptyset$, we write $\varphi^{-1}(k) = \{v_1^k, \cdots, v_{m(k)}^k\}$.

(b) The range of φ is either an interval $[0,\rho]$ in $\mathbb{N}$, if B is finite, or all of $\mathbb{N}$, if B is infinite.

(2) Two vertices v and w are adjacent only if $|\varphi(v)-\varphi(w)| = 1$. There are $\lambda_{i,j}^k$ edges joining v_j^k and v_i^{k+1}.

(3) If both the kth and $k+1^{st}$ floors are occupied (i.e., if $\varphi^{-1}(k) \neq \emptyset$ and $\varphi^{-1}(k+1) \neq \emptyset$) then each vertex on the kth floor is adjacent to at least one vertex on the $k+1^{st}$ floor, and each vertex on the $k+1^{st}$ floor is adjacent to at least one vertex

on the kth floor. That is, the m(k)–by–m(k+1) matrix $\Lambda^{(k)} = (\lambda^k_{i,j})$ is irredundant.

(4) Each vector v^k_i has a weight $\mu^k_i \in \{1,2,\cdots\}$ called its dimension. The dimensions $\{\mu^k_i\}$ and the "multiplicities" $\{\lambda^k_{i,j}\}$ satisfy

$$\sum_{j=1}^{m(k)} \lambda^k_{i,j}\mu^k_j = \mu^{k+1}_i.$$

Conversely, given a weighted multigraph B with properties (1)–(4) above, we can, by iterating the procedure of Proposition 2.3.9.b, construct a chain of multi–matrix algebras with Bratteli diagram B.

Proposition 2.3.10. *Suppose*

$$\mathbf{1} \in M_0 \subset M_1 \subset \cdots, \text{ and}$$
$$\mathbf{1} \in A_0 \subset A_1 \subset \cdots$$

are two chains of multi–matrix algebras with the same Bratteli diagram. Then there is an isomorphism ψ *of* $M_\infty = \bigcup_k M_k$ *onto* $A_\infty = \bigcup_k A_k$ *such that* $\psi(M_k) = A_k$ *for all* k.

Proof. We have to produce a sequence of isomorphisms $\psi_k : M_k \to A_k$ such that $\psi_{k+1}\big|_{M_k} = \psi_k$. Let $\psi_0 : M_0 \to A_0$ be any isomorphism. Suppose $\psi_0,\cdots,\psi_k$ have been defined. Then by Proposition 2.3.9.a, there is an isomorphism $\alpha_{k+1} : M_{k+1} \to A_{k+1}$ such that $\alpha_{k+1}(M_k) = A_k$, and by Proposition 2.3.3 there is an inner automorphism β_{k+1} of A_{k+1} extending $\psi_k \circ \alpha_{k+1}^{-1}\big|_{A_k}$. Thus we can set $\psi_{k+1} = \beta_{k+1} \circ \alpha_{k+1}$. #

2.3.11. A path model. Let B be a Bratteli diagram; we use paths on the diagram to construct a natural model for the chain of multi–matrix algebras associated to the diagram. We will suppose that B is infinite; it will be obvious how the construction must be modified for a finite diagram. First we produce an augmented diagram $\tilde{B}$ by adding a $(-1)^{st}$ story corresponding to the inclusion $\mathbf{K1} \subset M_0$; that is we append one vertex * with $\varphi(*) = -1$ and $\dim(*) = 1$, and we connect * to v^0_j by μ^0_j edges $(1 \le j \le m(0))$.

An oriented edge on any graph is an edge together with an ordering of its two vertices; we will call the first vertex of an oriented edge its start and the second its end. A path is a (possibly infinite) sequence (ξ_i) of oriented edges such that $\text{end}(\xi_i) = \text{start}(\xi_{i+1})$ for all i. A path $(\cdots,\xi_k)$ has end equal to $\mathrm{end}(\xi_k)$; a path $(\xi_0,\cdots)$ has start equal to $\mathrm{start}(\xi_0)$. If ξ and η are paths such that $\mathrm{end}(\xi) = \mathrm{start}(\eta)$ we define $\xi\circ\eta$ to be the path "first ξ, then η". A path ξ on $\tilde{B}$ is <u>monotone increasing</u> if $\varphi(\mathrm{end}(\xi_k)) = \varphi(\mathrm{start}(\xi_k)) + 1$ for all k.

We let Ω denote the set of infinite monotone increasing paths on $\tilde{B}$ starting at *; $\Omega_{[r}$ the set of infinite monotone increasing paths starting on the rth floor of $\tilde{B}$; $\Omega_{r]}$ the set of monotone increasing paths starting at * and ending on the rth floor; and $\Omega_{[r,s]}$ the set of monotone increasing paths starting on the rth floor and ending on the sth floor $(r < s)$. Given $\xi = (\xi_0,\xi_1,\cdots) \in \Omega$, set:

$$\xi_{r]} = (\xi_0,\cdots,\xi_r) \in \Omega_{r]} \qquad (0 \le r),$$

$$\xi_{[r,s]} = (\xi_{r+1},\cdots,\xi_s) \in \Omega_{[r,s]} \qquad (-1 \le r < s),$$

and

$$\xi_{[r} = (\xi_{r+1},\cdots) \in \Omega_{[r} \qquad (-1 \le r).$$

Also let $\xi_{[r]}$ be the vertex $\mathrm{end}(\xi_r) = \mathrm{start}(\xi_{r+1})$. Similarly if $\xi = (\xi_0,\cdots\xi_s) \in \Omega_{s]}$ and $r \le s$ we can define $\xi_{r]} = (\xi_0,\cdots,\xi_r) \in \Omega_{r]}$, and so forth.

Let $\mathbb{K}\Omega$ be the $\mathbb{K}$–vector space with basis Ω. For each $r \in \{0,1,2,\cdots\}$ we define an algebra $A_r \in \mathrm{End}_{\mathbb{K}}(\mathbb{K}\Omega)$ as follows. Let $R_r = \{(\xi,\eta) \in \Omega_{r]} \times \Omega_{r]} : \mathrm{end}(\xi) = \mathrm{end}(\eta)\}$. For $(\xi,\eta) \in R_r$ define $T_{\xi,\eta} \in \mathrm{End}_{\mathbb{K}}(\mathbb{K}\Omega)$ by

$$T_{\xi,\eta}\omega = \delta(\eta_{r]},\omega_{r]})\xi_{r]}\circ\omega_{[r} \quad (\omega \in \Omega).$$

Let A_r be the $\mathbb{K}$–linear span of $\{T_{\xi,\eta} : (\xi,\eta) \in R_r\}$ in $\mathrm{End}_{\mathbb{K}}(\mathbb{K}\Omega)$; since

$$\text{(2.3.11.1)} \qquad T_{\xi,\eta}T_{\xi',\eta'} = \delta(\eta,\xi')T_{\xi,\eta'}, \text{ and } \mathbf{1} = \sum_{\xi\in\Omega_{r]}} T_{\xi,\xi},$$

A_r is an algebra. Set

$$\Omega^i_{r]} = \{\xi \in \Omega_{r]} : \mathrm{end}(\xi) = v^r_i\} \quad (1 \le i \le m(r)),$$

so that $\Omega_{r]} = \biguplus_i \Omega^i_{r]}$ (disjoint union), and $R_r = \biguplus_i (\Omega^i_{r]} \times \Omega^i_{r]})$. It follows from the multiplication law (2.3.11.1) for the $T_{\xi,\eta}$ that

$$A^i_r = \mathrm{span}\{T_{\xi,\eta} : (\xi,\eta) \in \Omega^i_{r]}\times\Omega^i_{r]}\}$$

is an ideal of A_r and $A_r = \bigoplus_{i=1}^{m(r)} A_r^i$. There is an isomorphism of A_r^i onto $\mathrm{End}_{\mathbb{K}}(\mathbb{K}\Omega^i_{r]})$ defined by

$$T_{\xi,\eta}\omega = \delta(\eta,\omega)\xi \qquad (\xi,\eta,\omega \in \Omega^i_{r]}),$$

so that

$$A_r = \bigoplus_{i=1}^{m(r)} A_r^i \cong \bigoplus_{i=1}^{m(r)} \mathrm{End}_{\mathbb{K}}(\mathbb{K}\Omega^i_{r]}).$$

Note that the minimal central projections p_i^r in A_r have the form

$$p_i^r = \sum\{T_{\xi,\xi} : \xi \in \Omega^i_{r]}\}.$$

The cardinalities $\#(\Omega^i_{r]})$ satisfy

$$\#(\Omega^i_{0]}) = \mu_i^0, \text{ and } \#(\Omega^i_{r+1]}) = \sum_{j=1}^{m(r)} \lambda^r_{i,j}\, \#(\Omega^j_{r]}),$$

since each $\xi \in \Omega^j_{r]}$ can be extended in $\lambda^r_{i,j}$ ways, by adjunction of an edge λ in $\Omega_{[r,r+1]}$, to a path $\xi\circ\lambda$ in $\Omega^i_{r+1]}$. It follows from this and property (4) of the Bratteli diagram that $\#(\Omega^i_{r]}) = \mu_i^r$ for all r and i $(0 \le r, 1 \le i \le m(r))$. Thus

$$A_r \cong \bigoplus_{i=1}^{m(r)} \mathrm{Mat}_{\mu_i^r}(\mathbb{K}).$$

Finally $A_r \subset A_{r+1}$, because for $(\xi,\eta) \in R_r$,

$$T_{\xi,\eta} = \sum\{T_{\xi\circ\lambda,\eta\circ\lambda} : \lambda \in \Omega_{[r,r+1]}, \mathrm{end}(\xi) = \mathrm{start}(\lambda)\},$$

as operators on $\mathbb{K}\Omega$. If $(\xi,\eta) \in \Omega^j_{r]} \times \Omega^j_{r]}$, so $T_{\xi,\eta} \in A_r^j$, then

$$T_{\xi,\eta}p_i^{r+1} = \sum\{T_{\xi\circ\lambda,\eta\circ\lambda} : \lambda \in \Omega_{[r,r+1]}, \lambda_{[\tilde{r}]} = v_j^r, \lambda_{[r+1]} = v_i^{r+1}\}.$$

It follows that $\Lambda_{A_r}^{A_{r+1}} = (\lambda_{i,j}^r)$, and the Bratteli diagram for the chain $\mathbb{1} \in A_0 \subset A_1 \subset \cdots$ is B.

As an example of the utility of the path model, let us identify $C_{A_s}(A_r)$ for $r < s$. Let

$$R_{r,s} = \{(\xi,\eta) \in \Omega_{[r,s]} \times \Omega_{[r,s]} : \xi_{[r]} = \eta_{[r]} \text{ and } \xi_{[s]} = \eta_{[s]}\}.$$

For $(\xi,\eta) \in R_{[r,s]}$ define $T_{\xi,\eta} \in \mathrm{End}_{\mathbb{K}}(\mathbb{K}\Omega)$ by

$$T_{\xi,\eta}\omega = \delta(\eta_{[r,s]},\omega_{[r,s]})\omega_{r]} \circ \xi_{[r,s]} \circ \omega_{[s},$$

and let $A_{r,s} = \mathrm{span}_{\mathbb{K}}\{T_{\xi,\eta} : (\xi,\eta) \in R_{r,s}\}$. Then $A_{r,s}$ is an algebra, since again

$$T_{\xi,\eta}T_{\xi',\eta'} = \delta(\eta,\xi')T_{\xi,\eta'} \quad ((\xi,\eta),(\xi',\eta') \in R_{r,s}), \text{ and } \mathbb{1} = \sum\{T_{\xi,\xi} : \xi \in \Omega_{[r,s]}\}.$$

We have $A_{r,s} \subset A_s$, because if $(\xi,\eta) \in R_{r,s}$, then

$$T_{\xi,\eta} = \sum\{T_{\lambda\circ\xi,\lambda\circ\eta} : \lambda \in \Omega_{r]}, \lambda_{[r]} = \xi_{[r]} = \eta_{[r]}\},$$

as operators on $\mathbb{K}\Omega$. Clearly $A_{r,s} \subset C_{A_s}(A_r)$.

<u>Proposition 2.3.12.</u> $A_{r,s} = C_{A_s}(A_r)$.

<u>Proof.</u> For $x \in A_s$ define $P(x) =$

$$\sum\{T_{\lambda,\lambda'}xT_{\lambda',\lambda} : (\lambda,\lambda') \in R_r\}.$$

One verifies that P is a linear projection of A_s onto $C_{A_s}(A_r)$. But for $(\xi,\eta) \in R_s$,

$$\begin{aligned} P(T_{\xi,\eta}) &= \delta(\xi_{r]},\eta_{r]}) \sum\{T_{\lambda\circ\xi_{[r},\lambda\circ\eta_{[r}} : \lambda \in \Omega_{r]}, \mathrm{end}(\lambda) = \xi_{[r]}\} \\ &= \delta(\xi_{r]},\eta_{r]})T_{\xi_{[r},\eta_{[r}} \in A_{r,s}. \end{aligned}$$

Thus $C_{A_s}(A_r) \subset A_{r,s}$. #

It is an easy exercise to check that the factors of $A_{r,s}$ are in bijection with pairs of vertices (v,w), with v in floor r and w in floor s. The factor corresponding to a pair (v,w) is the algebra of endomorphisms of the free vector space over the set of paths from v to w.

Remarks. (1) The path model presented here is due to V.S. Sunder [Sun] and A. Ocneanu [Ocn]. Compare however [SV], in which a maximal abelian subalgebra of $A_\infty = \bigcup_k A_k$ is identified with $\mathbb{K}\Omega$.

(2) In case $\mathbb{K} = \mathbb{C}$, the action of the "path algebras" A_r on $\mathbb{C}\Omega$ extends to a representation on the Hilbert space $\ell^2(\Omega)$ with orthonormal basis Ω. It is evident that $T_{\xi,\eta}$ is then a rank-one partial isometry with adjoint $T^*_{\xi,\eta} = T_{\eta,\xi}$. So A_r is a C^*-subalgebra of $\mathcal{B}(\ell^2(\Omega))$.

2.4. The fundamental construction and towers for multi-matrix algebras.

We consider a pair of multi-matrix algebras $\mathbb{1} \in N \subset M$, and the associated tower of algebras

$$\mathbb{1} \in M_0 = N \subset M_1 = M \subset \cdots \subset M_k \subset M_{k+1} \subset \cdots$$

obtained by iterating the fundamental construction, as described in the chapter introduction. It turns out that all the M_k are then multi-matrix algebras:

Proposition 2.4.1. *Let* $N \subset M$ *be a pair of multi-matrix algebras and let* $M \subset \mathrm{End}^r_N(M)$ *be the pair obtained by the fundamental construction. Then*

(a) $\mathrm{End}^r_N(M)$ *is a multi-matrix algebra and its minimal central idempotents are of the form* $\rho(q)$, *where* q *is a minimal central idempotent in* N, *and* $\rho(q)$ *is right multiplication by* q.

(b) *The inclusion matrix for* $M \subset \mathrm{End}^r_N(M)$ *is the transpose of* Λ^M_N.

Proof. Set $F = \mathrm{End}_{\mathbb{K}}(M)$ and define maps $\lambda,\rho : M \longrightarrow F$ by $\lambda(x)(y) = xy$ and $\rho(x)(y) = yx$ for $x,y \in M$. The homomorphism λ is the composition of the inclusions $M \subset \mathrm{End}^r_N(M)$ and $\mathrm{End}^r_N(M) \subset F$; the map ρ is an algebra isomorphism from M^{opp} into F. As the pair $N \subset M$ is isomorphic to the pair $N^{opp} \subset M^{opp}$ by Corollary 2.3.4, it is also isomorphic to $\rho(N) \subset \rho(M)$. But $\mathrm{End}^r_N(M) = C_F(\rho(N))$ and $M = \lambda(M) = C_F(\rho(M))$. Consequently (a) follows from 2.2.3.a and (b) from 2.3.5. #

It follows at once by induction that all the algebras M_k in the tower generated by $N \subset M$ are also multi–matrix, and furthermore the inclusion matrix $\Lambda_{M_k}^{M_{k+1}}$ is Λ_N^M for k even and $(\Lambda_N^M)^t$ for k odd. Thus in the infinite Bratteli diagram for the tower, the $(k+1)^{st}$ story (for $M_{k+1} \subset M_{k+2}$) is the reflection of the kth story (for $M_k \subset M_{k+1}$).

We illustrate this with $N = \mathbb{C}[\mathfrak{S}_3]$ and $M = \mathbb{C}[\mathfrak{S}_4]$, as in Example 2.3.8. We have

$$\Lambda = \Lambda_N^M = \begin{bmatrix} 1 & 0 & 0 \\ 1 & 1 & 0 \\ 0 & 1 & 0 \\ 0 & 1 & 1 \\ 0 & 0 & 1 \end{bmatrix} \quad \text{and} \quad \vec{\nu} = \begin{bmatrix} 1 \\ 2 \\ 1 \end{bmatrix}.$$

These data determine the Bratteli diagram of the tower:

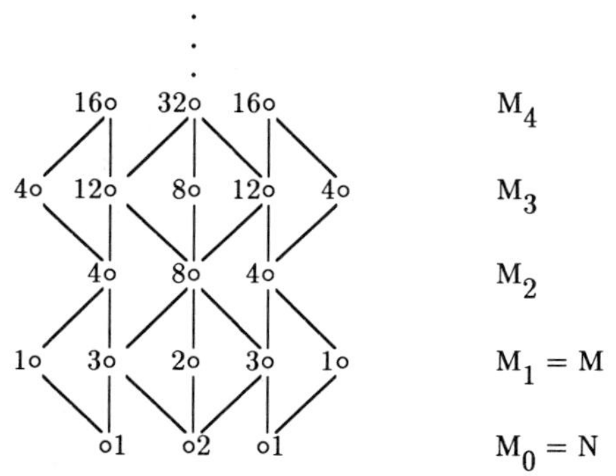

In particular $M_2 = \mathrm{End}_N^r(M) \cong \mathrm{Mat}_4(\mathbb{C}) \oplus \mathrm{Mat}_8(\mathbb{C}) \oplus \mathrm{Mat}_4(\mathbb{C})$.

An accidental feature of this example is that $\vec{\nu}$ is an eigenvector of $\Lambda^t\Lambda$ with eigenvalue 4. Consequently the vector of dimensions on the $k+2^{nd}$ floor is always 4 times the vector on the kth floor. To get a better idea of the general situation let us also consider the example with the same inclusion matrix Λ but with $N = \mathbb{C} \oplus \mathbb{C} \oplus \mathbb{C}$, i.e. $\vec{\nu} = (1,1,1)^t$. Then the Bratelli diagram is

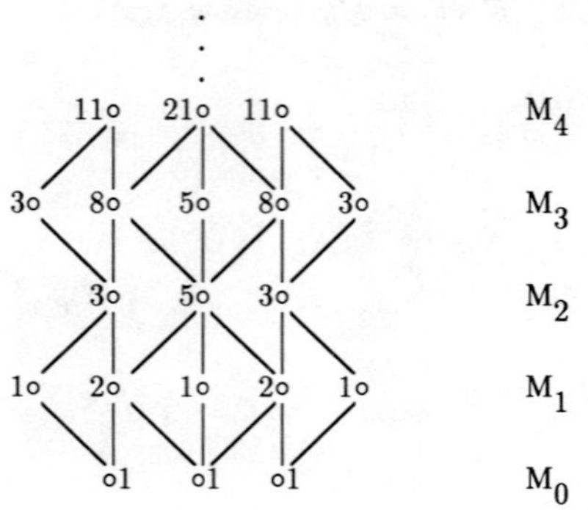

The reader might wish to continue the diagram for a few more levels and observe the *convergence of the ratio of dimensions* on the even floors to [1:2:1] and on the odd floors to [1:3:2:3:1], eigenvectors for $\Lambda^t\Lambda$ and $\Lambda\Lambda^t$ respectively. This observation is the key to the next proposition.

Proposition 2.4.2. *Let* $N \subset M$ *be a pair of multi–matrix algebras with inclusion matrix* Λ, *and let* $(M_k)_{k\geq 0}$ *be the associated tower. Then*

$$\lim_{k\to\infty} \{\dim_{\mathbb{K}} M_k\}^{1/k} = \|\Lambda\|^2.$$

Proof. It suffices to prove this in the case that $Z_M \cap Z_N = \mathbb{K}$. Then since $\Lambda^t\Lambda$ and $\Lambda\Lambda^t$ are irreducible and aperiodic (2.3.1.f and 1.3.2) and also positive semi–definite, it follows from Perron–Frobenius theory that

$$\lim_{k\to\infty} \|(\Lambda^t\Lambda)^k\xi\|^{1/k} = \lim_{k\to\infty} \|(\Lambda\Lambda^t)^k\xi\|^{1/k} = \|\Lambda\|^2,$$

for any non–zero $\xi \in \mathbb{R}^n$ with non–negative coordinates. (Set A equal to $\Lambda^t\Lambda$ or $\Lambda\Lambda^t$. Then $A = \|A\|E_0 + \sum \mu_i E_i$, where E_0 is the rank one orthogonal projection onto the span of the Perron–Frobenius eigenvector, E_i are the remaining spectral projections of A, and μ_i satisfy $0 \leq \mu_i < \|A\|$. Hence

$$\frac{A^k\xi}{\|A\|^k} = E_0\xi + \sum \frac{\mu_i^k}{\|A\|^k} E_i\xi,$$

which converges to $E_0\xi$. If z is the unique positive normalized ($\|z\| = 1$) Perron–Frobenius eigenvector, then $E_0\xi = (\sum x_i z_i)z$, which is non–zero because $\xi_i \geq 0$

and $z_i > 0$ for all i, and $\xi_j > 0$ for some j. It follows that

$$\lim \frac{\|\Lambda^k \xi\|}{\|\Lambda\|}^{1/k} = \lim \|E_0\xi\|^{1/k} = 1.)$$

The inclusion matrix for $M_0 \subset M_{2k}$ is $(\Lambda^t\Lambda)^k$, and that for $M_0 \subset M_{2k+1}$ is $(\Lambda\Lambda^t)^k\Lambda$. Thus $\dim_{\mathbb{K}} M_{2k} = \|(\Lambda^t\Lambda)^k\vec{\nu}\|^2$ and $\dim_{\mathbb{K}} M_{2k+1} = \|(\Lambda\Lambda^t)^k\Lambda\vec{\nu}\|^2$. Therefore

$$\lim_k (\dim_{\mathbb{K}} M_{2k})^{1/2k} = \lim_k (\dim_{\mathbb{K}} M_{2k+1})^{1/2k+1} = \|\Lambda\|^2. \ \#$$

Lemma 2.4.3. *Let* $N \subset M$ *be a pair of finite dimensional algebras over a field* $\mathbb{K}$, *and let* $(M_k)_{k \geq 0}$ *be the associated tower. Then*

$$[M:N] = \limsup_{k\to\infty} \{\dim_{\mathbb{K}} M_k\}^{1/k}.$$

Proof. As M_0 and M_k are finite dimensional $\mathbb{K}$-algebras, one has

$$\frac{\dim_{\mathbb{K}}(M_k)}{\dim_{\mathbb{K}}(M_0)} \leq \mathrm{rk}(M_k | M_0) \leq \dim_{\mathbb{K}}(M_k),$$

and therefore

$$[M : N] = \limsup_{k\to\infty} \{\mathrm{rk}(M_k | M_0)\}^{1/k} = \limsup_{k\to\infty} \{\dim_{\mathbb{K}} M_k\}^{1/k}. \qquad \#$$

Proposition 2.4.4. Let $\mathbb{1} \in N \subset M$ be a pair of finite dimensional algebras over a field $\mathbb{K}$, let $\mathbb{E}$ be any extension field of $\mathbb{K}$ and set

$$M^{\mathbb{E}} = M \otimes_{\mathbb{K}} \mathbb{E} \ \text{ and } \ N^{\mathbb{E}} = N \otimes_{\mathbb{K}} \mathbb{E}.$$

Then

(a) $\mathrm{End}_N^r(M) \otimes_{\mathbb{K}} \mathbb{E} \cong \mathrm{End}^r_{N^{\mathbb{E}}}(M^{\mathbb{E}})$.

(b) $[M : N] = [M^{\mathbb{E}} : N^{\mathbb{E}}]$.

Proof. (a) This is an example of a theorem on "change of rings in Hom"; see for example [R], p.24. We give a simple proof appropriate to the special case at hand.

Define

$$\alpha : \mathrm{End}_{\mathbb{K}}(M) \otimes_{\mathbb{K}} \mathbb{E} \to \mathrm{End}_{\mathbb{E}}(M^{\mathbb{E}})$$

by

$$\alpha(\varphi \otimes a)(x \otimes b) = \varphi(x) \otimes ba \quad (\varphi \in \mathrm{End}_{\mathbb{K}}(M),\ a,b \in \mathbb{E},\ x \in M).$$

Define also

$$\beta : \mathrm{End}_{\mathbb{E}}(M^{\mathbb{E}}) \to \mathrm{End}_{\mathbb{K}}(M) \otimes_{\mathbb{K}} \mathbb{E}$$

as follows. Let $\{a_i\}$ be a basis of $\mathbb{E}$ over $\mathbb{K}$. For each $\Phi \in \mathrm{End}_{\mathbb{E}}(M^{\mathbb{E}})$ and each i, there is a unique $\varphi_i \in \mathrm{End}_{\mathbb{K}}(M)$ such that

$$\Phi(x \otimes \mathbb{1}) = \sum \varphi_i(x) \otimes a_i \qquad (x \in M).$$

Only finitely many $\varphi_i(x)$ are non–zero for any particular x, and since M is finite dimensional over $\mathbb{K}$, only finitely many φ_i are non–zero altogether. Then β can be defined by

$$\beta(\Phi) = \sum \varphi_i \otimes a_i.$$

It is easy to check that α and β are isomorphisms of $\mathbb{E}$–algebras which are inverse to each other.

Next observe that

(2.4.4.1)
$$\alpha(\lambda(m) \otimes a) = \lambda(m \otimes a), \text{ and}$$
$$\alpha(\rho(n) \otimes a) = \rho(n \otimes a) \quad (m \in M,\ n \in N,\ a \in \mathbb{E}).$$

It follows from this that

$$\alpha(\mathrm{End}_N^r(M) \otimes \mathbb{E}) = \mathrm{End}^r_{N^{\mathbb{E}}}(M^{\mathbb{E}}).$$

(b) Let $(M_k)_{k \geq 0}$ be the tower of extensions generated by $N \subset M$ and let $(A_k)_{k \geq 0}$ be the tower generated by $N^{\mathbb{E}} \subset M^{\mathbb{E}}$. We produce a sequence of isomorphisms $\alpha_k : M_k \otimes_{\mathbb{K}} \mathbb{E} \to A_k$ such that $\alpha_{k+1}\big|_{M_k \otimes_{\mathbb{K}} \mathbb{E}} = \alpha_k$ for all k. Take α_0, α_1 to be the identity and α_2 to be the isomorphism defined in part (a); we have $\alpha_2\big|_M = \alpha_1$ by (2.4.4.1). Suppose $\alpha_1, \cdots, \alpha_k$ have been defined. Let

$$\delta_{k+1} : M_{k+1} \otimes \mathbb{E} = \mathrm{End}^{r}_{M_{k-1}}(M_k) \otimes \mathbb{E} \to \mathrm{End}^{r}_{M^{\mathbb{E}}_{k-1}}(M^{\mathbb{E}}_k)$$

be the isomorphism defined as in part (a), and let

$$\gamma_{k+1} : \mathrm{End}^{r}_{M^{\mathbb{E}}_{k-1}}(M^{\mathbb{E}}_k) \to A_{k+1} = \mathrm{End}^{r}_{A_{k-1}}(A_k)$$

be induced by the pair of isomorphisms

$$\begin{array}{ccc} M^{\mathbb{E}}_k & \xrightarrow{\alpha_k} & A_k \\ \cup & & \cup \\ M^{\mathbb{E}}_{k-1} & \xrightarrow{\alpha_{k-1}} & A_{k-1} \end{array}$$

Set $\alpha_{k+1} = \gamma_{k+1} \circ \delta_{k+1}$; this extends α_k because δ_{k+1} extends the identity on $M^{\mathbb{E}}_k$, and γ_{k+1} extends α_k.

Consequently, we have $\dim_{\mathbb{K}}(M_k) = \dim_{\mathbb{E}}(M^{\mathbb{E}}_k) = \dim_{\mathbb{E}}(A_k)$ for all k, and the equality $[M : N] = [M^{\mathbb{E}} : N^{\mathbb{E}}]$ follows from this and Lemma 2.4.3. #

Proof of Theorem 2.1.1 and Corollary 2.1.2.. Because of 2.4.4 and the definition of Λ^M_N for arbitrary semi-simple algebras (given in the chapter introduction), it suffices to consider the case that $\mathbb{K}$ is algebraically closed, so M and N are multi-matrix algebras. But then

$$[M : N] = \lim_k \{\dim_{\mathbb{K}} M_k\}^{1/k} = \|\Lambda^M_N\|^2,$$

by 2.4.2 and 2.4.3. The corollary follows from Kronecker's Theorem 1.1.1. #

Remark. The norm of a product of two matrices is not, in general, the product of their norms. It follows that, given a nested sequence $\mathbb{1} \in L \subset P \subset M$ of semi-simple algebras, the inequality

$$[M : L] \le [M : P][P : L]$$

is in general strict. However, even this inequality fails to hold for algebras with radicals, as we now show.

Example 2.4.5. Consider two integers $m', m'' \geq 1$ and set $m = m' + m''$. Let M be the factor $\mathrm{Mat}_m(\mathbb{C})$, let P be its "parabolic" subalgebra

$$P = \left\{ \begin{bmatrix} A & B \\ 0 & C \end{bmatrix} \in M : A \in \mathrm{Mat}_{m'}(\mathbb{C}),\ B \in \mathrm{Mat}_{m',m''}(\mathbb{C}),\ C \in \mathrm{Mat}_{m''}(\mathbb{C}) \right\}$$

and let L be the "Levi" subalgebra $\left\{ \begin{bmatrix} A & 0 \\ 0 & C \end{bmatrix} \right\}$ of P. Then L and M are semi-simple, and $[M : L] = 2$ as above, but P is of course not semi-simple.

We claim that $[M : P] = 1$. Indeed, from left multiplication $\begin{bmatrix} X & Y \\ Z & T \end{bmatrix} \mapsto \begin{bmatrix} A & B \\ 0 & C \end{bmatrix} \begin{bmatrix} X & Y \\ Z & T \end{bmatrix}$, one has the inclusion

$$\begin{cases} P & \to \mathrm{End}_{\mathbb{C}}(M) \approx M \otimes M^{opp} \\ \begin{bmatrix} A & B \\ 0 & C \end{bmatrix} & \mapsto \begin{bmatrix} \lambda_A & 0 & \lambda_B & 0 \\ 0 & \lambda_A & 0 & \lambda_B \\ 0 & 0 & \lambda_C & 0 \\ 0 & 0 & 0 & \lambda_C \end{bmatrix} \end{cases}$$

where λ_A is left-multiplication by A (and ρ_A below is right multiplication). As the commutant of P in M is reduced to the center $\mathbb{C}$ of M, the commutant of $\lambda(P)$ in $\mathrm{End}_{\mathbb{C}}(M)$ is isomorphic to M; moreover the natural morphism from M to $C_{\mathrm{End}(M)}(\lambda(P))$ is an isomorphism. Consequently the tower generated by $P \subset M$ is $P \subset M \subset M \subset \cdots$ and the index is 1.

We also claim that $[P : L] = 1$. From left multiplication $\begin{bmatrix} X & Y \\ 0 & Z \end{bmatrix} \mapsto \begin{bmatrix} A & 0 \\ 0 & B \end{bmatrix} \begin{bmatrix} X & Y \\ 0 & Z \end{bmatrix}$, one has the inclusion

$$\begin{cases} L & \to \quad \mathrm{End}\,(P) \\ (A,B) & \mapsto \begin{bmatrix} \lambda_A & 0 & 0 \\ 0 & \lambda_A & 0 \\ 0 & 0 & \lambda_B \end{bmatrix}. \end{cases}$$

Thus $C_{\mathrm{End}(P)}(L)$ is the subalgebra

$$N = \left\{ \begin{bmatrix} \rho_R & \rho_S & 0 \\ \rho_T & \rho_U & 0 \\ 0 & 0 & \rho_V \end{bmatrix} : \begin{bmatrix} R & S \\ T & U \end{bmatrix} \in \mathrm{Mat}_{2m'}(\mathbb{C}) \text{ and } V \in \mathrm{Mat}_{m''}(\mathbb{C}) \right\}$$

of $\mathrm{End}_{\mathbb{C}}(P)$, isomorphic to $(\mathrm{Mat}_{m'}(\mathbb{C}) \otimes \mathrm{Mat}_2(\mathbb{C})) \oplus \mathrm{Mat}_{m''}(\mathbb{C})$. As right multiplication $\begin{bmatrix} X & Y \\ 0 & Z \end{bmatrix} \mapsto \begin{bmatrix} X & Y \\ 0 & Z \end{bmatrix} \begin{bmatrix} A & B \\ 0 & C \end{bmatrix}$ is represented in $\mathrm{End}_{\mathbb{C}}(P)$ by the matrix

$$\begin{bmatrix} \rho_A & 0 & 0 \\ \rho_B & \rho_C & 0 \\ 0 & 0 & \rho_C \end{bmatrix}$$

the canonical morphism $P \longrightarrow N$ is given by

$$\begin{cases} P & \to N = (\mathrm{Mat}_{m'}(\mathbb{C}) \otimes \mathrm{Mat}_2(\mathbb{C})) \oplus \mathrm{Mat}_{m''}(\mathbb{C}) \\ \begin{bmatrix} A & B \\ 0 & C \end{bmatrix} & \mapsto \left[\begin{bmatrix} A & 0 \\ B & C \end{bmatrix}, \ C \right] \end{cases}$$

The argument used to show $[M : P] = 1$ shows also that the canonical construction applied to $P \subset N$ gives an algebra isomorphic to N. Finally, the tower generated by $L \subset P$ is $L \subset P \subset N \subset N \subset \cdots$ and the index is also 1.

2.4.6. A reprise of Proposition 2.4.1. Let $\mathbf{1} \in N \subset M$ be a pair of multi-matrix algebras with inclusion matrix Λ. Write $\{q_j : 1 \le j \le n\}$ and $\{p_i : 1 \le i \le m\}$ for the minimal central idempotents of N and M respectively. Let B be the two-story Bratteli diagram whose 0th story is $B(N \subset M)$ and whose 1st story is the reflection of $B(N \subset M)$; that is $\Lambda^{(0)} = \Lambda$ and $\Lambda^{(1)} = \Lambda^t$. Let $\tilde{B}$ be the augmented diagram, as in 2.3.11. For example for $\mathbb{C}\mathfrak{S}_3 \subset \mathbb{C}\mathfrak{S}_4$, $\tilde{B}$ is

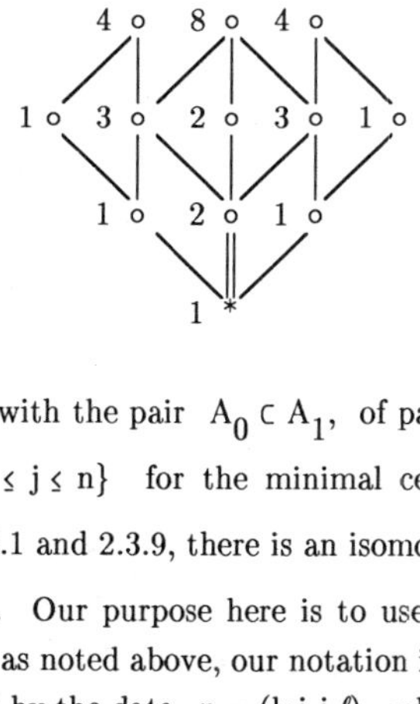

We identify the pair $N \subset M$ with the pair $A_0 \subset A_1$, of path algebras associated with $\tilde{B}$. (See 2.3.11.) Write $\{\tilde{q}_j : 1 \le j \le n\}$ for the minimal central idempotents of the path algebra A_2. According to 2.4.1 and 2.3.9, there is an isomorphism of $\mathrm{End}_N^r(M)$ onto A_2 which takes $\lambda(M)$ onto M. Our purpose here is to use the path model to provide an explicit isomorphism. Except as noted above, our notation is as in 2.3.11.

An edge on $\tilde{B}$ is specified by the data $\eta = (k;i,j,\ell)$, where k is the story on which η lies, v_j^k and v_i^{k+1} are the two vertices of η, and the index ℓ distinguishes among the $\lambda_{i,j}^k$ edges joining v_j^k and v_i^{k+1}. Define an involution $*$ of $\Omega_{[0,1]} \cup \Omega_{[1,2]}$ by

$$(k;i,j,\ell)^* = (1-k;j,i,\ell).$$

Thus $*$ is the reflection through the first floor. (Nevertheless we regard the reflection of an upward oriented edge to be upward oriented.)

Let $V_j = \mathbb{K}\Omega^j_{2]} \otimes \mathbb{K}\Omega^j_{0]}$ and $V = \bigoplus_{j=1}^{n} V_j$. Define a linear map U from M to V by requiring

$$U(T_{\xi,\eta}) = (\xi_0,\xi_1,\eta_1^*) \otimes \eta_0 \qquad ((\xi,\eta) \in R_1).$$

U is a linear isomorphism, its inverse F being determined by

$$F(\xi' \otimes \eta_0) = T_{(\xi'_0,\xi'_1),(\eta_0,\xi'^*_2)}$$
$$(\xi' \in \Omega^j_{2]} \text{ and } \eta_0 \in \Omega^j_{0]};\ 1 \le j \le n).$$

Note that U breaks and unfolds the round trip path $\xi \circ \eta^{-1}$, while F folds and joins the pair (ξ',η_0). For example:

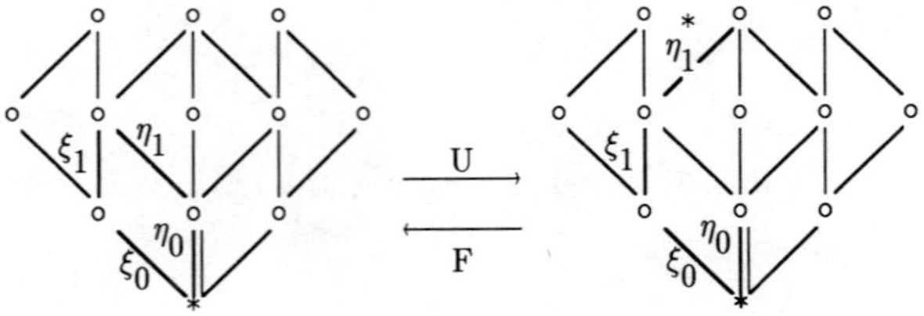

V carries both a right action of N and a left action of A_2, arising from the right action of N on $\mathbb{K}\Omega_{0]}$ and the left action of A_2 on $\mathbb{K}\Omega_{2]}$:

$$\rho(n)(\xi \otimes \eta) = \xi \otimes \rho(n)\eta$$
$$x(\xi \otimes \eta) \quad = x\xi \otimes \eta.$$

It is easy to check that A_2 is in fact the commutant of $\rho(N)$ in $\mathrm{End}_{\mathbb{K}}(V)$, and that U intertwines the right actions of N on M and V. Hence

$$\alpha : \varphi \mapsto U \circ \varphi \circ F$$

is an isomorphism from $\mathrm{End}^r_N(M) = C_{\mathrm{End}_{\mathbb{K}}(M)}(\rho(N))$ to $C_{\mathrm{End}_{\mathbb{K}}(V)}(\rho(N)) = A_2$.

Let $(\xi,\eta) \in R_2$ and $(\sigma,\tau) \in R_1$ (so $T_{\xi,\eta} \in A_2$ and $T_{\sigma,\tau} \in M$). One checks that

$$\alpha^{-1}(T_{\xi,\eta})(T_{\sigma,\tau}) = \delta(\eta,\sigma\circ\tau_1^*)T_{(\xi_0,\xi_1),(\tau_0,\xi_2^*)}.$$

It follows that $\alpha^{-1}(x) = \lambda(x)$ for $x \in M \subset A_2$. Also

$$\alpha^{-1}(\tilde{q}_j) = \alpha^{-1}(\sum_{\xi \in \Omega^j_{2]}} T_{\xi,\xi}) = \rho(q_j),$$

as required by Proposition 2.4.1.

Remark. Later we will want to modify the definition of U somewhat. If $c : \Omega_{[0,1]} \longrightarrow \mathbb{K}^*$ is any function and we instead define U by

$$U(T_{\xi,\eta}) = c(\eta_1)(\xi_0,\xi_1,\eta_1^*) \otimes \eta_0,$$

then $\varphi \mapsto U \circ \varphi \circ U^{-1}$ is another isomorphism of $\mathrm{End}^r_N(M)$ onto A_2.

2.5. Traces.

A $\mathbb{K}$-linear map φ from $\mathbb{K}$-algebra M to a $\mathbb{K}$-vector space V is said to be faithful if the corresponding bilinear map

$$(x,y) \mapsto \varphi(xy)$$

is non-degenerate; that is for each non-zero $x \in M$ there is a $y \in M$ such that $\varphi(xy) \neq 0$. This is a one-sided notion, but if M is finite dimensional and $\varphi : M \longrightarrow \mathbb{K}$ is linear, then φ is faithful on one side if and only if it is faithful on the other. Furthermore, in this case, for each linear $\psi : M \longrightarrow \mathbb{K}$, there is an $a \in M$ such that $\psi(x) = \varphi(xa)$ for all $x \in M$.

A trace on M is a linear map $\mathrm{tr} : M \longrightarrow \mathbb{K}$ such that $\mathrm{tr}(xy) = \mathrm{tr}(yx)$ for all $x,y \in M$. On a factor, any non-zero trace is faithful, and any two traces are proportional. In fact a trace on $\mathrm{Mat}_d(\mathbb{K})$ satisfies $\mathrm{tr}(e_{i,j}) = \delta_i^j \, \mathrm{tr}(e_{1,1})$, where $\{e_{i,j}\}$ are the standard matrix units.

Let M be a multi-matrix algebra over $\mathbb{K}$, written as before as $M = \bigoplus_{i=1}^{m} p_i M$, with $p_i M \cong \mathrm{Mat}_{\mu_i}(\mathbb{K})$. We associate to a trace tr on M the row-vector

$$\overleftarrow{s} = (\mathrm{tr}(e_1),\cdots,\mathrm{tr}(e_m)) \in \mathbb{K}^m,$$

where e_i is a minimal idempotent in Mp_i. For example, the trace $\sigma^{(i)}$ defined by $\sigma^{(i)}(e_i) = 1$ and $\sigma^{(i)}{}_{|p_kM} = 0$ for $k \neq i$ corresponds to the ith vector of the canonical basis of $\mathbb{K}^m$. Any row vector $\overleftarrow{s} \in \mathbb{K}^m$ determines a unique trace

$$\mathrm{tr}_s = \sum_{i=1}^{m} s_i\sigma^{(i)} : M \to \mathbb{K}$$

with associated vector $\overleftarrow{s}$.

A trace tr on M is faithful if and only if the associated vector $\overleftarrow{s}$ has no zero entries. When the characteristic of $\mathbb{K}$ is zero, we say that tr is <u>positive</u> if $s_i \geq 0$ for all i. (There is an ambiguity here; if $\mathbb{K}$ is given as an extension of the reals, the meaning of $s_i \geq 0$ is clear. Otherwise we take $s_i \geq 0$ to mean that there is an imbedding of $\mathbb{Q}(s_1,\cdots,s_m)$ in $\mathbb{C}$ such that $s_i \geq 0$ for all i.) A positive trace is faithful if $s_i > 0$ for all i.

<u>Proposition 2.5.1.</u> *Let* $\mathbb{1} \in N \subset M$ *be a pair of multi–matrix algebras with*

$$N = \bigoplus_{j=1}^{n} q_jN\ ,\ M = \bigoplus_{i=1}^{m} p_iM$$

and with inclusion matrix Λ_N^M.

(a) *Let* σ *be a trace on* M *corresponding to* $\overleftarrow{s} \in \mathbb{K}^m$ *and let* τ *be a trace on* N *corresponding to* $\overleftarrow{t} \in \mathbb{K}^n$. *Then* σ *extends* τ *if and only if* $\overleftarrow{t} = \overleftarrow{s}\Lambda_N^M$.

(b) *If* $\mathrm{char}(\mathbb{K}) = 0$, *then there exists a faithful trace on* M *with faithful restriction to* N. *If* $\mathrm{char}(\mathbb{K}) = p > 0$, *then a sufficient condition for the existence of a faithful trace on* M *with faithful restriction to* N *is that for all* j, *the sum* $\sum_i \lambda_{i,j}$ *is not divisible by* p.

<u>Proof</u>. (a) If f_j is a minimal idempotent in q_jN, then f_jp_i is the sum of $\lambda_{i,j}$ minimal idempotents in p_iM. Hence the restriction of σ to N is described by the vector $\overleftarrow{t}'$ with components

$$t'_j = \sigma(f_j) = \sum_{i=1}^{m} \sigma(f_jp_i) = \sum_{i=1}^{m} s_i\lambda_{i,j} = (\overleftarrow{s}\Lambda_N^M)_j.$$

(b) Define the trace on M with weights $s_i = 1$ for all i. The restriction to N has weights $t_j = \sum_i \lambda_{i,j}$ (mod char($\mathbb{K}$)). If char($\mathbb{K}$) = 0, or if the column sums of Λ_N^M are not divisible by characteristic, then the weights t_j are non–zero, and the restricted trace is faithful. #

Remarks

(1) With the notation of the proposition, one has, when σ extends τ,

$$(\overleftarrow{s}, \vec{\mu}) = \sum_{i=1}^{m} s_i \mu_i = \sigma(1) = \tau(1) = (\overleftarrow{t}, \vec{\nu}).$$

By Propositions 2.3.1.b and 2.5.1.a, this implies

$$(\overleftarrow{s}, \Lambda_N^M \vec{\nu}) = (\overleftarrow{s}\Lambda_N^M, \vec{\nu}),$$

which is, of course, obvious!

(2) A faithful trace on M may have zero restriction to N. Consider for example

$$N = \mathbb{C}[\mathfrak{S}_2] \subset M = \mathbb{C}[\mathfrak{S}_3] \cong \mathbb{C} \oplus M_2(\mathbb{C}) \oplus \mathbb{C}$$

as in Example 2.3.7, and the trace on M associated to the vector (1,–1,1) $\in \mathbb{C}^3$. Or consider the two element field $\mathbb{F}_2$ and the pair $\mathbb{F}_2 \subset \mathrm{Mat}_2(\mathbb{F}_2)$ (with inclusion matrix [2]); any trace on $\mathrm{Mat}_2(\mathbb{F}_2)$ has zero restriction to the center $\mathbb{F}_2$! One may thus say about traces on multi–matrix algebras, that positivity is hereditary, but faithfulness is not.

(3) The assignment of a vector $\overleftarrow{s} \in \mathbb{K}^m$ to a trace tr : M $\to \mathbb{K}$ has been defined above via the values of tr on (classes of) minimal idempotents of M. In Chapter 3, we shall consider a new situation, where M is a finite direct sum of continuous (type II_1) von Neumann factors; since no minimal idempotents are present in this situation, we shall describe a trace tr by the vector $\tilde{s} = (\mathrm{tr}(p_1), \cdots, \mathrm{tr}(p_m))$ of values of tr on minimal *central* projections of M. In principle, the description of tr via $\tilde{s}$ is also possible for multi–matrix algebras, but this causes complications which would be out of place in the present chapter.

(4) Given a Bratteli diagram representing a sequence of inclusions of multi–matrix algebras, $A_k \subset A_{k+1}$, and a trace on $\bigcup_k A_k$, it is sometimes convenient to record all the data on the diagram by marking each vertex with the weight of the trace on the corresponding factor, that is the value of the trace on a minimal idempotent in the factor,

as well as with the dimensions. Thus, in the situation of Proposition 2.5.1, and for $\Lambda = \begin{bmatrix} 1 & 1 \\ 2 & 1 \\ 0 & 1 \end{bmatrix}$, the diagram is

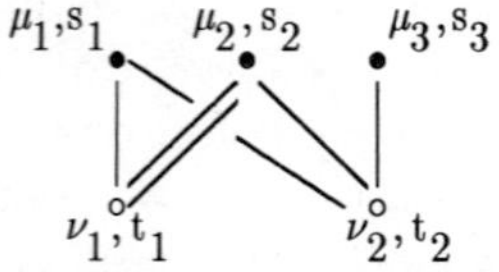

(Note that on any Bratteli diagram it is actually superfluous to record the dimensions except on the first floor. Similarly on a finite Bratteli diagram it is superfluous to record the weights of a trace except on the top floor, but on an infinite Bratteli diagram it is not (in general) superfluous to record the traces, since the traces on the higher floors are not determined by those on the floors below.)

(5) Suppose $\mathbb{K} \supset \mathbb{R}$, and $N \subset M$ is a pair of multi-matrix algebras over $\mathbb{K}$ with $Z_N \cap Z_M = \mathbb{K}$ and with inclusion matrix Λ. Let $(M_k)_{k\geq 0}$ be the tower obtained by iterating the fundamental construction. Then it follows from Perron–Frobenius theory that there is a unique positive normalized $(\mathrm{tr}(\mathbb{1}) = 1)$ trace on $M_\infty = \bigcup_k M_k$.

In fact, let $\overleftarrow{t}^{(0)}$ be the Perron–Frobenius eigenvector for $\Lambda^t\Lambda$, normalized by $\sum t_j^{(0)}\nu_j = 1$. Define $\overleftarrow{t}^{(2k)} = \|\Lambda\|^{-2k}\,\overleftarrow{t}^{(0)}$ and $\overleftarrow{t}^{(2k-1)} = \overleftarrow{t}^{(2k)}\Lambda^t$ $(k \geq 1)$. Then $\{\overleftarrow{t}^{(k)}\}_{k\geq 1}$ defines a consistent family of traces on the algebras M_k, since $\overleftarrow{t}^{(k+1)}\Lambda_{M^k}^{M^{k+1}} = \overleftarrow{t}^{(k)}$ for all k.

An argument similar to one given in the proof of 2.4.2. shows that $\bigcap_{r\geq 0} (\Lambda^t\Lambda)^r(\mathbb{R}_+^n)$ consists of Perron–Frobenius eigenvectors for $\Lambda^t\Lambda$. Suppose tr is any positive normalized trace on M_∞ and $\overleftarrow{s}^{(k)}$ is the vector determining tr on M_k. Then for all k and r,

$$\overleftarrow{s}^{(2k+2r)}(\Lambda^t\Lambda)^r = \overleftarrow{s}^{(2k)},$$

whence $\overleftarrow{s}^{(2k)}$ is a Perron–Frobenius eigenvector for $\Lambda^t\Lambda$. Since $\overleftarrow{s}^{(2k)}(\Lambda^t\Lambda)^k = \overleftarrow{s}^{(0)}$, we have $\overleftarrow{s}^{(2k)} = \|\Lambda\|^{-2k}\,\overleftarrow{s}^{(0)}$.

2.6. Conditional expectations.

We are primarily interested in the following situation:

(1) $N \subset M$ is a pair of multi-matrix algebras.

(2) M has a faithful trace with faithful restriction to N.

(3) $E : M \to N$ is the orthogonal projection of M onto N with respect to the inner product determined by the trace.

However, to clarify somewhat the roles played by semi– simplicity, the pair of faithful traces, and the conditional expectation E, we begin in a more general setting.

A <u>conditional</u> <u>expectation</u> from a $\mathbb{K}$–algebra M onto a subalgebra N is an (N,N)–linear map whose restriction to N is the identity. Recall that such a map E is faithful if for each non–zero $x \in M$ there is a $y \in M$ such that $E(xy) \neq 0$. For example, if M is a factor, $M = \mathrm{Mat}_{\mu}(\mathbb{K})$, where $\mathbb{K}$ has charactersitic 0 or μ is relatively prime to $\mathrm{char}(\mathbb{K})$, then the trace on M normalized by $\mathrm{trace}(\mathbb{1}) = 1$ is a faithful conditional expectation of M onto $\mathbb{K}$.

Consider $\mathrm{Hom}_N^r(M,N)$, the set of right N–linear maps from M to N, with its left N–module structure defined by $(x\varphi)(y) = x\varphi(y)$ $(x \in N,\ y \in M,\ \varphi \in \mathrm{Hom}_N^r(M,N))$. We associate to a conditional expectation $E : M \to N$ the left N–linear map $E^b : M \to \mathrm{Hom}_N^r(M,N)$ defined by $E^b(x)(y) = E(xy)$ for $x,y \in M$. Then E is faithful if and only if E^b is injective. We say that E is <u>very faithful</u> if E^b is an isomorphism.

<u>Lemma 2.6.1</u>. *Let* $N \subset M$ *be a pair of finite dimensional* $\mathbb{K}$*–algebras. Suppose* N *has a faithful* $\mathbb{K}$*–linear functional. Then any faithful expectation* E *from* M *to* N *is very faithful.*

<u>Proof</u>. Choose a faithful functional $\tau : N \longrightarrow \mathbb{K}$ and set $\sigma = \tau \circ E$. If $x \in M$ is such that $\sigma(xx') = 0$ for all $x' \in M$, then $\sigma(xyz) = \tau(E(xy)z) = 0$ for all $y \in M$ and for all $z \in N$, so that $E(xy) = 0$ by faithfulness of τ and $x = 0$ by that of E. Thus σ is faithful. It follows that any $\mathbb{K}$–linear map $M \longrightarrow \mathbb{K}$ is of the form $x \mapsto \sigma(ax)$ for some $a \in M$, since M is finite dimensional.

Consider a right N–linear map $\varphi : M \to N$. There exists $a \in M$ with $\tau\varphi(x) = \sigma(ax)$ for all $x \in M$. Define $\psi : M \to N$ by $\psi = E^b(a)$; i.e., $\psi(x) = E(ax)$. We claim that $\psi = \varphi$. It is enough to check that $\lambda\psi = \lambda\varphi$ for any $\mathbb{K}$–linear $\lambda : N \to \mathbb{K}$. But as τ is faithful, such a λ is given by $y \mapsto \tau(yb)$ for some $b \in N$. Now one has for all $x \in M$

$$\lambda\psi(x) = \tau(E(ax)b) = \tau E(axb), \text{ and}$$

$$\lambda\varphi(x) = \tau(\varphi(x)b) = \tau(\varphi(xb)) = \sigma(axb) = \tau E(axb). \ \#$$

<u>Remarks</u>. (1) If N is a multi–matrix algebra, then N has a faithful $\mathbb{K}$–linear functional.

(2) Let V be a $\mathbb{K}$–vector space and define a multiplication on $A = \mathbb{K} \oplus V$ by $(\lambda,v)(\lambda',v') = (\lambda\lambda', \lambda v' + \lambda' v)$. The result is a $\mathbb{K}$–algebra for which any subspace of $0 \oplus V$ is an ideal. Suppose $\dim V \geq 2$. If $\varphi : A \to \mathbb{K}$ is any $\mathbb{K}$–linear functional, then $\ker(\varphi) \cap V$ is a non–zero ideal in $\ker(\varphi)$. So A has no faithful linear functional.

But if V is one-dimensional, spanned by v, then the functional $(a,bv) \mapsto a + b$ is faithful on A.

The next proposition concerns the existence of faithful conditional expectations.

Proposition 2.6.2. *Let* $N \subset M$ *be a pair of* $\mathbb{K}$*-algebras with* N *finite dimensional, and let* $tr : M \to \mathbb{K}$ *be a faithful trace with faithful restriction* to N. *Then there exists a unique* $\mathbb{K}$*-linear map* $E : M \to N$ *such that*

(i) $tr(E(x)) = tr(x) \qquad x \in M$

(ii) $E(y) = y \qquad y \in N$

(iii) $E(xy) = E(x)y \qquad x \in M,\ y \in N.$

Moreover E *is a faithful conditional expectation from* M *to* N, *namely*

(iv) $E(yx) = yE(x) \qquad x \in M,\ y \in N$

(v) $E(xy) = 0$ *for all* y *implies* $x = 0$.

If M *is finite dimensional, then* E *is very faithful; that is*

(vi) $E^{\flat} : M \to \mathrm{Hom}_N^r(M,N)$ *defined by* $a \mapsto (x \mapsto (E(ax))$ *is an isomorphism.*

Proof. We consider M together with the nondegenerate symmetric $\mathbb{K}$-bilinear form $(x,z) \mapsto tr(xz)$ and with the associated orthogonality relation. As tr and $tr|_N$ are faithful one has $M = N \oplus N^{\perp}$.

We begin by checking uniqueness. Let $E : M \to N$ satisfy (i) to (iii). As E is defined on N by (ii), it is enough to check that $E = 0$ on $N^{\perp}$. Let $t \in N^{\perp}$. For any $y \in N$ one has by (iii) and (i)

$$tr(E(t)y) = tr(E(ty)) = tr(ty) = 0$$

so that $E(t) \perp N$. But $E(t)$ is also in N, so that $E(t) = 0$.

To prove existence, define E to be the projection of M onto N along $N^{\perp}$. It is obvious that (ii) holds. For $x \in M$, one has $E(x)-x$ orthogonal to N and hence to $\mathbf{1}$, so (i) holds.

Note that $N^{\perp}$ is a right N-module because of the trace property of tr. Namely if $y,y' \in N$ and $z \in N^{\perp}$. Then

$$tr(y'(zy)) = tr((yy')z) = 0,$$

so $zy \in N^{\perp}$. Now $xy - E(xy)$ and $x - E(x)$ are in $N^{\perp}$, and hence also $xy - E(x)y \in N^{\perp}$. The difference

$$(xy-E(xy)) - (xy-E(x)y) = E(x)y-E(xy)$$

is in $N^{\perp} \cap N = (0)$, which proves (iii). One obtains (iv) similarly.

Since $tr = tr \circ E$, the faithfulness of E follows from that of tr. Finally, if M is finite dimensional, then E is very faithful by Lemma 2.6.1. #

Remark. Conditions (i)–(iii) are equivalent to the single condition

$$tr(E(x)y) = tr(xy) \quad \text{for } x \in M \text{ and } y \in N,$$

as the reader may verify.

The relevance of conditional expectations for the *fundamental construction* comes from the following fact.

Proposition 2.6.3. *Let* M, N *be* $\mathbb{K}$*–algebras with* $1 \in N \subset M$; *set* $L = End_N^r(M)$ *and let* $\lambda: M \to L$ *denote the inclusion. Assume moreover that*

(i) *the right* N–*module* M *is projective of finite type, and*

(ii) *there exists a very faithful conditional expectation* E *from* M *to* N.

Then L *is generated by* M *and* E (*viewed as a map from* M *to* M). *More precisely,* L *is generated as a* $\mathbb{K}$–*vector space by elements of the form* $\lambda(x)E\lambda(y)$ *with* $x,y \in M$. *Furthermore, the map* $x \otimes y \mapsto \lambda(x)E\lambda(y)$ *from* $M \otimes_N M$ *to* $End_N^r(M)$ *is an isomorphism.*

Proof. Hypothesis (ii) says that $E^b : M \to M^* = Hom_N^r(M,N)$ is an isomorphism. As projective modules of finite type are flat (see [BAC 1], page 28), the $\mathbb{K}$–linear map

$$id_M \otimes E^b : M \otimes_N M \longrightarrow M \otimes_N M^*$$

is an isomorphism. Let

$$\theta \quad \begin{cases} M \otimes_N M^* & \to L \\ x \otimes x^* & \mapsto (z \mapsto xx^*(z)) \end{cases}$$

be the canonical homomorphism. By (i), it is an isomorphism (see, e.g., [BA 2], page 111). Consequently, the composition

$$\Phi = \theta(id_M \otimes E^b) : M \otimes_N M \to L$$

is an isomorphism. Routine computations show that

$$\Phi(x \otimes y) = \lambda(x)E\lambda(y) \qquad x,y \in M$$

$$\Phi(x \otimes y)\Phi(z \otimes t) = \Phi(xE(yz) \otimes t) \quad x.y,z,t \in M.$$

The proposition follows from the first of these. #

Remarks.

(1) It could be that M is projective of finite type as a right N–module but not as a left N–module, as observed in [BA 8], page 53.

(2) In the situation of the previous proposition can we conclude that L is projective of finite type over M (as a *right* $\lambda(M)$–module)?

For pairs of multi–matrix algebras, the situation regarding pairs of faithful traces and conditional expectations is the following:

(1) If $\operatorname{char} \mathbb{K} = 0$, then for any pair of multi–matrix algebras $N \subset M$ over $\mathbb{K}$, there exist faithful traces on M with faithful restriction to N (2.5.1), hence faithful conditional expectations $E : M \to N$ (2.6.2).

(2) Whenever $E : M \to N$ is a faithful conditional expectation, it is very faithful, since N always has a faithful functional (2.6.1).

(3) If $\operatorname{char} \mathbb{K} > 0$, M need not have a faithful trace with faithful restriction to N. For example there is no pair of faithful traces for $\mathbb{F}_2 \subset \mathrm{Mat}_2(\mathbb{F}_2)$. Note that nevertheless $\begin{bmatrix} a & b \\ c & d \end{bmatrix} \mapsto a + b + c$ defines a faithful conditional expectation $\mathrm{Mat}_2(\mathbb{F}_2) \to \mathbb{F}_2$.

Corollary 2.6.4. *Consider a pair of multi–matrix algebras*

$$\mathbf{1} \in N = \bigoplus_{j=1}^{n} q_j N \subset M = \bigoplus_{i=1}^{m} p_i M$$

as well as

$$L = \bigoplus_{j=1}^{n} \rho(q_j)L = \mathrm{End}_N^r(M).$$

Suppose there is a faithful conditional expectation $E : M \to N$. *Then*

(a) L *is generated as a* $\mathbb{K}$*–vector space by elements* $\lambda(x)E\lambda(y)$ *for* $x,y \in M$;

(b) *The* $\mathbb{K}$*–linear map* $\varphi\colon N \to ELE$ *defined by* $\varphi(x) = \lambda(x)E$ *is an isomorphism of algebras.*

(c) If f_j *is a minimal idempotent in the factor* q_jN, *then* $\lambda(f_j)E$ *is a minimal idempotent in the factor* $\rho(q_j)L$.

Proof. (a) Condition (i) of Proposition 2.6.3 is fulfilled because any module over a semi–simple algebra is projective, and condition (ii) is fulfilled by Lemma 2.6.1.

To prove (b), first note that φ is a morphism because E is an idempotent which commutes with $\lambda(x)$ for all $x \in N$. If $x \in N$ and $\varphi(x) = 0$, then also $x = \varphi(x)(\mathbf{1}) = 0$, so φ is injective. Finally φ is surjective by part (a).

For $j \in \{1,\cdots,n\}$, the idempotent $\rho(q_j)E = \lambda(q_j)E$ is not zero and lies in the factor $\rho(q_j)L$. The resulting reduced factor is $\rho(q_j)ELE = \rho(q_j)\lambda(N)E$. As φ is an isomorphism, its restriction φ_j to q_jN is also an isomorphism onto $\rho(q_j)ELE$. It follows that the idempotent $\varphi_j(f_j) = \lambda(f_j)E$ is minimal in the factor $\rho(q_j)ELE$. But if $e \in L$ is an non–zero idempotent in L dominated by $\lambda(f_j)E$, and thus also by $\lambda(q_j)E = \rho(q_j)E$, then $e = \rho(q_j)E\,e\,\rho(q_j)E \in \rho(q_j)ELE$, and therefore $e = \lambda(f_j)E$. In other words, $\lambda(f_j)E$ is also minimal in L. #

Remark: The following instructive proof of 2.6.4.a was given by Wenzl [Wen3]. First note that the map φ of 2.6.4.b is an injective homomorphism. Now consider the subalgebra A of L generated by $\lambda(M)$ and E, and note that

$$A = \{\ \lambda(y_0) + \sum_i \lambda(x_i)E\lambda(y_i) :\ x_i, y_i \in M\},\ \text{and}$$

$$EAE = \varphi(N) \cong N$$

If ψ is a non–zero element of rad(A), then there exist $x, y \in M$ such that $E(y\psi(x)) \neq 0$ (using the faithfulness of E). But then $E\lambda(y)\psi\lambda(x)E = \lambda(E(y\psi(x))E = \varphi(E(y\psi(x)))$ is a non–zero element of $\mathrm{rad}(A) \cap EAE = \mathrm{rad}(EAE)$, a contradiction since EAE is isomorphic to the semi–simple algebra N. Thus A is semi–simple. Note that $A' = \lambda(M)' \cap \{E\}' = \rho(N)$, so $A'' = \rho(N)' = L$, where primes denote centralizers in $\mathrm{End}_K(M)$. Since A is semi–simple, $A = A'' = L$. Finally observe that $\lambda(M)E\lambda(M) = \{\sum_i \lambda(x_i)E\lambda(y_i) :\ x_i, y_i \in M\}$ is an ideal in L, and if ψ is a central projection in L orthogonal to this ideal, then for all $x, y \in M$,

$$0 = (E\lambda(y)\psi)(x) = E(y\psi(x)).$$

Hence $\psi = 0$ by faithfulness of E, so $L = \lambda(M)E\lambda(M)$. #

2.6.5 Reprise of 2.6.4 using the path model. Let N, M, and L be as in 2.6.4. Suppose tr is a faithful trace on M with faithful restriction to N, and let $E : M \to N$ be the conditonal expectation determined by tr, as in 2.6.2. Let B be the Bratteli diagram for $N \subset M \subset L$, and let $\tilde{B}$ be the augmented diagram, as in 2.3.11 and 2.4.6. We identify $N \subset M$ with the pair of path algebras $A_0 \subset A_1$, but we distinguish between $L = \mathrm{End}_N^r(M)$ and the isomorphic path algebra A_2. Let $\overleftarrow{s}$ and $\overleftarrow{t}$ be the vectors determing the trace tr on M and N. We also regard $\overleftarrow{s}$ and $\overleftarrow{t}$ as functions of vertices on the 0th and 1st floors respectively: $t(v_j^0) = t_j$ and $s(v_i^1) = s_i$. Recall the *–operation which reflects edges through the first floor. We define the reflection on vertices as well:

$$(v_i^k)^* = v_i^{2-k} \quad (0 \le k \le 2 \ \ 1 \le i \le m(k))$$

We first give a formula for $E \in \mathrm{End}_N^r(M)$. Recall that E is determined by the requirement $\mathrm{tr}(E(z)x) = \mathrm{tr}(zx)$, for $z \in M$ and $x \in N$. If $(\xi,\eta) \in R_1$ and $(\alpha,\beta) \in R_0$, so that $T_{\xi,\eta} \in M$ and $T_{\alpha,\beta} \in N$, then one verifies that

$$\mathrm{tr}(T_{\xi,\eta}T_{\alpha,\beta}) = \delta(\eta_0,\alpha_0)\delta(\xi_0,\beta_0)\delta(\xi_1,\eta_1)s(\xi_{[1]}),$$

while

$$\mathrm{tr}(\delta(\xi_1,\eta_1)T_{\xi_0,\eta_0}T_{\alpha,\beta}) = \delta(\eta_0,\alpha_0)\delta(\xi_0,\beta_0)\delta(\xi_1,\eta_1)t(\xi_{[0]}).$$

Hence

$$E(T_{\xi,\eta}) = \frac{s(\xi_{[1]})}{t(\xi_{[0]})}\delta(\xi_1,\eta_1)T_{\xi_0,\eta_0} \qquad ((\xi,\eta) \in R_1). \tag{2.6.5.1}$$

(Remark that $\xi_{[1]} = \eta_{[1]}$ and $\xi_{[0]} = \eta_{[0]}$ if $E(T_{\xi,\eta}) \neq 0$, so the expression is not so asymmetric as it may first appear.)

Let

$$U(T_{\xi,\eta}) = c(\eta_1)(\xi_0,\xi_1,\eta_1^*) \otimes \eta_0 \qquad ((\xi,\eta) \in R_1),$$

and $F = U^{-1}$, as in 2.4.6. Next we compute $e = U \circ E \circ F$, the image of E in A_2. For $\xi \otimes \eta_0$, an elementary tensor in V_j for some j,

$$\begin{aligned} e\xi \otimes \eta_0 = e(\xi \otimes \eta_0) &= \frac{1}{c(\xi_2^*)} U(E(T_{(\xi_0,\xi_1),(\eta_0,\xi_2^*)})) \\ &= \frac{1}{c(\xi_2^*)} \frac{s(\xi_{[1]})}{t(\xi_{[0]})} \delta(\xi_1,\xi_2^*)\, U(T_{\xi_0,\eta_0}) \\ &= \frac{1}{c(\xi_1)} \frac{s(\xi_{[1]})}{t(\xi_{[0]})} \delta(\xi_1,\xi_2^*)\, U\Big[\sum_{\substack{\lambda \\ \lambda_{[0]}=\xi_{[0]}}} T_{(\xi_0,\lambda),(\eta_0\circ\lambda)} \Big] \end{aligned}$$

$$= \left[\delta(\xi_1,\xi_2^*)\frac{s(\xi_{[1]})}{t(\xi_{[0]})} \sum_{\substack{\lambda \\ \lambda_{[0]}=\xi_{[0]}}} \frac{c(\lambda)}{c(\xi_1)}(\xi_0,\lambda,\lambda^*)\right] \otimes \eta_0.$$

It follows that

$$e = \sum_{\substack{\xi,\lambda\in\Omega_{[0,1]} \\ \xi_{[0]}=\lambda_{[0]}}} \frac{c(\lambda)}{c(\xi)}\frac{s(\xi_{[1]})}{t(\xi_{[0]})} T_{(\lambda,\lambda^*),(\xi,\xi^*)}. \tag{2.6.5.2}$$

Remark. If $\mathbb{K} = \mathbb{C}$, and the trace tr is positive, we prefer to use the inner product $\langle x,y\rangle = tr(xy^*)$ on M, where $*$ is the natural $*$ operation on the path algebra M, rather than the bilinear form $(x,y) \mapsto tr(xy)$. (The orthogonal projection $E : M \to N$ is unaffected by the change.) We give V the inner product for which $\biguplus_j \Omega^j_{2]} \otimes \Omega^j_{0]}$ is an orthonormal basis. Then the choice

$$U(T_{\xi,\eta}) = \sqrt{s(\xi_{[1]})}\,(\xi_0,\xi_1,\eta_1^*) \otimes \eta_0 \tag{2.6.5.3}$$

makes U into a unitary operator from M onto V. In this case e is given by

$$e = \sum_{\substack{\xi,\lambda\in\Omega_{[0,1]} \\ \xi_{[0]}=\lambda_{[0]}}} \frac{\sqrt{s(\xi_{[1]})}\,\sqrt{s(\lambda_{[1]})}}{t(\xi_{[0]})} T_{(\lambda,\lambda^*),(\xi,\xi^*)}. \tag{2.6.5.4}$$

Then e is a self-adjoint projection in the C^*–algebra A_2. This formula for e is due to Sunder [Sun] and Ocneanu [Ocn]. The formulae (2.6.5.3) and (2.6.5.4) are also sensible if $\mathbb{K}$ is any quadratically closed field.

We know from 2.6.4 that any $\varphi \in End^r_N(M)$ has a decomposition $\varphi = \sum_i \lambda(x_i)\, E\, \lambda(y_i)$ where $x_i,y_i \in M$, but so far we have not considered how to compute such a decomposition. Since the isomorphism $\alpha : \varphi \mapsto U\circ\varphi\circ F$ of $End^r_N(M)$ onto A_2 takes $\lambda(x)$ to x ($x \in M$), it suffices to decompose $z \in A_2$ into a sum $z = \sum_i x_i e y_i$ with $x_i,y_i \in M$. For (α,γ) and $(\delta,\beta) \in R_1$ (so $T_{\alpha,\gamma}$ and $T_{\delta,\beta} \in M$) one computes from (2.6.5.2) that

$$T_{\alpha,\gamma} e T_{\delta,\beta} = \delta(\gamma_0,\delta_0)\frac{c(\gamma_1)}{c(\delta_1)}\frac{s(\delta_{[1]})}{t(\delta_{[0]})} T_{(\alpha_0,\alpha_1,\gamma_1^*),(\beta_0,\beta_1,\delta_1^*)}.$$

Hence for $(\alpha,\beta) \in R_2$,

$$(2.6.5.5) \qquad T_{\alpha,\beta} = \frac{c(\beta_2^*)t(\beta_{[2]}^*)}{c(\alpha_2^*)s(\beta_{[1]})} T_{(\alpha_0,\alpha_1),(\gamma_0,\alpha_2^*)}\, e\, T_{(\gamma_0,\beta_2^*),(\beta_0,\beta_1)},$$

where γ_0 is an arbitrary edge in $\Omega_{0]}$ with $\gamma_{[0]} = \text{end}(\beta)^* = \text{start}(\beta_2^*)$. In particular if we use the convention (2.6.5.3), and formula (2.6.5.4) we get

$$(2.6.5.6) \qquad T_{\alpha,\beta} = \frac{t(\beta_{[2]}^*)}{\sqrt{s(\alpha_{[1]})}\sqrt{s(\beta_{[1]})}} T_{(\alpha_0,\alpha_1),(\gamma_0,\alpha_2^*)}\, e\, T_{(\gamma_0,\beta_2^*),(\beta_0,\beta_1)}.$$

Another way to write this is

$$T_{\alpha,\beta} = t(\text{end}(\beta)^*)F(\alpha \otimes \gamma_0)\, e\, F(\beta \otimes \gamma_0)^*.$$

As an exercise in using (2.6.5.6) we compute a decomposition for the minimal central idempotent $\rho(q_j)$ in $\text{End}_N^r(M)$. We have

$$\alpha(\rho(q_j)) = \tilde{q}_j = \sum_{\xi\in\Omega_{2]}^j} T_{\xi,\xi}$$

$$= \sum_{\xi\in\Omega_{2]}^j} \frac{t_j}{s(\xi_{[1]})} T_{(\xi_0,\xi_1),(\gamma_0,\xi_2^*)}\, e\, T_{(\gamma_0,\xi_2^*),(\xi_0,\xi_1)},$$

for any $\gamma_0 \in \Omega_{0]}^j$. Taking the average over the ν_j elements of $\Omega_{0]}^j$, we arrive at

$$\rho(q_j) = \frac{t_j}{\nu_j} \sum_{\substack{(\xi,\eta)\in R_1 \\ \eta_{[0]}=v_j^0}} \frac{1}{s(\xi_{[1]})} \lambda(T_{\xi,\eta})\, E\lambda(T_{\eta,\xi}).$$

In the remainder of this section we discuss, following [Wen3] and [BW], the notion of an extension of an algebra with respect to a conditional expectation. This type of structure appears frequently in Chapter 4.

Definition 2.6.6. Let $N \subset M$ be a pair of algebras over a field $\mathbb{K}$, and $E : M \to N$ a faithful conditional expectation. An E-extension of M is a pair (L,f), where L is an algebra containing M, $f \in L$, and

(i) L is generated as an algebra by M and f.

(ii) $f^2 = f$.

(iii) $fyf = E(y)f = fE(y)$ for all $y \in M$.

(iv) The morphism $\begin{cases} x \mapsto xf \text{ is injective.} \\ N \to L \end{cases}$

The model example of an E-extension is the fundamental construction $(\mathrm{End}_N^r(M),E)$, when E is very faithful and M is projective of finite type as a right N-module.

Lemma 2.6.7. *Let* (L,f) *be an* E*-extension of* M.

(i) *Any element of* L *has the form* $y_0 + \sum_{j=1}^{n} y_j' f y_j''$, *with* $y_0, y_j', y_j'' \in M$. *In particular* MfM *is an ideal of* L.

(ii) *There is a unique conditional expectation* $\overline{E} : L \to N$ *extending* E *and satisfying* $\overline{E}(x)f = fxf$ *for* $x \in L$. *Moreover* $\overline{E}(x) = \overline{E}(xf) = \overline{E}(fx)$ *for all* $x \in L$.

(iii) *For* $x \in L$ *there exist unique* $b_1, b_2 \in M$ *with* $xf = b_1 f$ *and* $fx = fb_2$.

Proof. (i) is immediate from the definition 2.6.6.

(ii) Let ψ denote the isomorphism $x \mapsto xf$ from N to L, whose range is exactly fLf. Then $\overline{E} : x \mapsto \psi^{-1}(fxf)$ has the desired properties.

(iii) If $x = y_0 + \sum_j y_j' f y_j''$, then $b_1 = y_0 + \sum_j y_j' E(y_j'')$ satisfies $xf = b_1 f$. If $b \in M$ and $bf = 0$, then for all $y \in M$, $0 = fybf = E(yb)f = \psi \circ E(yb)$. Since E is faithful and ψ injective, $b = 0$. This proves the existence and uniqueness of b_1. Proceed similarly for b_2. #

Remarks 2.6.8. (1) If $N \subsetneq M$, then $\overline{E}$ is never faithful since $f \neq \mathbf{1}$ and $\overline{E}((\mathbf{1}-f)x) = 0$ for all $x \in L$.

(2) Let $x \in L$. One has $\overline{E}(xy) = 0$ for all $y \in M$ if, and only if, $fx = 0$. Similarly $\overline{E}(yx) = 0$ for all $y \in M$ if, and only if, $xf = 0$.

Let us check the first assertion. Suppose $\overline{E}(xy) = 0$ for all $y \in M$. Then for all y,

$$0 = \overline{E}(xy) = \overline{E}(fxy) = \overline{E}(fb_2 y) = \overline{E}(b_2 y) = E(b_2 y).$$

Since E is faithful, $b_2 = 0$ and $fx = fb_2 = 0$.

(3) If N, M and L are $*$-algebras, $E = E^*$, and $f = f^*$, then $\overline{E}$ is self adjoint, because ψ is a $*$-morphism.

(4) If N and M are C^*-algebras, L is a $*$-subalgebra of a C^*-algebra, $E = E^*$ and $f = f^*$, then $\overline{E}$ is positive. Indeed $x \mapsto fxf$ is positive and ψ^{-1} is positive.

Proposition 2.6.9. *Assume that* M *is projective of finite type as a right* N*-module and* E *is very faithful. Let* (L,f) *be an* E*-extension of* M. *Then*

$$\sigma \begin{cases} \mathrm{End}_N^r(M) \longrightarrow L \\ \sum_j y_j' E y_j'' \mapsto \sum y_j' f y_j'' \end{cases}$$

defines a (non-unital) isomorphism of $\mathrm{End}_N^r(M)$ *onto the ideal* MfM *of* L. *Moreover there is a morphism of algebras* $\varphi : L \longrightarrow \mathrm{End}_N^r(M)$ *such that* $L = MfM \oplus \ker \varphi$ *(direct sum of algebras).*

Proof. Identify M with its image in $\mathrm{End}_N^r(M)$. Since by 2.6.3, $\sum y_j' \otimes y_j'' \mapsto \sum y_j' E y_j''$ is an isomorphism of $M \otimes_N M$ onto $\mathrm{End}_N^r(M)$, the map σ is well-defined, and it is an algebra morphism with image MfM, by definition 2.6.6. We set

$$\varphi \begin{cases} L \to \mathrm{End}_N^r(M) \\ y_0 + \sum_j y_j' f y_f'' \mapsto y_0 + \sum y_j'' E y_j'' . \end{cases}$$

We have to check that φ is well-defined. Let $x = y_0 + \sum_j y_j' f y_j''$ and $a = y_0 + \sum_j y_j' E y_j''$ with $y_0, y_j', y_j'' \in M$. Then for all $y', y'' \in M$

$$fy' x y'' f = \{E(y' y_0 y'') + \sum_j E(y' y_j') E(y_j'' y'')\} f,$$

while

$$Ey' a y'' E = \{E(y' y_0 y'') + \sum_j E(y' y_j') E(y_j'' y'')\} E.$$

If $x = 0$, then $Ey' a y'' E = 0$ for all $y', y'' \in M$, so $MEMaMEM = 0$; but $\mathrm{End}_N^r(M) = MEM$ by 2.6.3, and since this algebra has a unit, $a = 0$.

It is clear that φ is a surjective algebra morphism (indeed $\varphi(MfM) = \mathrm{End}_N^r(M)$), and that $\varphi \circ \sigma$ is the identity. Hence σ is injective and $L = MfM \oplus \ker \varphi$ as vector spaces. Since both MfM and $\ker \varphi$ are ideals in L, this is actually a direct sum of algebras. #

2.7. Markov traces on pairs of multi–matrix algebras.

Let $N \subset M$ be a pair of multi–matrix algebras and let $\lambda : M \to L = \mathrm{End}_N^r(M)$ be the pair obtained by the fundamental construction. If $E : M \to N$ is a faithful conditional expectation, we know from Corollary 2.6.4 that L is generated as a vector space by elements of the form $\lambda(x)E\lambda(y)$ with $x,y \in M$. Any trace $\mathrm{Tr} : L \to \mathbb{K}$ satisfies

$$\mathrm{Tr}(\lambda(x)E\lambda(y)) = \mathrm{Tr}(\lambda(yx)E) = \mathrm{Tr}(E\lambda(yx)E) = \mathrm{Tr}(\lambda(E(yx))E),$$

for all $x,y \in M$, and hence Tr is determined by its values on elements of the form $\lambda(x)E$ for $x \in N$.

Let tr be a faithful trace on M with faithful restriction to N and let $E : M \to N$ be the conditional expectation defined in Proposition 2.6.2. Let $\beta \in \mathbb{K}$. Define tr to be a <u>Markov trace of modulus β</u> if there exists a trace $\mathrm{Tr} : L \to \mathbb{K}$ such that

$$\left.\begin{aligned} \mathrm{Tr}(\lambda(x)) &= \mathrm{tr}(x) \\ \beta\,\mathrm{Tr}(\lambda(x)E) &= \mathrm{tr}(x) \end{aligned}\right\} \text{ for all } x \in M.$$

Observe that this relation implies $\beta \neq 0$, because tr is faithful. If such a Tr exists, it is unique in the following strong sense.

<u>Lemma 2.7.1</u>. *Let $N \subset M$ be a pair of multi–matrix algebras and let $\beta \in \mathbb{K}^*$. Let* tr *and* E *be as above. Then there exists at most one trace* Tr *on* L *such that*

$$\beta\,\mathrm{Tr}(\lambda(y)E) = \mathrm{tr}(y) \qquad \textit{for all } y \in N.$$

If such a Tr *exists, then it is faithful and satisfies*

$$\beta\,\mathrm{Tr}(\lambda(x)E) = \mathrm{tr}(x) \qquad \textit{for all } x \in M.$$

If $\overleftarrow{r}$ *is the vector describing* Tr *and* $\overleftarrow{t}$ *the vector describing* $\mathrm{tr}|_N$, *then* $\overleftarrow{r}\beta = \overleftarrow{t}$.

<u>Proof</u>. We use the notation of Corollary 2.6.4. If such a trace Tr exists, then for each j,

$$\begin{aligned} \beta\, r_j &= \beta \mathrm{Tr}(\lambda(f_j)E) && \text{(by 2.6.4.c)} \\ &= \mathrm{tr}(f_j) = t_j, && j = 1,,\cdots,,n, \end{aligned}$$

so that $\overleftarrow{r}\beta = \overleftarrow{t}$. Uniqueness and faithfulness of Tr follow. Finally

$$\beta \operatorname{Tr}(\lambda(x)E) = \beta \operatorname{Tr}(E\lambda(x)E) = \beta \operatorname{Tr}(\lambda(E(x))E) = \operatorname{tr}(E(x)) = \operatorname{tr}(x)$$

for all $x \in M$. #

Proposition 2.7.2. *Let $\beta \in \mathbb{K}^*$, let $N \subset M$ be a multi-matrix algebra pair with inclusion matrix Λ and let $\lambda : M \to L$ be the pair obtained by the fundamental construction. Let the decompositions into factors be*

$$N = \bigoplus_{j=1}^{n} q_j N \subset M = \bigoplus_{i=1}^{m} p_i M \xrightarrow{\lambda} L = \bigoplus_{j=1}^{n} \rho(q_j) L$$

$$q_j N \cong \operatorname{Mat}_{\nu_j}(\mathbb{K}) \qquad p_i M \cong \operatorname{Mat}_{\mu_i}(\mathbb{K}) \qquad \rho(q_j) L \cong \operatorname{Mat}_{\kappa_j}(\mathbb{K}),$$

$$\vec{\nu} = (\nu_1, \cdots, \nu_n) \qquad \vec{\mu} = (\mu_1, \cdots, \mu_m) \qquad \vec{\kappa} = (\kappa_1, \cdots, \kappa_n)$$

so that in particular

$$\Lambda\vec{\nu} = \vec{\mu} \qquad \Lambda^t\vec{\mu} = \vec{\kappa}$$

Let tr *be a faithful trace on* M *with faithful restriction* to N *and associated conditional expectation* $E : M \to N$. *Let $\overleftarrow{s} \in \mathbb{K}^m$ and $\overleftarrow{t} \in \mathbb{K}^n$ be the corresponding vectors, so that in particular $\overleftarrow{t} = \overleftarrow{s}\Lambda$. Finally, let $\beta \in \mathbb{K}^*$.*

Then the following are equivalent.

(i) tr *is a Markov trace of modulus* β.

(ii) $\overleftarrow{s}(\Lambda\Lambda^t) = \beta\overleftarrow{s}$ *and* $\overleftarrow{t}(\Lambda^t\Lambda) = \beta\overleftarrow{t}$.

In particular, if $\operatorname{char}(\mathbb{K}) = 0$ *and if* β *is the modulus of some Markov trace on* M, *then* β *is a totally positive algebraic number; that is,* $\beta > 0$ *for any imbedding of* $\mathbb{Q}(\beta)$ *in* $\mathbb{C}$.

Proof. (i) $\Rightarrow$ (ii) Let Tr be as in the definition of a Markov trace, and let $\overleftarrow{r} \in \mathbb{K}^n$ be the corresponding vector. Then $\overleftarrow{t} = \overleftarrow{r}\Lambda^t\Lambda$ because Tr extends tr, and $\overleftarrow{t} = \beta\overleftarrow{r}$ by the previous lemma, so that $\beta\overleftarrow{t} = \overleftarrow{t}\Lambda^t\Lambda$. One has also $\overleftarrow{s} = \overleftarrow{r}\Lambda^t$, so that

$$\overleftarrow{s}\Lambda\Lambda^t = \overleftarrow{r}\Lambda^t\Lambda\Lambda^t = \beta\,\overleftarrow{r}\Lambda^t = \beta\,\overleftarrow{s}.$$

(ii) $\Rightarrow$ (i) Set $\overleftarrow{r} = \beta^{-1}\overleftarrow{t}$ and let $\operatorname{Tr} : L \to \mathbb{K}$ be the corresponding trace. Then Tr extends tr because

$$\overleftarrow{r}\Lambda^t = \beta^{-1}\overleftarrow{t}\Lambda^t = \beta^{-1}\overleftarrow{s}\Lambda\Lambda^t = \overleftarrow{s}.$$

Consider the linear map $\tilde{\tau} : N \to \mathbb{K}$ defined by $\tilde{\tau}(y) = \beta \operatorname{Tr}(\lambda(y)E)$; it is a trace, because E is N–linear and idempotent. If f_j denotes some minimal idemotent in q_jN, one has

$$\tilde{\tau}(f_j) = \beta \operatorname{Tr}(\lambda(f_j)E) = \beta r_j = t_j, \quad j = 1,, \cdots, n$$

by Corollary 2.6.4.c and the definition of $\overleftarrow{r}$, so that $\tilde{\tau} = tr|_N$. Thus Tr satisfies the Markov condition $\beta \operatorname{Tr}(\lambda(x)E) = tr(x)$ for all $x \in M$ by the previous lemma.

Finally matrices of the form $\Lambda^t \Lambda$ have totally positive eigenvalues, when $\operatorname{char}(\mathbb{K}) = 0$. #

Remarks.

(1) Take $\Lambda = \begin{bmatrix} 1 & 1 \\ 1 & 1 \end{bmatrix}$ and $\overleftarrow{s} = (3,1)$, so that $\overleftarrow{t} = (4,4)$. Then $\overleftarrow{t}\Lambda^t\Lambda = 4\overleftarrow{t}$, but $\overleftarrow{s}\Lambda\Lambda^t$ is not a scalar multiple of $\overleftarrow{s}$. This shows that one cannot delete the first equality in condition (ii). Although $\overleftarrow{t}\Lambda^t\Lambda = \overleftarrow{t}\beta$ follows from $\overleftarrow{s}\Lambda\Lambda^t = \overleftarrow{s}\beta$ (because $\overleftarrow{t} = \overleftarrow{s}\Lambda$), we prefer to state (ii) in a symmetric form.

(2) *We stress that* $\beta > 0$ *holds without any positivity assumption on* tr, *in case* $\operatorname{char}(\mathbb{K}) = 0$.

Theorem 2.7.3. *Let* $\mathbb{K}$ *be a field extension of* $\mathbb{R}$. *Let* $N \subset M$ *be a pair of multi–matrix algebras over* $\mathbb{K}$ *with inclusion matrix* Λ, *and with* $Z_M \cap Z_N = \mathbb{K}$. *Let* $\beta \in \mathbb{K}^*$.

A necessary and sufficient condition for the existence of a positive Markov trace of modulus β *on* M *is that* $\beta = [M{:}N] = \|\Lambda\|^2$. *Any two positive Markov traces on* M *are proportional.*

Proof. Since $Z_M \cap Z_N = \mathbb{K}$, it follows that Λ is indecomposable and $\Lambda\Lambda^t$ is irreducible (2.3.1f and 1.3.2b). Recall that $[M{:}N] = \|\Lambda\|^2$ by Theorem 2.1.1.

If tr is a positive Markov trace of modulus β on M, then $\beta = \|\Lambda\Lambda^t\| = [M{:}N]$ by the previous proposition and Perron–Frobenius theory.

Conversely, set $\beta = [M{:}N]$. Let $\overleftarrow{s}$ be a Perron–Frobenius vector such that $\overleftarrow{s}\,\Lambda\Lambda^t = \beta\overleftarrow{s}$. Let $\overleftarrow{t} = \overleftarrow{s}\Lambda$; it follows as in remark (1) above that $\overleftarrow{t}\Lambda^t\Lambda = \beta\,\overleftarrow{t}$. Hence if tr is the (positive) trace corresponding to the vector $\overleftarrow{s}$, then tr is a Markov trace of modulus β by 2.7.2.

The final statement follows from the uniqueness of the Perron–Frobenius eigenvector for $\Lambda\Lambda^t$. #

A crucial property of a Markov trace tr on a pair $N \subset M$ is that the trace Tr on L $= \operatorname{End}^r_N(M)$ entering the definition of the Markov property is again a Markov trace on $M \subset L$. More precisely:

Proposition 2.7.4. *Let* tr *be a Markov trace of modulus* β *on a multi-matrix pair* $N \subset M$, *set* $L = \mathrm{End}_N^r(M)$ *as usual, let* $\mathrm{Tr} : L \to \mathbb{K}$ *be the extension of* tr *to a trace on* L *as in Lemma* 2.7.1, *and let* $D : L \to \lambda(M)$ *be the conditional expectation defined by* Tr *and* tr. *Then*

(a) Tr *is a Markov trace of modulus* β *(with respect to* $\lambda : M \to L$);

(b) $\beta\, D(E) = \mathbf{1}$;

(c) $\beta\, D\overline{\lambda}(E)D = D$, where $\overline{\lambda}(\cdot)$ means left multiplication on L;

(d) $\beta\overline{\lambda}(E)D\overline{\lambda}(E) = \overline{\lambda}(E)$.

Proof. (a) Let $\overleftarrow{s}$ and $\overleftarrow{t}$ be the vectors defining the trace tr on M and N respectively. As tr is a Markov trace of modulus β, one has

$$\overleftarrow{s}\Lambda\Lambda^t = \beta\overleftarrow{s}, \quad \overleftarrow{t}\Lambda^t\Lambda = \beta\overleftarrow{t}$$

by Proposition 2.7.2. From the proof of 2.7.1, we know that Tr is described by $\overleftarrow{r} = \beta^{-1}\overleftarrow{t}$. Consequently

$$\overleftarrow{r}\Lambda^t\Lambda = \beta\overleftarrow{r}, \quad \overleftarrow{s}\Lambda\Lambda^t = \beta\overleftarrow{s}$$

and (a) now follows from 2.7.2.

(b) The bilinear form $(u,v) \mapsto \mathrm{Tr}(uv)$ is nondegenerate on L and its restriction to $\lambda(M)$ is nondegenerate; thus $L = \lambda(M) \oplus \lambda(M)^{\perp}$, where orthogonality is meant with respect to this bilinear form. For all $x \in M$ one has

$$\mathrm{Tr}(\beta\, E\lambda(x) - \lambda(x)) = \beta\,\mathrm{Tr}(\lambda(x)E) - \mathrm{Tr}(\lambda(x)) = \mathrm{tr}(x) - \mathrm{tr}(x) = 0$$

so that $\beta\, E - \mathbf{1} \in \lambda(M)^{\perp}$. As D is the orthogonal projection of L onto $\lambda(M)$, this implies $D(\beta E) = \mathbf{1}$.

(c) By M–linearity of D one has $D\overline{\lambda}(E)D = \overline{\lambda}(D(E))D$, so (c) follows from (b).

(d) Choose $x,y \in M$ and set $u = \lambda(x)E\lambda(y) \in L$. The maps from M to M,

$$E\lambda(x)E\lambda(y) : z \mapsto E(xE(yz))$$

$$E\lambda(E(x))\lambda(y) : z \mapsto E(E(x)yz)$$

$$\lambda(E(x))E\lambda(y) : z \mapsto E(x)E(yz)$$

are equal by (N,N)–linearity of E. By (M,M)–linearity of D one has

$$\overline{\lambda}(E)D\overline{\lambda}(E)u = ED(E\lambda(x)E\lambda(y)) = ED(\lambda(E(x))E\lambda(y))$$

$$= E\lambda(E(x))D(E)\lambda(y).$$

Consequently, using (a),

$$\beta\,\overline{\lambda}(E)D\overline{\lambda}(E)u = E\lambda(E(x))\lambda(y) = Eu = \overline{\lambda}(E)u,$$

which proves (d). #

This completes the proof of Theorems 2.1.3 *and* 2.1.4.

We now analyze the role of Markov traces for towers. Changing our notation slightly, we consider a multi–matrix pair $M_0 \subset M_1$, the tower $(M_k)_{k\geq 0}$ it generates, and a trace $tr = tr_1$ on M_1, which is a Markov trace of modulus β on the pair $M_0 \subset M_1$. We denote by tr_2 the extension of the trace to M_2 denoted previously by Tr, and by

$$E_1 = E : M_1 \to M_0,\ E_1 \in M_2$$
$$E_2 = D : M_2 \to M_1,\ E_2 \in M_3$$

the associated conditional expectations. According to Proposition 2.7.4, the process of extending a Markov trace on M_k to M_{k+1} iterates; namely, if

> E_k: $M_k \to M_{k-1}$ is the conditional expectation associated to tr_k and tr_{k-1}, and $tr_{k+1} : M_{k+1} \to \mathbb{K}$ is the unique extension of tr_k satisfying $\beta\, tr_{k+1}(xE_k) = tr_k(x)$ for all $x \in M_k$ (see 2.7.1),

then tr_{k+1} is also a Markov trace, and the process can continue. Note that M_{k+1} is the algebra generated by M_k and E_k, for short $M_{k+1} = \langle M_k, E_k\rangle$. Denote by M_∞ the inductive limit (union) of the nested sequence

$$M_0 \subset M_1 \subset \cdots \subset M_k \subset M_{k+1} \subset \cdots.$$

This is a $\mathbb{K}$–algebra with unit which is the union of its finite dimensional semi–simple subalgebras, and which has a finite dimensional center isomorphic to $Z_M \cap Z_N$. The union of the tr_k's constitutes a trace $tr : M_\infty \to \mathbb{K}$ which is nondegenerate (namely, $tr(xy) = 0$ for all $y \in M_\infty$ implies $x = 0$). If $\mathbb{K} \supset \mathbb{R}$ and $tr = tr_1$ is positive, then tr is also positive in the sense that $tr(\epsilon) > 0$ for any non zero idempotent ϵ in M_∞. If this holds, and if moreover $Z_M \cap Z_N \cong \mathbb{K}$, then tr is the unique positive trace on M_∞, up to normalization; see Remark (5) at the end of Section 2.5.

Proposition 2.7.5. *Let* $M_0 \subset M_1$ *be a pair of multi–matrix algebras and let* $tr : M_1 \to \mathbb{K}$ *be a Markov trace of modulus* β. *With the notation above one has*

(a) $\beta E_i E_j E_i = E_i$ *for* $i,j \geq 1$ *with* $|i-l| = 1$;

(b) $E_i E_j = E_j E_i$ *for* $i,j \geq 1$ *with* $|i-j| \geq 2$;

(c) $\beta\, tr(wE_k) = tr(w)$ *for all* $w \in M_k$. *In particular, if* tr *is normalized by* $tr(\mathbf{1}) = 1$, *then* $tr(E_k) = \beta^{-1}$ *for all* $k \geq 1$.

Proof. Statements (a) and (c) follow from (a), (c) and (d) of Proposition 2.7.4. If $j \geq i+2$, then $E_i \in M_{j-1}$, and (b) follows because E_j is M_{j-1}–linear. #

Observe that this Proposition contains Theorem 2.1.6.

2.7.6. The path model for M_∞ and the idempotents E_i. Let $M_0 \subset M_1$ be a pair of multi–matrix algebras and let

$$M_0 \subset M_1 \subset \cdots \subset M_k \subset M_{k+1} \subset \cdots$$

be the tower generated by iterating the fundamental construction. Let $\tilde{B}$ be the augmented Bratteli diagram of the tower and

$$A_0 \subset A_1 \subset \cdots \subset A_k \subset A_{k+1} \subset \cdots$$

the chain of path algebras associated to $\tilde{B}$ as in 2.3.11. Having identified $M_0 \subset M_1$ with $A_0 \subset A_1$, we can obtain an explicit sequence of isomorphisms $\alpha_k : M_k \to A_k$ with $\alpha_{k+1}\big|_{M_k} = \alpha_k$ for all k, by iterating the procedure of 2.4.6.

If tr is a Markov trace of modulus β on M_1, then tr extends uniquely to a trace on M_∞ which is faithful on each M_k and which has the Markov property: if $E_k : M_k \to M_{k-1}$ is the conditional expectation determined by the trace, then $\beta tr(E_k x) = tr(x)$ for all $x \in M_k$. If $t_j^{(k)}$ denotes the weights of the trace on the kth floor of B, then $t_j^{(k)} = \beta^{-1} t_j^{(k-2)}$ for all k and j. We also write tr for the corresponding trace on $A_\infty = \bigcup_k A_k$.

Assuming (just for the sake of having definite formulae) that $\mathbb{K}$ is quadratically closed, we can choose the isomorphisms $\{\alpha_k\}$ so that $e_k = \alpha_k(E_k) =$

$$\sum_{\substack{\xi,\eta\in\Omega_{[k-1,k]}\\ \xi_{[k-1]}=\eta_{[k-1]}}} \frac{\sqrt{t^{(k)}(\xi_{[k]})}\sqrt{t^{(k)}(\eta_{[k]})}}{t^{(k-1)}(\xi_{[k-1]})} T_{(\xi,S_k\xi),(\eta,S_k\eta)},$$

where S_k denotes reflection of an edge through the kth floor of B. In fact we know that this choice determines $\{\alpha_k\}$ completely because of the decomposition 2.6.4.(a). Then $\{e_k\}$ is a sequence of idempotents (self–adjoint projections on $\ell^2(\Omega)$ in case $\mathbb{K}=\mathbb{C}$ and tr is positive) satisfying (a)–(c) of 2.7.5.

Iterating the decomposition (2.6.5.6), we can write any matrix unit $T_{\alpha,\beta}$ in A_k as a monomial in the matrix units of A_1 and the idempotents $e_1,\cdots,e_{k-1}$. For example for $(\alpha,\beta)\in R_3$ $(T_{\alpha,\beta}\in A_3)$ one finds

$$T_{\alpha,\beta}=C(\alpha,\beta)T_{\alpha_{1]},(\gamma^1,\tilde{\alpha}_2)}e_1e_2\,T_{(\gamma^1,\tilde{\alpha}_3),(\gamma^2,\tilde{\beta}_3)}e_1\,T_{(\gamma^2,\tilde{\beta}_2),\beta_{1]}},$$

where $\tilde{\alpha}_i$ denotes the edge in $B(M_0\subset M_1)$ directly below the edge α_i, and γ^1,γ^2 are arbitrary edges in $\Omega_{0]}$ with the appropriate endpoints. The constant $C(\alpha,\beta)$ can be evaluated by computing $tr(T_{\alpha,\beta}\cdot T_{\beta,\alpha})$, using the fact that $e_kxe_k=E_k(x)e_k=e_kE_k(x)$ $(x\in A_k)$ and the Markov property of tr.

Let $A_{tr,k}(M_0\subset M_1)$ be the subalgebra of M_k generated by $\mathbb{1},E_1,\cdots,E_{k-1}$. Our next goal is to understand the structure of these algebras. We shall see that, when the modulus β of the Markov trace tr lies in a certain generic set, these algebras depend only on β and k, and not on any other data pertaining to the inclusion $M_0\subset M_1$ or the trace tr. For β in this generic set, $A_{tr,k}(M_0\subset M_1)$ is isomorphic to an abstractly defined algebra $\mathcal{A}_{\beta,k}$, whose structure we describe in detail in the next section. For non–generic β, new phenomena can occur, and our knowledge is much less satisfactory in this case; see Section 2.9. The following two sections borrow heavily from [Jo 1].

2.8 – The algebras $\mathcal{A}_{\beta,k}$ for generic β

For any integer $k\geq 1$ and for any number $\beta\neq 0$ in the basic field $\mathbb{K}$, let $\mathcal{A}_{\beta,k}$ be the algebra abstractly defined (as an associative algebra over $\mathbb{K}$) by the generators $\epsilon_1, \epsilon_2, \cdots, \epsilon_{k-1}$ and the unit $\mathbf{1}$

the relations $\epsilon_i^2 = \epsilon_i$

$\beta \epsilon_i \epsilon_j \epsilon_i = \epsilon_i$ if $|i-j| = 1$

$\epsilon_i \epsilon_j = \epsilon_j \epsilon_i$ if $|i-j| \geq 2$

(Observe k indexes the algebra generated by idempotents up to k–1; this agrees with the usual convention for Artin's braid groups, but is *not* as in [Jo1] or [Jo2].)

A monomial in $\mathcal{A}_{\beta,k}$ is a product $\epsilon_{i_1}\epsilon_{i_2}\cdots\epsilon_{i_q}$ where each ϵ_{i_j} is one of $\epsilon_1, \cdots, \epsilon_{k-1}$; the unit $\mathbf{1}$ of $\mathcal{A}_{\beta,k}$ is a monomial (the empty product).

Proposition 2.8.1. *Any monomial* $w \in \mathcal{A}_{\beta,k}$ *may be written in one reduced form*

$$\beta^{-r}(\epsilon_{i_1}\epsilon_{i_1-1}\cdots\epsilon_{j_1})(\epsilon_{i_2}\epsilon_{i_2-1}\cdots\epsilon_{j_2})\cdots(\epsilon_{i_p}\epsilon_{i_p-1}\cdots\epsilon_{j_p})$$

where $r \in \mathbb{N}$ *is an appropriate integer and where*

$$1 \leq i_1 < i_2 < \cdots < i_p \leq k-1$$
$$1 \leq j_1 < j_2 < \cdots < j_p \leq k-1$$
$$i_1 \geq j_1, i_2 \geq j_2, \cdots, i_p \geq j_p$$
$$0 \leq p \leq k-1.$$

Moreover $\dim_{\mathbb{K}} \mathcal{A}_{\beta,k} \leq \frac{1}{k+1}\begin{bmatrix} 2k \\ k \end{bmatrix}$.

Proof. Consider an integer m with $0 \leq m \leq k-1$; we prove the first part of the lemma by induction on m for a monomial w in $\{\epsilon_1, \cdots, \epsilon_m\}$. As this is obvious for monomials with $m \leq 1$, we may assume that $m \geq 2$ and that the claim holds for m–1.

Suppose w is a monomial in which ϵ_m appears at least twice. Then w has one of the forms

$$w = w_1 \epsilon_m a \epsilon_m w_2$$

or

$$w = w_1 \epsilon_m a \epsilon_{m-1} b \epsilon_m w_2,$$

where a,b are monomials in $\{\epsilon_1, \cdots, \epsilon_{m-2}\}$. As ϵ_m commutes with these, w equals either $$w_1\epsilon_m a w_2$$

or $$w_1 a\beta^{-1}\epsilon_m b w_2,$$

and the number of ϵ_m's has been reduced. Consequently we may assume that w involves exactly one ϵ_m.

Let $w = w_1\epsilon_m w_2$ with w_1 and w_2 monomials in $\{\epsilon_1,\cdots,\epsilon_{m-1}\}$. Using first the induction hypothesis on w_2 and then the commutation $\epsilon_m\epsilon_j = \epsilon_j\epsilon_m$ for $j \leq m-2$, we can reduce to the case that $w = w_1\epsilon_m\epsilon_{m-1}\cdots\epsilon_n$, with w_1 a reduced monomial finishing, say, with ϵ_ℓ. If $\ell \geq n$ one has

$$\begin{aligned}\epsilon_\ell\epsilon_m\epsilon_{m-1}\cdots\epsilon_n &= \epsilon_m\cdots\epsilon_{\ell+2}(\epsilon_\ell\epsilon_{\ell+1}\epsilon_\ell)\epsilon_{\ell-1}\cdots\epsilon_n \\ &= \beta^{-1}\epsilon_\ell\epsilon_{\ell-1}\cdots\epsilon_n\epsilon_m\epsilon_{m-1}\cdots\epsilon_{\ell+2}.\end{aligned}$$

Consequently we may assume that $\ell < n$, so that w is of the form

$$w = \beta\epsilon_{i_1}\cdots\epsilon_{j_1}\epsilon_{i_2}\cdots\epsilon_{j_2}\cdots\epsilon_{i_p}\cdots\epsilon_{j_p} \quad (i_p = m,\ j_p = n)$$

with all desired relations for the i's and the j's. This ends the induction argument.

We now count the number of reduced monomials, following Chapter III in [Fel]. By a path in the lattice $\mathbb{Z}^2$, we mean here an oriented connected polygonal line with vertices at integral points and with edges being either horizontal and directed to the right or vertical and directed upwards. A path starting at (a,b) and ending at (c,d) has c–a + d–b unit edges, c–a horizontal ones and d–b vertical ones. The number of these paths is consequently the binomial coefficient

$$N\binom{c,d}{a,b} = \begin{bmatrix} c-a+d-b \\ c-a \end{bmatrix}.$$

Assume first $a > b$ and $c > d$. To each of these paths touching the main diagonal, associate the following "reflected" path: if (j,j) is the diagonal point on the path with smallest j, replace the subpath from (a,b) to (j,j) by the reflected path (with respect to the diagonal) from (b,a) to (j,j) and leave the subpath from (j,j) to (c,d) unchanged. This defines a bijection between the set of paths from (a,b) to (c,d) which touch the diagonal and the set of paths from (b,a) to (c,d). Thus the number of paths from (a,b) to (c,d) which do <u>not</u> touch the main diagonal is $N\binom{c,d}{a,b} - N\binom{c,d}{b,a}$.

Assume now $a = b$ and $c = d = a+n$ for some $n > 0$. Consider the paths from (a,a) to (a+n,a+n) whose vertices are on or below the main diagonal. These are in bijection with paths from (a+1,a) to (a+n+1,a+n) which do not touch the main diagonal, and their number is

$$\begin{bmatrix}2n\\n\end{bmatrix} - \begin{bmatrix}2n\\n+1\end{bmatrix} = \frac{1}{n+1}\begin{bmatrix}2n\\n\end{bmatrix}.$$

Consider finally a sequence $(i_1,j_1,\cdots,i_p,j_p)$ corresponding to a reduced monomial in $\mathcal{A}_{\beta,k}$. We may associate to this sequence the following path from (0,0) to (k,k), and any path from (0,0) to (k,k) which remains on or below the diagonal can be obtained in this way.

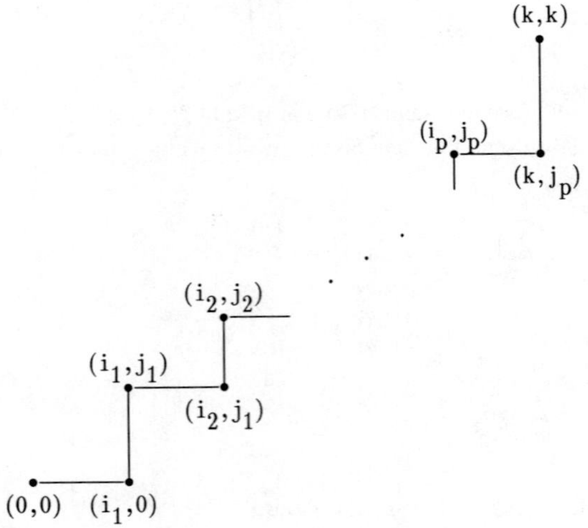

it follows that the number of reduced monomials is $\frac{1}{k+1}\begin{bmatrix}2k\\k\end{bmatrix}$. #

<u>Remark</u>. The Catalan numbers may be defined by

$$\tfrac{1}{2}(1-\sqrt{1-4\lambda}) = \sum_{n\geq 1} C_n\lambda^n = \sum_{n\geq 1}\frac{1}{n}\begin{bmatrix}2n-2\\n-1\end{bmatrix}\lambda^n.$$

With this notation, dim $\mathcal{A}_{\beta,k} \leq C_{k+1}$. See e.g. n^0 2.7.3 (page 111) of [GJ].

We shall also need the following computation. We agree that a binomial coefficient $\begin{bmatrix}a\\b\end{bmatrix}$ is zero if the integers a,b satisfy $b < 0$ or $b > a$.

Lemma 2.8.2. *Let* $k \geq 1$ *be an integer and set* $m = [k/2]$, *the greatest integer less than or equal to* $k/2$. *Then*

$$\sum_{j=0}^{m} \left(\binom{k}{j} - \binom{k}{j-1} \right)^2 = \frac{1}{k+1} \binom{2k}{k}.$$

Proof. By comparison of the coefficients of t^c on both sides of $(1+t)^a(1+t)^b = (1+t)^{a+b}$, one has

$$\sum_{j=0}^{c} \binom{a}{j} \binom{b}{c-j} = \binom{a+b}{c} \qquad (*)$$

for any integers $a,b,c \geq 0$. (See for example Section II.12 in [Fel].)

Assume first that k is even: $k = 2m$. Setting $a = b = c = k$ in $(*)$, one obtains

$$\sum_{j=0}^{m} \binom{k}{j}^2 = \frac{1}{2} \sum_{j=0}^{k} \binom{k}{j}^2 + \frac{1}{2} \binom{k}{m}^2$$

$$= \frac{1}{2} \binom{2k}{k} + \frac{1}{2} \binom{k}{m}^2, \text{ and}$$

$$\sum_{j=0}^{m} \binom{k}{j-1}^2 = \frac{1}{2} \binom{2k}{k} - \frac{1}{2} \binom{k}{m}^2.$$

Setting $a = b = k$ and $c = k + 1$ in $(*)$, one obtains

$$2 \sum_{j=0}^{m} \binom{k}{j} \binom{k}{j-1} = \sum_{j=0}^{k} \binom{k}{j} \binom{k}{j-1} = \binom{2k}{k+1}$$

Thus

$$\sum_{j=0}^{m} \left(\binom{k}{j} - \binom{k}{j-1} \right)^2 = \binom{2k}{k} - \binom{2k}{k+1} = \frac{1}{k+1} \binom{2k}{k}.$$

For k odd $(k = 2m+1)$, one obtains similarly

$$\sum_{j=0}^{m} \begin{bmatrix} k \\ j \end{bmatrix}^2 = \frac{1}{2}\sum_{j=0}^{k} \begin{bmatrix} k \\ j \end{bmatrix}^2 = \frac{1}{2}\begin{bmatrix} 2k \\ k \end{bmatrix},$$

$$\sum_{j=0}^{m} \begin{bmatrix} k \\ j-1 \end{bmatrix}^2 = \frac{1}{2}\begin{bmatrix} 2k \\ k \end{bmatrix} - \begin{bmatrix} k \\ m \end{bmatrix}^2, \text{ and}$$

$$2\sum_{j=0}^{m} \begin{bmatrix} k \\ j \end{bmatrix}\begin{bmatrix} k \\ j-1 \end{bmatrix} = \sum_{j=0}^{k} \begin{bmatrix} k \\ j \end{bmatrix}\begin{bmatrix} k \\ j-1 \end{bmatrix} - \begin{bmatrix} k \\ m+1 \end{bmatrix}\begin{bmatrix} k \\ m \end{bmatrix}$$

$$= \begin{bmatrix} 2k \\ k+1 \end{bmatrix} - \begin{bmatrix} k \\ m \end{bmatrix}^2;$$

so the conclusion follows for k odd as well. #

Define now a sequence $(P_k)_{k\geq 0}$ of polynomials in $\mathbb{Z}[\lambda]$ by

$$P_0 = 1, \ P_1 = 1, \ P_{k+1}(\lambda) = P_k(\lambda) - \lambda P_{k-1}(\lambda) \ \ (k \geq 1)$$

so that in particular

$$P_2(\lambda) = 1-\lambda \quad P_4(\lambda) = 1-3\lambda + \lambda^2 \quad P_6(\lambda) = 1-5\lambda + 6\lambda^2 - \lambda^3$$
$$P_3(\lambda) = 1-2\lambda \quad P_5(\lambda) = 1-4\lambda + 3\lambda^2 \quad P_7(\lambda) = 1-6\lambda + 10\lambda^2 - 4\lambda^3.$$

(Observe P_k here is as in [Wen1], but as P_{k-1} in [Jo1].)

Proposition 2.8.3. *Consider an integer* $k \geq 0$ *and set* $m = \left[\frac{k}{2}\right]$. *Then*

(i) *The polynomial* P_k *is of degree* m. *Its leading coefficient is* $(-1)^m$ *if* $k = 2m$ *is even and* $(-1)^m(m+1)$ *if* $k = 2m+1$ *is odd.*

(ii) P_k *has* m *distinct roots which are given by* $\dfrac{1}{4\cos^2\left[\frac{j\pi}{k+1}\right]}$ for $j = 1,2,\cdots,m$.

(iii) *Assume* $k \geq 1$. *Let* λ *be a real number with*

$$\frac{1}{4\cos^2\left[\frac{\pi}{k+2}\right]} < \lambda < \frac{1}{4\cos^2\left[\frac{\pi}{k+1}\right]}.$$

Then $P_1(\lambda)>0, P_2(\lambda)>0,\cdots,P_k(\lambda)>0, P_{k+1}(\lambda)<0.$

(iv) *Set* $Q_k(\lambda) = P_k(\lambda(\lambda+1)^{-2})$. *Then*

$$Q_k(\lambda) = \frac{\lambda^{k+1}-1}{(\lambda-1)(\lambda+1)^k}.$$

Proof. Claims (i) and (iv) are easily checked by induction.

For (ii), we compute in the ring $\mathbb{Q}[\lambda,\sqrt{1-4\lambda}\,]$ and proceed as in the proof of I.2.2. The difference equation for the P_k's has an indicial equation $\mu^2 - \mu + \lambda = 0$ with roots

$$\mu_1 = \tfrac{1}{2}(1+\sqrt{1-4\lambda}) \quad \mu_2 = \tfrac{1}{2}(1-\sqrt{1-4\lambda})$$

so that $P_k = C\mu_1^k + D\mu_2^k$. By adjustment of the constants C,D to fit P_0,P_1 we find $P_k = (\mu_1-\mu_2)^{-1}(\mu_1^{k+1}-\mu_2^{k+1})$ for each $k \geq 0$. Consider now a real number θ with $0 < \theta < \pi/2$ and set $\lambda = \frac{1}{4\cos^2\theta}$, so that $\mu_1 = \frac{e^{i\theta}}{2\cos\theta}$ and $\mu_2 = \frac{e^{-i\theta}}{2\cos\theta}$. Then

$$P_k(\lambda) = \frac{\sin((k+1)\theta)}{2^k \cos^k(\theta)\sin\theta},$$

which vanishes when $\theta = \frac{j\pi}{k+1}$ with $j = 1,2,\cdots,m$.

Claim (iii) is obvious for $k = 1$, and we may assume $k \geq 2$. For $\ell \in \{2,\cdots,k\}$, the smallest root of P_ℓ is $\frac{1}{4\cos^2\left[\frac{\pi}{\ell+1}\right]}$, and $P_\ell(\lambda) > 0$ for $\frac{1}{4\cos^2\left[\frac{\pi}{\ell+1}\right]} > \lambda$. As $\frac{1}{4\cos^2\left[\frac{\pi}{\ell+1}\right]} > \frac{1}{4\cos^2\left[\frac{\pi}{k+1}\right]} > \lambda$, one has $P_\ell(\lambda) > 0$. The two smallest roots of P_{k+1} are

$$\lambda_1 = \frac{1}{4\cos^2\left[\frac{\pi}{k+2}\right]} \quad \lambda_2 = \frac{1}{4\cos^2\left[\frac{2\pi}{k+2}\right]},$$

and $P_{k+1} < 0$ on $]\lambda_1,\lambda_2[$. As $\frac{1}{4\cos^2\left[\frac{\pi}{k+1}\right]} < \lambda_2$ one has in particular $P_{k+1}(\lambda) < 0$. #

Since the polynomials P_k have coefficients in $\mathbb{Z}$, it makes sense to evaluate them at any number in our reference field $\mathbb{K}$. Given an integer $k \geq 1$, we define $\beta \in \mathbb{K}^*$ to be k-generic if

$$P_1(\beta^{-1}) \neq 0, P_2(\beta^{-1}) \neq 0,\cdots,P_k(\beta^{-1}) \neq 0.$$

Say that β is generic if it is k-generic for all k.

For example, any $\beta \in \mathbb{K}^*$ is 1-generic, and β is 2-generic if and only if $\beta \neq 1$.

If $\mathbb{K}$ is not algebraic over its prime field, transcendental numbers are obviously generic. If $\mathbb{K}$ contains the reals, Proposition 2.8.3 (ii) shows also that any β outside the interval $]0,4[$ is generic.

For $\beta \in \mathbb{K}^*$, let q be a number distinct from 0 and -1, in $\mathbb{K}$ or possibly in some quadratic extension of $\mathbb{K}$, such that $\beta = q^{-1}(q+1)^2$. Claim (iv) of Proposition 2.8.3 shows that β is not generic if and only if $\sum_{j=0}^{m} q^j = 0$ for some integer $m \geq 2$. In particular, if $\mathbb{K}$ is a finite field, no β is generic.

For generic $\beta \in \mathbb{K}^*$, we shall define inductively a nested sequence $(B_{\beta,k})_{k\geq 1}$ of associative $\mathbb{K}$-algebras with unit, and a normalized trace on each of these.

Set $B_{\beta,1} = \mathbb{K}$ and denote by tr_1 the tautological trace on $B_{\beta,1}$. Set $B_{\beta,2} = \mathbb{K}e_1 \oplus \mathbb{K}(1-e_1)$ where e_1 is an idempotent, not zero. Define $tr_2 : B_{\beta,2} \to \mathbb{K}$ by $tr_2(e_1) = \beta^{-1}$ and $tr_2(1-e_1) = 1 - \beta^{-1}$. Identify $B_{\beta,1}$ with the multiples of the identity in $B_{\beta,2}$. The Bratteli diagram of the pair $B_{\beta,1} \subset B_{\beta,2}$ is

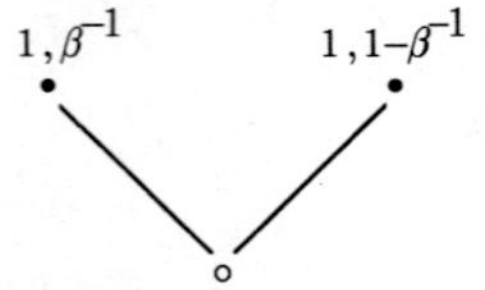

(see the end of Section 2.5 for the notation).

In the next lemma, we set

$$\left\{ {k \atop i} \right\} = \left[{k \atop i} \right] - \left[{k \atop i-1} \right]$$

Lemma 2.8.4. *Consider an integer $n \geq 2$, and assume $\beta \in \mathbb{K}^*$ is n-generic. Suppose there is given a nested sequence $(B_{\beta,k})_{1\leq k\leq n}$ of $\mathbb{K}$-algebras, together with traces $tr_k : B_{\beta,k} \to \mathbb{K}$ extending one another, such that the following hold for $k \in \{2,\cdots,n\}$:*

(i) $B_{\beta,k}$ *is generated by its unit, by elements $e_1,\cdots,e_{k-2}$ (all in $B_{\beta,k-1}$) and by e_{k-1}. Denote by $B'_{\beta,k}$ the two-sided ideal in $B_{\beta,k}$ generated by $e_1,\cdots,e_{k-1}$.*

(ii) *The generators satisfy the relations*

$$\begin{aligned} e_i^2 &= e_i, & \\ \beta\, e_i e_j e_i &= e_i & \text{if } |i-j| = 1, \\ e_i e_j &= e_j e_i & \text{if } |i-j| \geq 2 \end{aligned}$$

for all $i, j \in \{1,\cdots k-1\}$..

(iii) $B_{\beta,k}$ *is a direct sum of $\left[\frac{k}{2}\right] + 1$ factors Q_j^k, with Q_j^k isomorphic to the algebra of matrices of order $\left\{ {k \atop j} \right\}$, for $j = 0,1,\cdots,\left[\frac{k}{2}\right]$. One has $B'_{\beta,k} = \bigoplus_{j>0} Q_j^k$. Denote by d_k*

the (unique) nonzero idempotent in Q_0^k.

(iv) *The inclusion* $B_{\beta,k-1} \subset B_{\beta,k}$ *is described by the Bratteli diagram:*

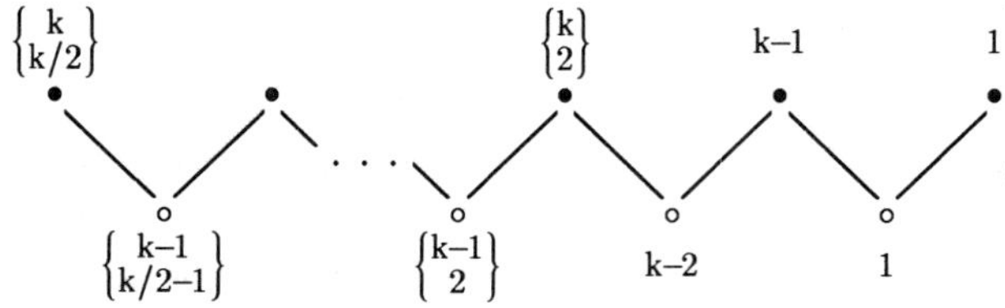

(k even),

or,

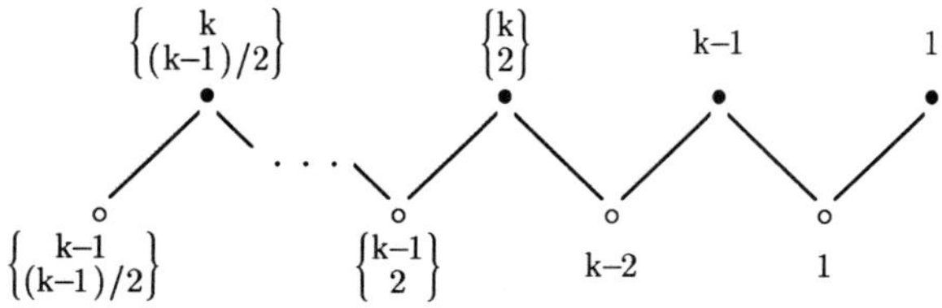

(k odd).

The index j *of the subfactors increases from right to left, so the white [resp. black] vertex on the extreme right represents* Q_0^{k-1} *[resp.* Q_0^k*].*

(v) *The dimension of* $B_{\beta,k}$ *is* $\frac{1}{k+1}\begin{bmatrix}2k\\ k\end{bmatrix}$.

(vi) *The trace satisfies* $tr_k(\mathbb{1}) = 1$ *and* $\beta tr_k(we_j) = tr_k(w)$ *for all* $w \in B_{\beta,j}$, $1 \le j \le k-1$.

(vii) *The value of* tr_k *on a minimal idempotent of* Q_j^k *is* $\beta^{-j}P_{k-2j}(\beta^{-1})$ *for* $j = 0,1,\cdots,\left[\frac{k}{2}\right]$, *so that in particular:*

(viii) tr_k *is faithful.*

Then one may define an algebra $B_{\beta,n+1}$ *containing* $B_{\beta,n}$ *and a trace* tr_{n+1} *on it, such that* (i) *to* (vii) *hold for* $k = n+1$. *Moreover, if* β *is* $(n+1)$*–generic, then* (viii) *holds also for* $k = n + 1$.

Suppose in addition that $\mathbb{K} = \mathbb{C}$, *that each* $B_{\beta,k}$ $(k \le n)$ *has a* C^**–algebra structure making the idempotents* e_i *self–adjoint projections, that* $\beta > 0$, *and that* $P_k(\beta^{-1}) > 0$ *for* $1 \le k \le n+1$. *Then* $B_{\beta,n+1}$ *also has a* C^**–algebra structure making* e_n *a self–adjoint projection, and the trace* tr_{n+1} *is faithful and positive.*

Proof (see §5.1 in [Jo1]). During the proof, we write B_k for $B_{\beta,k}$.

Both tr_n and it restriction tr_{n-1} to B_{n-1} are nondegenerate by (vii), since β is n-generic. Let $E_n : B_n \to B_{n-1}$ be the conditional expectation associated to tr_n as in 2.6.2. Set $\tau = \beta^{-1}$.

Then $E_n e_{n-1} E_n = \tau E_n$. Indeed, for all $y \in B_n$ and all $z \in B_{n-1}$, one has, first by 2.6.2(i) and then by (vi)

$$tr_{n-1}(\{E_n(e_{n-1}E_n(y)) - \tau E_n(y)\}z) = tr_n(e_{n-1}E_n(yz)) - \tau tr_{n-1}(E_n(yz)) = 0.$$

Thus $E_n(e_{n-1}E_n(y)) = \tau E_n(y)$ because tr_{n-1} is non-degenerate, and in particular $E_n(e_{n-1}) = \tau \mathbf{1}$..

Next we claim that

$$e_{n-1}E_n(e_{n-1}x) = \tau e_{n-1}x \qquad (*)$$

for all $x \in B_n$.

Obviously (*) holds for $x = \mathbf{1}$ because $E_n(e_{n-1}) = \tau \mathbf{1}$ by the previous claim.

Next we check that (*) holds if $x = ye_{n-1}$ for some $y \in B_{n-1}$. First, if $y = y_1 e_{n-2} y_2$ with $y_1, y_2 \in B_{n-2}$, then $e_{n-1}x = \tau y_1 e_{n-1} y_2 = \tau e_{n-1} y_1 y_2$, and

$$e_{n-1}E_n(e_{n-1}x) = \tau e_{n-1}E_n(e_{n-1})y_1y_2 = \tau^2 e_{n-1}y_1y_2 = \tau e_{n-1}x,$$

by B_{n-2}-linearity of E_n. If $y \in B_{n-2}$, then $e_{n-1}x = e_{n-1}y$, and again

$$e_{n-1}E_n(e_{n-1}x) = \tau e_{n-1}x.$$

Thus (*) holds when $x = ye_{n-1}$, for any $y \in B_{n-2} + B_{n-2}e_{n-2}B_{n-2}$, namely for all $y \in B_{n-1}$.

Now using the B_{n-1}-linearity of E_n, we see finally that (*) holds for all $x \in B_{n-1} + B_{n-1}e_{n-1}B_{n-1}$, namely for all $x \in B_n$.

Define B'_{n+1} to be the algebra obtained from the pair $B_{n-1} \subset B_n$ by the fundamental construction, and set

$$B_{n+1} = B'_{n+1} \oplus \mathbb{K} d_{n+1}$$

where d_{n+1} is a central idempotent. By Corollary 2.6.4, the two-sided ideal B'_{n+1} is generated by B_n and E_n. From now on, we write e_n (an element in B_{n+1}) rather than E_n (a mapping from B_n onto B_{n-1}). Then B_{n+1} is a multi-matrix algebra by 2.4.1 in which $e_1,\cdots,e_n$ generate B'_{n+1}, so that $\mathbf{1},e_1,\cdots,e_n$ generate B_{n+1}. We have checked (i) and (ii).

Define a map $J : B_n \to B'_{n+1} \oplus \mathbb{K}d_{n+1}$ by $J(x) = (x,0)$ if $x \in B'_n$ and $J(d_n) = (d_n,d_{n+1})$. (This is of course an abuse of notation: the first component of $J(d_n)$ is the element of $B'_{n+1} = \mathrm{End}_{B_{n-1}}(B_n)$ which is left (or right) multiplication by d_n!) Then J is obviously an injective morphism, so that we may (and we shall) identify B_n with a subalgebra of B_{n+1}. Now the shape of the diagram in (iv) follows from the induction hypothesis and Proposition 2.4.1b, and the dimensions from the relations

$$\left\{{n+1 \atop (n+1)/2}\right\} = \left\{{n \atop (n+1)/2-1}\right\} \qquad \text{(n odd)},$$

and

$$\left\{{n+1 \atop j}\right\} = \left\{{n \atop j}\right\} + \left\{{n \atop j-1}\right\} \qquad \text{(all n and j)}.$$

This shows (iv), and consequently also (iii). Now (v) follows from Lemma 2.8.2.

Define the trace $tr_{n+1} : B_{n+1} \to \mathbb{K}$ by assigning the weight $\tau^j P_{n+1-2j}(\tau)$ to the factor Q_j^{n+1}, as desired for (vii). Let f_j^k denote a minimal idempotent in Q_j^k. When n is even and $j = n/2$ we have

$$tr_{n+1}(f^n_{n/2}) = tr_{n+1}(f^{n+1}_{n/2}) = \tau^{n/2}P_1(\tau) = \tau^{n/2},$$

while

$$tr_n(f^n_{n/2}) = \tau^{n/2}P_0(\tau) = \tau^{n/2}.$$

In all other cases we have

$$\begin{aligned} tr_{n+1}(f^n_j) &= tr_{n+1}(f^{n+1}_j) + tr_{n+1}(f^{n+1}_{j+1}) \\ &= \tau^j\{P_{n-2j+1}(\tau) + \tau P_{n-2j-1}(\tau)\} \\ &= \tau^j P_{n-2j}(\tau), \end{aligned}$$

by the three term recursion for the P's. Consequently tr_{n+1} extends tr_n, and in particular $tr_{n+1}(\mathbf{1}) = 1$. (This point shows precisely why the factor $Q_0^{n+1} = \mathbb{K}d_{n+1}$ had to be introduced in B_{n+1}!) Incidentally, this gives the relation

$$\sum_{j=0}^{\left[\frac{k}{2}\right]} \begin{Bmatrix} k \\ j \end{Bmatrix} \tau^j P_{k-2j}(\tau) = 1,$$

which could also be checked directly.

We next verify the relation

$$\beta \mathrm{tr}_{n+1}(we_n) = \mathrm{tr}_{n+1}(w) \quad (**)$$

for all $w \in B_n$.

We check this first for $w \in B_{n-1}$. We may then as well assume that w is some minimal idempotent f_j^{n-1} of Q_j^{n-1}, where j is an integer with $0 \le j \le \left[\frac{n-1}{2}\right]$. But then we know from Corollary 2.6.4c that $f_j^{n-1} e_n$ is a minimal idempotent in Q_{j+1}^{n+1}. Consequently

$$\begin{aligned} \mathrm{tr}_{n+1}(f_j^{n-1} e_n) &= \tau^{j+1} P_{n+1-2(j+1)}(\tau) \\ &= \tau\{\tau^j P_{n-1-2j}(\tau)\} = \tau \mathrm{tr}_{n-1}(f_j^{n-1}) \end{aligned}$$

and (**) follows because tr_{n+1} extends tr_{n-1}.

We now set $w = xe_{n-1}y$ for some $x,y \in B_{n-1}$. Then $e_n w e_n = xe_n e_{n-1} e_n y = \tau x e_n y$ by (ii) and, using the case of (**) already checked

$$\mathrm{tr}_{n+1}(we_n) = \tau \mathrm{tr}_{n+1}(yxe_n) = \tau^2 \mathrm{tr}_{n+1}(yx).$$

On the other hand, by the induction hypothesis

$$\mathrm{tr}_{n+1}(w) = \mathrm{tr}_n(yxe_{n-1}) = \tau \mathrm{tr}_n(yx).$$

Thus (**) holds for $w = xe_{n-1}y$.

Consequently (**) holds for all w in $B_{n-1} + B_{n-1}e_{n-1}B_{n-1}$, namely for all $w \in B_n$. This proves (vi) and (vii). If β is $(n+1)$–generic, then (viii) follows from (vii).

Finally, if $\mathbb{K} = \mathbb{C}$, and the B_k are C^*–algebras for $k \le n$, then B'_{n+1} also may be given a C^*–structure making the idempotent e_n self–adjoint; see the discussion in Appendix IIa, or the remark under 2.6.5. Clearly B_{n+1} also has a C^*–structure. Moreover the weights of the trace on B_{n+1} are strictly positive by (vii). #

Theorem 2.8.5. *Consider an integer* $k \geq 1$ *and a number* $\beta \in \mathbb{K}^*$ *such that* $P_j(\beta^{-1}) \neq 0$ *for* $j \leq k-1$, *where* $(P_j)_{j\geq 1}$ *are the polynomials of Proposition 2.8.3.*

(a) $\mathcal{A}_{\beta,k}$ *is a multi-matrix algebra of dimension* $\frac{1}{k+1}\begin{bmatrix}2k\\k\end{bmatrix}$, *isomorphic to* $\bigoplus_{j=0}^{m} M_{\left\{k \atop j\right\}}(\mathbb{K})$, *where* $m = \left[\frac{k}{2}\right]$ *and* $\left\{k \atop j\right\} = \begin{bmatrix}k\\j\end{bmatrix} - \begin{bmatrix}k\\j-1\end{bmatrix}$.

(b) *There exists a unique normalized trace* $tr_k : \mathcal{A}_{\beta,k} \to \mathbb{K}$ *such that*

$$\beta tr_k(w\epsilon_j) = tr_k(w)$$

whenever $1 \leq j \leq k-1$ *and* w *is in the subalgebra generated by* $\mathbf{1}, \epsilon_1, \cdots, \epsilon_{j-1}$. *Moreover* tr_k *is faithful if* $P_k(\beta^{-1}) \neq 0$.

(c) *The natural map* $\mathcal{A}_{\beta,k-1} \to \mathcal{A}_{\beta,k}$ *is injective and* tr_k *extends* tr_{k-1}.

(d) *If* $B_{\beta,k}$ *is as in Lemma 2.8.4, the assignment* $\epsilon_j \mapsto e_j$ $(1 \leq j \leq k-1)$ *extends to an isomorphism from* $\mathcal{A}_{\beta,k}$ *onto* $B_{\beta,k}$.

(e) *The trace* tr_k *on* $\mathcal{A}_{\beta,k}$ *also satisfies*

$$\beta\, tr_k(\epsilon_j w) = tr_k(w)$$

whenever $1 \leq j \leq k-2$ *and* w *is a word in* $\{\epsilon_{j+1}, \cdots, \epsilon_{k-1}\}$. *More generally we have*

$$tr_k(uv) = tr_k(u) tr_k(v)$$

whenever u *is a word in* $\{\epsilon_1, \cdots, \epsilon_j\}$ *and* w *is a word in* $\{\epsilon_{j+1}, \cdots, \epsilon_{k-1}\}$.

(f) *The map* $\epsilon_j \longmapsto \epsilon_{k-j}$ *extends to a trace preserving automorphism* σ_k *of* $\mathcal{A}_{\beta,k}$. *Furthermore* σ_k *is inner in case* $\mathbb{K}$ *contains a solution* q *of the equation* $q^{-1}(q+1)^2 = \beta$.

Proof. Claims (i) and (ii) of the previous lemma show that the map of (d) is a morphism onto. Claim (v) of the lemma and Proposition 2.8.1 show that this morphism is injective. Consequently, assertions (a) and (c) and the existence of tr_k in (b) follow from the lemma. But the relation in (b) together with the normalization $tr_k(\mathbf{1}) = 1$ and the trace property $tr_k(xy) = tr_k(yx)$ suffice to compute the trace on any word in the generators $\{\epsilon_i\}$ of $\mathcal{A}_{\beta,k}$, so the trace is unique.

We prove by induction on m $(j+1 \leq m \leq k-1)$ that the formula of (e) holds for $u \in \text{alg}\{\mathbf{1}, \epsilon_1, \cdots, \epsilon_j\}$ and $w \in \text{alg}\{\mathbf{1}, \epsilon_{j+1}, \cdots, \epsilon_m\}$. The case $m = j+1$ is clear from (b).

Suppose that $m > j+1$ and that the result is verified for elements of $\text{alg}\{\mathbf{1},\epsilon_{j+1},\cdots,\epsilon_{m-1}\}$. It suffices then to deal with a reduced word $w = x\epsilon_m y$ where x and y are words in $\{\epsilon_{j+1},\cdots,\epsilon_{m-1}\}$. Then $tr_k(w) = \beta^{-1}tr_k(xy)$, and $tr_k(uw) = tr_k(yux\epsilon_m) = \beta^{-1}tr_k(uxy) = tr_k(u)\,\beta^{-1}tr_k(xy)$, where the last step follows from the induction hypothesis.

Let q be an element of $\mathbb{K}$, or of a quadratic extension of $\mathbb{K}$, satisfying $q^{-1}(1+q)^2 = \beta$. Define elements

$$\gamma_i = (q+1)\epsilon_i - \mathbf{1} \qquad \text{and}$$
$$c_i = (\gamma_1\gamma_2\cdots\gamma_{j-1})\cdots(\gamma_1\gamma_2)\gamma_1$$

in $\mathcal{A}_{\beta,k}\otimes_{\mathbb{K}}\mathbb{K}(q)$ for $1 \le i \le k-1$. These are invertible, with $\gamma_i^{-1} = (q^{-1}+1)\epsilon_i - \mathbf{1}$, and one verifies by induction that $c_j\gamma_i c_j^{-1} = \gamma_{j-1}$ and $c_j\epsilon_i c_j^{-1} = \epsilon_{j-1}$ for $i \le j-1$. In particular, $\sigma_k\colon x \mapsto c_k x c_k^{-1}$ is the automorphism of part (f). This automorphism is trace preserving, because the trace tr_k extends uniquely to $\mathcal{A}_{\beta,k}\otimes_{\mathbb{K}}\mathbb{K}(q)$. #

Corollary 2.8.6. *Consider an integer* $k \ge 1$ *and an arbitrary number* $\beta \in \mathbb{K}^*$. *Let* φ *be the homomorphism* $\mathcal{A}_{\beta,k} \to \mathcal{A}_{\beta,k+1}$ *which, for* $j \le k-1$, *maps* ϵ_j *viewed as a generator of* $\mathcal{A}_{\beta,k}$ *to* ϵ_j *(sic) viewed as a generator of* $\mathcal{A}_{\beta,k+1}$.

(a) $\mathcal{A}_{\beta,k}$ *is of dimension* $\frac{1}{k+1}\begin{bmatrix}2k\\k\end{bmatrix}$.

(b) φ *is an injection and any element* $x \in \mathcal{A}_{\beta,k+1}$ *can be written as* $x = \varphi(u) + \sum\varphi(u_i)\epsilon_k\varphi(w_i)$, where u, v_i, and w_i are elements of $\mathcal{A}_{\beta,k}$.

(c) *There is a sequence of traces* $tr_\ell\colon \mathcal{A}_{\beta,\ell} \to \mathbb{K}$ $(1 \le \ell \le k)$ *such that* $tr_\ell(\mathbf{1}) = 1$, and $tr_{\ell+1}(\varphi(u) + \sum\varphi(u_i)\epsilon_\ell\,\varphi(w_i)) = tr_\ell(u) + \beta^{-1}\sum tr_\ell(v_i w_i)$ for all u, u_i, $w_i \in \mathcal{A}_{\beta,\ell}$

Proof. It is enough to prove the corollary for any extension of the field $\mathbb{K}$, so that we may assume $\mathbb{K}$ to contain infinitely many generic numbers.

Assume first that β is generic. Then $\mathcal{A}_{\beta,k}$ has a basis over $\mathbb{K}$ made of the $\frac{1}{k+1}\begin{bmatrix}2k\\k\end{bmatrix}$ reduced monomials (see 2.8.1 and 2.8.5a), say $(\epsilon_\sigma)_{\sigma\in S}$. The structure constants are defined by

$$\epsilon_\sigma\epsilon_\tau = \sum_{\mu\in S} c^\mu_{\sigma,\tau}\epsilon_\mu. \quad (*)$$

Proposition 2.8.1 shows that, for any given pair (σ,τ), all but one of the $c^\mu_{\sigma,\tau}$ vanish and the one non-zero $c^{\mu}_{\sigma,\tau}$ is a power of β^{-1} depending on σ and τ. In particular there are monomials $c^{\mu}_{\sigma,\tau}(t) \in \mathbb{K}[t]$ such that $c^{\mu}_{\sigma,\tau}$ as above is just $c^{\mu}_{\sigma,\tau}(\beta^{-1})$ for any $\sigma,\tau,\mu \in S$.

Define now the "generic" algebra $\mathcal{A}_{gen,k}$ over the polynomial ring $\mathbb{K}[t]$ as the free $\mathbb{K}[t]$-module over S, with canonical basis denoted again by $(\epsilon_\sigma)_{\sigma\in S}$, and with multiplication defined by

$$\epsilon_\sigma\epsilon_\tau = \sum_{\mu\in S} c^{\mu}_{\sigma,\tau}(t)\epsilon_\mu \qquad \sigma,\tau \in S.$$

The relations which express that this multiplication is associative are polynomial, and they hold when t is specialized at β^{-1} for any generic $\beta \in \mathbb{K}^*$, by Theorem 2.8.5. Hence they hold identically, and $\mathcal{A}_{gen,k}$ is a well-defined associative algebra. Indeed, it is the algebra with unit over $\mathbb{K}[t]$ abstractly defined by generators $\epsilon_1,\cdots,\epsilon_{k-1}$ and relations

$$\begin{aligned} &\epsilon_i^2 = \epsilon_i \\ &\epsilon_i\epsilon_j\epsilon_i = t\epsilon_i \qquad \text{if } |i-j| = 1 \\ &\epsilon_i\epsilon_j = \epsilon_j\epsilon_i \qquad \text{if } |i-j| \geq 2 \end{aligned}$$

Consider finally an arbitrary $\beta \in \mathbb{K}^*$. Then $\mathcal{A}_{\beta,k}$ is isomorphic to $\mathcal{A}_{gen,k} \otimes_{\mathbb{K}[t]} \mathbb{K}$, where $\mathbb{K}$ is made a $\mathbb{K}[t]$-module by $c(t)\lambda = c(\beta^{-1})\lambda$ for $c(t) \in \mathbb{K}[t]$ and $\lambda \in \mathbb{K}$. This shows claim (a). That φ is an injection follows similarly. As observed in the proof of (a), there exist bases of $\mathcal{A}_{\beta,k}$ and $\mathcal{A}_{\beta,k+1}$ consisting of the reduced monomials of 2.8.1, and claim (b) follows from this. We leave the details of part (c) to the reader; compare, however, 2.9.6. #

Remark: In general the traces tr_ℓ of claim (c) are not faithful; see Theorem 2.9.6.d.

Consider now the situation at the end of Section 2.7: One has a multi-matrix pair $M_0 \subset M_1$ and a Markov trace $tr : M_1 \to \mathbb{K}$ of modulus β; these generate a tower, and the conditional expectations $E_j : M_j \to M_{j-1}$ for $j = 1,\cdots,k-1$ generate (together with $\mathbb{1}$) a subalgebra $A_{tr,k}(M_0 \subset M_1)$ of M_k.

Proposition 2.8.7. *Suppose that* $\beta \in \mathbb{K}^*$ *satisfies* $P_j(\beta^{-1}) \neq 0$ *for* $i \leq j \leq k$.

(a) *Suppose that* $\chi : \mathcal{A}_{\beta,k} \to C$ *is a surjective homomorphism of* $\mathbb{K}$*-algebras and that* C *has a trace* tr *satisfying* $\beta\, tr(w\chi(\epsilon_j)) = tr(w)$ *for* $i \leq j \leq k-1$ *and* $w \in \chi(\mathcal{A}_{\beta,j})$. *Then* χ *is an isomorphism and* tr *is non-degenerate.*

(b) *In particular, with the notation above, the map* $\chi : \epsilon_j \mapsto E_j$ *extends to an isomorphism of* $\mathcal{A}_{\beta,k}$ *onto* $A_{tr,k}(M_0 \subset M_1)$, *and the restriction to* $A_{tr,k}(M_0 \subset M_1)$ *of the Markov trace* $tr : M_k \to \mathbb{K}$ *is non-degenerate.*

<u>Proof.</u> (a) It follows from 2.8.5(b) that $tr \circ \chi = tr_k$. Hence if $x \in \ker(\chi)$, then for all $y \in \mathcal{A}_{\beta,k}$ one has $tr_k(xy) = tr(\chi(x)\chi(y)) = 0$, so that $x = 0$, by the non-degeneracy of tr_k. Thus χ is an isomorphism and tr is non-degenerate.

(b) By 2.7.5, the map χ extends to an homomorphism of $\mathcal{A}_{\beta,k}$ onto $A_{tr,k}(M_0 \subset M_1)$, and 2.7.5 together with 2.8.5(b) imply that $tr \circ \chi = tr_k$. Thus (b) follows from (a). #

Suppose $\beta \in \mathbb{K}^*$ is generic. The following picture sums up the structure of the traced algebras introduced in this section (with $\tau = \beta^{-1}$).

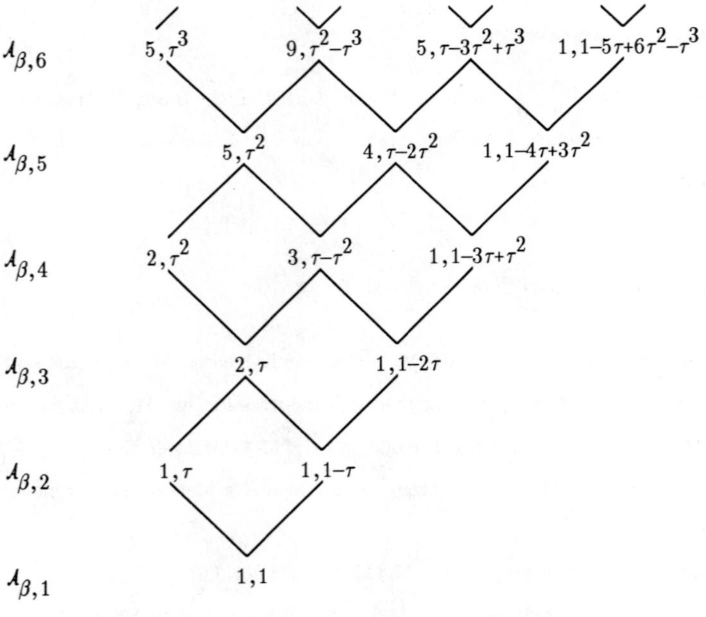

2.9. An approach to the non–generic case.

If $\beta \in \mathbb{K}^*$ is non–generic, then

(1) The algebra $\mathcal{A}_{\beta,k}$ defined by generators and relations as in Section 2.8 need not be semi–simple.

(2) Given a multi–matrix pair $M_0 \subset M_1$ and a Markov trace tr of modulus β on M_1, the restriction of tr to $A_{tr,k}(M_0 \subset M_1)$, the algebra generated by $\{\mathbf{1}, E_1, \cdots, E_{k-1}\}$ in M_k, need not be faithful.

(3) Given a second such pair $\tilde{M}_0 \subset \tilde{M}_1$ and a Markov trace $\tilde{tr}$ of modulus β on $\tilde{M}_1$, the algebras $A_{tr,k}(M_0 \subset M_1)$ and $A_{\tilde{tr},k}(\tilde{M}_0 \subset \tilde{M}_1)$ need not be isomorphic.

All this contrasts with the generic case described in 2.8.5 and 2.8.6. The modulus $\beta = 1$ illustrates these phenomena.

Example 2.9.1. The algebra $\mathcal{A}_{1,3}$ is not semi–simple. (This is a particular case of Theorem II.10 in Appendix II.c.)

Proof. Let $T = \left\{ \begin{bmatrix} * & * \\ 0 & * \end{bmatrix} \right\}$ be the algebra of 2–by–2 upper triangular matrices over $\mathbb{K}$. As T is not semi–simple, it suffices to show that T is a quotient of $\mathcal{A}_{1,3}$. But the assignment

$$1 \mapsto \begin{bmatrix} 1 & 0 \\ 0 & 1 \end{bmatrix} \quad \epsilon_1 \mapsto \begin{bmatrix} 1 & 0 \\ 0 & 0 \end{bmatrix} \quad \epsilon_2 \mapsto \begin{bmatrix} 1 & 1 \\ 0 & 0 \end{bmatrix}$$

extends to a homomorphism from $\mathcal{A}_{1,3}$ onto T. #

Example 2.9.2. Consider the pair $M_0 = \mathbb{C} \oplus \mathbb{C}$ imbedded in $M_1 = \mathrm{Mat}_3(\mathbb{C}) \oplus \mathrm{Mat}_3(\mathbb{C})$ with inclusion matrix $\Lambda = \begin{bmatrix} 1 & 2 \\ 2 & 1 \end{bmatrix}$, together with the trace tr on M_1 with weight vector (1,–1). Then tr is a Markov trace of modulus 1 on M_1. Consider also a pair $\tilde{M}_0 = \tilde{M}_1$ with any faithful trace $\tilde{tr}$ on $\tilde{M}_1$; then $\tilde{tr}$ is evidently a Markov trace of modulus 1 on $\tilde{M}_1$. We have

(2) The restriction of tr to $A_{tr,1}(M_0 \subset M_1)$ is not faithful.

(3) $A_{tr,k}(M_0 \subset M_1)$ and $A_{\tilde{tr},k}(\tilde{M}_0 \subset \tilde{M}_1)$ are non–isomorphic for all $k \geq 2$.

Proof. The matrix $\Lambda\Lambda^t = \Lambda^2 = \begin{bmatrix} 5 & 4 \\ 4 & 5 \end{bmatrix}$ has eigenvectors (1,–1) and (1,1) with eigenvalues 1 and 9 respectively. The Perron–Frobenius eigenvalue 9 is also the index $[M_1:M_0]$. But the other eigenvector (1,–1) also defines a Markov trace tr on M_1 with modulus $\beta = 1$.

Let $M_0 \subset M_1 \subset M_2 \cdots$ be the tower generated by $M_0 \subset M_1$. Since M_k is generated as an algebra by M_1 and $A_{tr,k}(M_0 \subset M_1)$, if for some n_0 the algebras A_{tr,n_0} and A_{tr,n_0+1} were equal, then $M_{n_0} = M_{n_0+1}$ as well, and therefore $M_k = M_{n_0}$ for all $k \geq n_0$. But $\dim_{\mathbb{C}} M_k$ increases as $[M_1:M_0]^k = 9^k$, by Proposition 2.4.2. Hence

$$A_{tr,k}(M_0 \subset M_1) \subsetneq A_{tr,k+1}(M_0 \subset M_1)$$

for all k. On the other hand $\tilde{M}_1 = \tilde{M}_n$ for all k and $A_{\tilde{tr},k}(\tilde{M}_0 \subset \tilde{M}_1) \cong \mathbb{C}$ for all k. This proves (3).

The algebra $A_{tr,2}(M_0 \subset M_1)$ is spanned by $\mathbb{1}$ and E_1, and is of dimension 2, since $A_{tr,2} \supsetneq A_{tr,1} \cong \mathbb{C}$. The trace tr on M_2 restricted to $A_{tr,2}$ is given by

$$tr(a+bE_1) = a + b \qquad (a,b \in \mathbb{C}).$$

It is not faithful because

$$tr((\mathbb{1}-E_1)(a+bE_1)) = a\, tr(\mathbb{1}-E_1) = 0$$

for all $a,b \in \mathbb{C}$. #

We do not intend to make a detailed study of the algebras $A_{tr,k}(M_0 \subset M_1)$ when β is not generic. But we want to describe the structure of the unique quotient of $\mathcal{A}_{\beta,k}$ on which the usual rules $tr(\mathbb{1}) = 1$ and $\beta\, tr(we_i) = tr(w)$ for $w \in alg\{\mathbb{1},e_1,\cdots,e_{i-1}\}$ defines a faithful normalized trace. (Here e_i denotes the image of ϵ_i in the quotient.)

<u>The algebras $B_{\beta,k}$</u>. For the rest of this section we fix a $\beta \in \mathbb{K}^*$ which is n–generic but not (n+1)–generic for some $n \geq 1$. That is $P_k(\beta^{-1}) \neq 0$ for $k \leq n$, but $P_{n+1}(\beta^{-1}) = 0$. We again define a nested sequence $(B_{\beta,k})_{k\geq 1}$ of multi–matrix algebras over $\mathbb{K}$, and a consistent family of normalized faithful traces tr_k on these algebras. For $k \leq n$, define $B_{\beta,k}$ and tr_k exactly as in Lemma 2.8.4; since β is n–generic there is no problem in doing so. For $k \geq n$ define $B_{\beta,k+1}$ to be the algebra obtained by applying the fundamental construction to the pair $B_{\beta,k-1} \subset B_{\beta,k}$. Observe that $B_{\beta,n+1}$ is the same as $B'_{\beta,n+1}$ in 2.8.4. For $k \leq n+1$, define tr_k as in Lemma 2.8.4; then tr_{n+1} is also faithful because P_{n+1} does not appear in the computation of the weights of

the trace on $B_{\beta,n+1} = \underset{j>0}{\oplus} Q_j^{n+1}$. Also since $P_{n+1}(\tau) = 0$, the trace on $B_{\beta,n+1} = B'_{\beta,n+1}$ extends that on $B_{\beta,n}$; it thus follows from 2.8.4(vi) (with $k = n+1$) that tr_n is a Markov trace of modulus β on $B_{\beta,n-1} \subset B_{\beta,n}$. For $k \geq n+1$, we define tr_k as in Proposition 2.7.4. Thus tr_k is a Markov trace on $B_{\beta,k-1} \subset B_{\beta,k}$ for $k \geq n$, but *not* for $k < n$. Note that $B_{\beta,k}$ is a multi–matrix algebra generated by the identity and idempotents $\{e_1,\cdots,e_{k-1}\}$ satisfying the relations 2.8.4(ii); in fact these relations hold for $\{e_1,\cdots,e_n\}$ by 2.8.4 and for $\{e_n,e_{n+1},\cdots\}$ by 2.7.5. For $k \geq n+1$ the identity is contained in the algebra generated by $\{e_1,\cdots,e_{k-1}\}$, in contrast to the case of generic β; this follows from 2.6.4.

Note that if $\mathbb{K} = \mathbb{C}$ and $\beta = 4\cos^2(\pi/(n+2))$, then the algebras $B_{\beta,k}$ can be given a C^*–structure such that the generators $\{e_i\}$ are self–adjoint projections, and the trace is faithful and positive. This is shown in 2.8.4 for $k \leq n$. The assertion for $k \geq n+1$ follows, because the tower construction for a pair of finite dimensional C^*–algebras with a positive Markov trace produces a chain of C^*–algebras with a positive trace, and self–adjoint projections e_i; see the discussion in Appendix IIa.

Example 2.9.3. Let $\beta = 1$, so that $n = 1$. The definitions above (*cum grano salis*) give $B_{\beta,k} = B_{\beta,1} = \mathbb{K}$ for all $k \geq 1$.

Example 2.9.4. Assume that the characteristic of $\mathbb{K}$ is not 2 and let $\beta = 2$, so that $n = 2$. The structure of the algebras $B_{\beta,k}$ and of the traces tr_k is shown in figure 2.9.4 below.

Example 2.9.5. Assume that $\mathbb{K}$ contains $\mathbb{Q}(\sqrt{5})$ and choose $\beta \in \{4\cos^2(\pi/5), 4\cos^2(2\pi/5)\}$, so that $n = 3$. The picture (with $\tau = \beta^{-1}$ satisfying $P_4(\tau) = \tau^2 - 3\tau + 1 = 0$) is given below in figure 2.9.5.

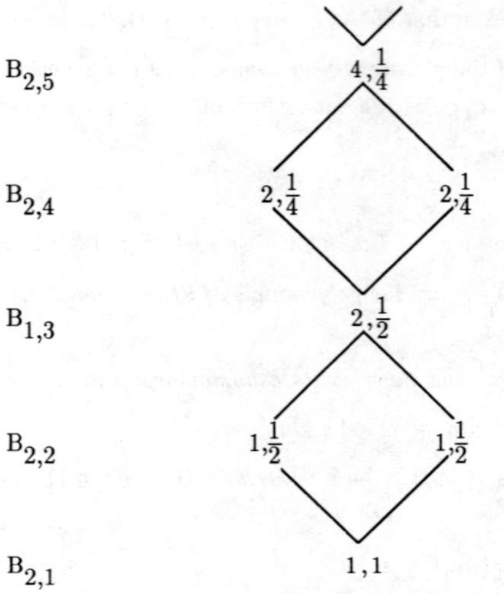

Figure 2.9.4.

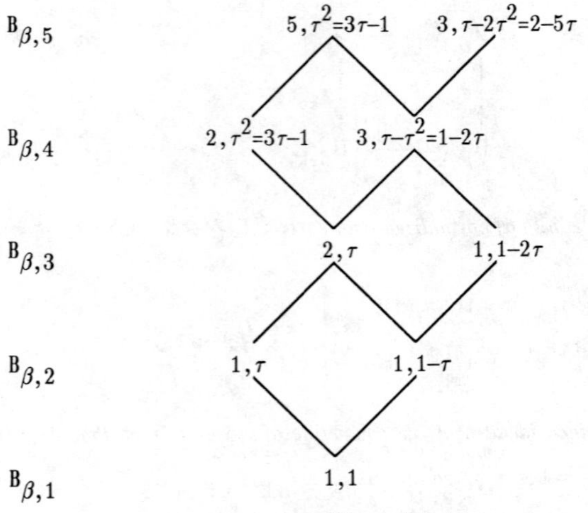

Figure 2.9.5.

In general, the picture for the $B_{\beta,k}$'s is obtained from that of the $\mathcal{A}_{\beta,k}$'s at the end of Section 2.8 by deleting the factor Q_0^{n+1} (represented as the extreme right point in the (n+1)st row) as well as all factors above and to the right.

Theorem 2.8.5 gives a complete description of $\mathcal{A}_{\beta,k}$ when β is (k–1)–generic. The following theorem indicates how part of the picture changes when β is not generic. Recall that we may (and do) always identify $\mathcal{A}_{\beta,k}$ with a subalgebra of $\mathcal{A}_{\beta,k+1}$ (see Corollary 2.8.6) and that $B_{\beta,k}$ is also a subalgebra of $B_{\beta,k+1}$.

Theorem 2.9.6. *Consider an integer* $n \geq 2$. *Let* $\beta \in \mathbb{K}^*$ *be such that* $P_j(\beta^{-1}) \neq 0$ *for* $j \leq n$ *and* $P_{n+1}(\beta^{-1}) = 0$, *where* $(P_j)_{j\geq 1}$ *are the polynomials of Proposition 2.8.3. Then one has for all* $k \geq 1$,

(a) $B_{\beta,k}$ *is a multi–matrix algebra, and there exists a homomorphism* π_k *of* $\mathcal{A}_{\beta,k}$ *onto* $B_{\beta,k}$ *mapping each generator* ϵ_j *onto* e_j $(1 \leq j \leq k-1)$.

(b) *There exists a normalized trace* $tr_k : B_{\beta,k} \to \mathbb{K}$ *such that, for any* $j \in \{1,\cdots,k-1\}$

$$\beta tr_k(we_j) = tr_k(w)$$

whenever w *is in the subalgebra* $B_{\beta,j}$ *of* $B_{\beta,k}$. *Moreover* tr_k *is faithful and the restriction of* tr_k *to* $B_{\beta,j}$ *is* tr_j *for* $j \leq k$.

(c) *For* $k \geq 2$ *the following diagram commutes.*

$$\begin{array}{ccc} \mathcal{A}_{\beta,k-1} & \longrightarrow & \mathcal{A}_{\beta,k} \\ \downarrow{\scriptstyle \pi_{k-1}} & & \downarrow{\scriptstyle \pi_k} \\ B_{\beta,k-1} & \longrightarrow & B_{\beta,k} \end{array}$$

(d) *There is a unique family of normalized traces* $tr_k : \mathcal{A}_{\beta,k} \to \mathbb{K}$ *such that*

$$(*) \qquad \left.\begin{aligned} tr_k(x) &= tr_{k-1}(x) \\ \beta\, tr_k(x\epsilon_{k-1}) &= tr_{k-1}(x) \end{aligned}\right\} \quad (x \in \mathcal{A}_{\beta,k-1}).$$

If $I_{\beta,k}$ *denotes the two sided ideal in* $\mathcal{A}_{\beta,k}$ *consisting of those* x *such that* $tr_k(xy) = 0$ *for all* $y \in \mathcal{A}_{\beta,k}$, then $I_{\beta,k} = \ker(\pi_k)$, *so that* $B_{\beta,k} \cong \mathcal{A}_{\beta,k}/I_{\beta,k}$.

(e) *Suppose* $(C_k)_{k\geq 1}$ *is an increasing sequence of* $\mathbb{K}$–*algebras and* $\psi_k : \mathcal{A}_{\beta,k} \to C_k$ *are surjective homomorphisms such that* $\psi_k\big|_{\mathcal{A}_{\beta,k-1}} = \psi_{k-1}$ *for all* k. *Suppose further that each* C_k *has a faithful normalized trace* $tr_k : C_k \to \mathbb{K}$ *satisfying*

$$\mathrm{tr}_k\Big|_{C_{k-1}} = \mathrm{tr}_{k-1}, \textit{ and}$$
$$\beta \mathrm{tr}_k(w\psi_k(\epsilon_{k-1})) = \mathrm{tr}_{k-1}(w)$$

for $w \in C_{k-1}$. *Then* $C_k \cong B_{\beta,k} \cong \mathcal{A}_{\beta,k}/I_{\beta,k}$.

(f) *The trace* tr_k *on* $\mathcal{A}_{\beta,k}$ *also satisfies*

$$\beta\, \mathrm{tr}_k(\epsilon_j w) = \mathrm{tr}(w)$$

whenever $1 \leq j \leq k-2$ *and* w *is an element of* $\mathrm{alg}\,\{\mathbb{1}, e_{j+1}, \cdots, e_{k-1}\}$. *More generally, we have*

$$\mathrm{tr}_k(uw) = \mathrm{tr}_k(u)\, \mathrm{tr}_k(w)$$

whenever $u \in \mathrm{alg}\,\{\mathbb{1}, e_1, \cdots, e_j\}$ *and* $w \in \mathrm{alg}\,\{\mathbb{1}, e_{j+1}, \cdots, e_{k-1}\}$.

(g) *The map* $\epsilon_j \mapsto \epsilon_{k-j}$ *extends to a trace preserving automorphism* σ_k *of* $\mathcal{A}_{\beta,k}$, *and* $e_j \mapsto e_{k-j}$ *extends to a trace preserving automorphism* $\overline{\sigma}_k$ *of* $B_{\beta,k}$. *These automorphisms are inner in case* $\mathbb{K}$ *contains an element* q *satisfying* $q^{-1}(q+1)^2 = \beta$.

Proof. Claims (a) to (c) follow from the construction of the $B_{\beta,k}$ above. The traces $\mathrm{tr}_k|B_{\beta,k} \circ \pi_k$ on $\mathcal{A}_{\beta,k}$ satisfy (∗). The uniqueness statements in (b) and (d) are proved as in 2.8.5(b). We have $\mathrm{tr}_k(xy) = \mathrm{tr}_k(\pi_k(x)\pi_k(y))$, so that if $x \in \ker(\pi_k)$, then $x \in I_{\beta,k}$. Conversely if $x \in I_{\beta,k}$, then $\pi_k(x) = 0$ by faithfulness of tr_k on $B_{\beta,k}$. This proves (d), and (e) follows similarly. Statement (f) is proved as 2.8.5(e), and statement (g) as 2.8.5(f). #

Corollary 2.9.7. *Suppose that* $\mathbb{K} \supset \mathbb{R}$, *that* $M_0 \subset M_1$ *is a pair of multi–matrix algebras over* $\mathbb{K}$, *and that* tr *is a* *positive* *Markov trace on* M_1 *of modulus* $\beta = [M_1:M_0]$. *Then* $A_{\mathrm{tr},k}(M_0 \subset M_1)$ *is isomorphic to* $B_{\beta,k}$ *for all* $k \geq 1$.

Proof. This follows from 2.8.5 and 2.8.7 when β is generic, so we suppose that β is non–generic. Let $(M_k)_{k\geq 1}$ be the tower of algebras generated by $M_0 \subset M_1$, and tr the extension of the trace to $\bigcup_k M_k$, as described in Section 2.7. Both $\beta = \|\Lambda_{M_0}^{M_1}\|^2$ and the weights of the trace are real and positive; see 2.7.3. Using the path model (2.4.6 and 2.6.5), we see that it is possible to choose a system of matrix units $T_{\xi,\eta}$ for the algebra M_k so that the idempotents E_i $(1 \le i \le k-1)$ are positive linear combinations of certain minimal idempotents $T_{\xi,\xi'}$; see especially 2.6.5.2 and 2.6.5.4. Let $M_k^{\mathbb{R}}$ be the $\mathbb{R}$–linear span of the matrix units generating M_k. Thus $M_k^{\mathbb{R}}$ is a multi–matrix algebra over $\mathbb{R}$, and $M_k = M_k^{\mathbb{R}} \otimes_{\mathbb{R}} \mathbb{K}$. Let $A_k^{\mathbb{R}}$ be the $\mathbb{R}$–subalgebra of $M_k^{\mathbb{R}}$ generated by $\{\mathbf{1},E_1,\cdots,E_{k-1}\}$. The trace tr restricts to a positive $\mathbb{R}$–valued trace on $M_k^{\mathbb{R}}$. Note that $A_k^{\mathbb{R}}$ is closed under the $\mathbb{R}$–linear involution $*$ of $M_k^{\mathbb{R}}$ defined by $T^*_{\xi,\eta} = T_{\eta,\xi}$. Positivity of the trace implies that $tr(x^*x) > 0$ for all non–zero $x \in M_k$, and as this holds in particular for $x \in A_k^{\mathbb{R}}$, we conclude that $tr|_{A_k^{\mathbb{R}}}$ is faithful. It follows by linear algebra that tr is also faithful on $A_{tr,k}(M_0 \subset M_1) = A_k^{\mathbb{R}} \otimes_{\mathbb{R}} \mathbb{K}$, and therefore 2.9.6(e) implies the conclusion. #

The proof of Theorem 2.1.8 is now complete.

Theorem 2.9.8. ([Jo2]). *Let* $n \ge 2$ *be an integer and suppose that* $\beta \in \mathbb{K}^*$ *is* n–*generic but not* (n+1)–*generic. Then the generating function* $f_n(x)$ *for* $(\dim_{\mathbb{K}}(B_{\beta,k+1})_{k\ge 0}$ *is*

$$f_n(z) = \sum_{k=0}^{\infty} \dim_{\mathbb{K}}(B_{\beta,k+1})z^k = \frac{P_{n-1}(z)}{P_{n+1}(z)},$$

where the P_j *are the polynomials of Proposition 2.8.3.*

Proof. Set $\Lambda_n = \Lambda_{B_{\beta,n-1}}^{B_{\beta,n}}$ and $b_k^n = \dim_{\mathbb{K}}(B_{\beta,k})$. Also let $\xi^{(n,k)}$ be the vector of dimensions of the multi–matrix algebra $B_{\beta,k}$. Note that the Bratteli diagram for $B_{\beta,n-1} \subset B_{\beta,n}$ is the Coxeter graph A_{n+1}, with a particular bicoloration and labelling of the vertices. (See 2.8.4(iv) for the picture, substituting n for k.) Thus for n odd Λ_n is the $\frac{n+1}{2}$–by–$\frac{n+1}{2}$ Jordan block

$$\begin{bmatrix} 1 & 0 & & & \\ 1 & 1 & \ddots & & \\ 0 & \ddots & \ddots & \ddots & \\ & \ddots & \ddots & 1 & 0 \\ & & 0 & 1 & 1 \end{bmatrix},$$

while for n even Λ_n is the $(\frac{n}{2}+1)$–by–$\frac{n}{2}$ matrix

$$\begin{bmatrix} 1 & 0 & & & & \\ 1 & 1 & \ddots & & & \\ 0 & 1 & \ddots & \ddots & & \\ & \ddots & \ddots & 1 & 0 & \\ & & \ddots & 1 & 1 & \\ & & & 0 & 1 & \end{bmatrix}.$$

In order to accomodate vectors and matrices of different sizes, we adopt the convention that $\mathbb{R}^d$ imbeds in $\mathbb{R}^{d+1}$ via

$$(\xi_1,\cdots,\xi_d)^t \mapsto (0,\xi_1,\cdots,\xi_d)^t.$$

With this convention we have <u>for n odd</u>

$$\xi^{(n,k)} = \begin{cases} (\Lambda_n\Lambda_n^t)^{(k-1)/2}\xi & \text{for } k \text{ odd} \\ \Lambda_n^t(\Lambda_n\Lambda_n^t)^{k/2-1}\xi & \text{for } k \text{ even,} \end{cases}$$

where $\xi = (0,0,\cdots 0,1)^t$. Hence

$$b_k^n = \|\xi^{(n,k)}\|^2 = \langle(\Lambda_n\Lambda_n^t)^{k-1}\xi\,|\,\xi\rangle \quad (n \text{ odd}). \tag{2.9.8.1}$$

The corresponding formulae *for* n *even* are

$$\xi^{(n,k)} = \begin{cases} (\Lambda_n^t\Lambda_n)^{(k-1)/2}\xi & \text{for } k \text{ odd} \\ \Lambda_n(\Lambda_n^t\Lambda_n)^{k/2-1}\xi & \text{for } k \text{ even.} \end{cases}$$

Hence

$$b_k^n = \langle(\Lambda_n^t\Lambda_n)^{k-1}\xi\,|\,\xi\rangle \quad (n \text{ even}). \tag{2.9.8.2}$$

One can visualize these results quite easily by adding to the Bratteli diagram of the chain $(B_{\beta,k})_{k\geq 1}$ some "phantom" vertices with zero dimension. The picture for $n = 5$, for example, is

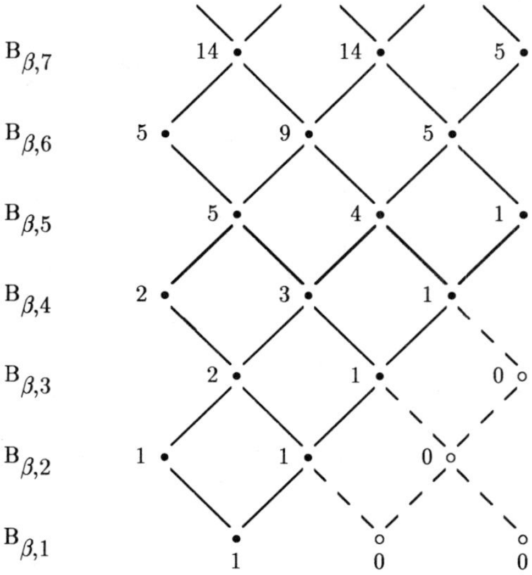

Recall also that our labelling of the vertices on each floor increases from right to left. Since $\Lambda_n\xi = \xi$ (n odd) and $\Lambda_n^t\xi = \xi$ (n even), (2.9.8.1) and (2.9.8.2) give

(2.9.8.3) $$b_k^n = \langle(\Lambda_n^t\Lambda_n)^k\xi\,|\,\xi\rangle \quad (\text{n odd}), \text{ and}$$

(2.9.8.4) $$b_k^n = \langle(\Lambda_n\Lambda_n^t)^k\xi\,|\,\xi\rangle \quad (\text{n even}).$$

Finally one verifies that

(2.9.8.5) $$\Lambda_{n+1}^t\Lambda_{n+1} - \Lambda_n^t\Lambda_n = E \quad (\text{n odd}), \text{ and}$$

(2.9.8.6) $$\Lambda_{n+1}\Lambda_{n+1}^t - \Lambda_n\Lambda_n^t = E \quad (\text{n even}),$$

where E is the orthogonal projection onto $\mathbb{R}\xi$, in the Euclidean space of the appropriate dimension.

We claim that the functions $(f_n(z))_{n\geq 2}$ satisfy the first order difference equation

$$f_{n+1}(z) - [zf_n(z)+1] = zf_{n+1}(z)[zf_n(z)+1].$$

First consider the case that n is odd. Then $zf_n(z) + 1 =$

$$= \sum_{k=0}^{\infty} b_{k+1}^{n} z^{k+1} + 1 = \sum_{k=1}^{\infty} b_{k}^{n} z^{k} + 1$$

$$= \sum_{k=1}^{\infty} \langle (\Lambda_n^t \Lambda_n)^k \xi \,|\, \xi \rangle z^k + \langle \xi \,|\, \xi \rangle$$

$$= \sum_{k=0}^{\infty} \langle (\Lambda_n^t \Lambda_n)^k z^k \xi \,|\, \xi \rangle,$$

using 2.9.8.3. Setting $B = \Lambda_n^t \Lambda_n$ we have

$$z f_n(z) + 1 = \langle (\mathbb{1}-Bz)^{-1} \xi \,|\, \xi \rangle.$$

Similarly using 2.9.8.2 and setting $A = \Lambda_{n+1}^t \Lambda_{n+1}$, we have

$$f_{n+1}(z) = \sum_{k=0}^{\infty} b_{k+1}^{n+1} z^k = \sum_{k=0}^{\infty} \langle (\Lambda_{n+1}^t \Lambda_{n+1})^k z^k \xi \,|\, \xi \rangle$$

$$= \langle (\mathbb{1}-Az)^{-1} \xi \,|\, \xi \rangle.$$

The difference $f_{n+1}(z) - [z f_n(z)+1]$ is computed using 2.9.8.5, and the resolvent identity:

$$\begin{aligned} f_{n+1}(z) - [z f_n(z)+1] &= \langle [(\mathbb{1}-Az)^{-1} - (\mathbb{1}-Bz)^{-1}] \xi \,|\, \xi \rangle \\ &= \langle (\mathbb{1}-Az)^{-1} z (\Lambda_{n+1}^t \Lambda_{n+1} - \Lambda_n^t \Lambda_n)(\mathbb{1}-Bz)^{-1} \xi \,|\, \xi \rangle \\ &= z \langle (\mathbb{1}-Az)^{-1} E (\mathbb{1}-Bz)^{-1} \xi \,|\, \xi \rangle \\ &= z \langle (\mathbb{1}-Az)^{-1} \xi \,|\, \xi \rangle \langle (\mathbb{1}-Bz)^{-1} \xi \,|\, \xi \rangle \\ &= z f_{n+1}(z) [z f_n(z)+1]. \end{aligned}$$

The case n even is entirely similar.

Next we observe that the functions $s_n(z) = \dfrac{P_{n-1}(z)}{P_{n+1}(z)}$ satisfy the same difference equation. First note that

$$z s_n + 1 = \frac{z P_{n-1} + P_{n+1}}{P_{n+1}} = \frac{P_n}{P_{n+1}},$$

by the second–order difference equation for the P_j. Hence

$$s_{n+1} - [zs_n+1] = \frac{P_n}{P_{n+2}} - \frac{P_n}{P_{n+1}}$$

$$= \frac{P_n}{P_{n+2}P_{n+1}}(P_{n+1}-P_{n+2}) = \frac{zP_n^2}{P_{n+2}P_{n+1}},$$

using the defining relation for the P_j again. But this last expression is $zs_{n+1}[zs_n+1]$.

Since $(f_n)_{n\geq 1}$ and $(s_n)_{n\geq 1}$ satisfy the same first order difference equation, it suffices now to check that $f_2 = s_2$. But $b_{k+1}^2 = 2^k$ for all k, so $f_2(z) = \sum_{k=0}^{\infty} 2^k z^k = \frac{1}{1-2z}$, while $s_2(z) = \frac{P_1(z)}{P_3(z)} = \frac{1}{1-2z}$. #

2.10. A digression on Hecke algebras.

As a general reference for this section, we use [BLie], especially exercises ≥ 22 in §IV.2. See also [CR], §11D. For the origin of the term "Hecke algebra", see p. xi in [Lus].

2.10.a – The complex Hecke algebra defined by $GL_n(q)$ and its Borel subgroup.

If G is a finite group and G_0 is a subgroup, the complex Hecke algebra $H(G,G_0)$ of the pair $G_0 \subset G$ is the commutant of the natural representation of G on the complex vector space $\mathbb{C}[G/G_0]$ of functions from G/G_0 to $\mathbb{C}$.

We denote by $\mathbb{C}[G]$ the algebra of complex functions on G, with the convolution product. We identify $\mathbb{C}[G/G_0]$ with the subspace of this algebra consisting of functions φ with $\varphi(gh) = \varphi(g)$ for $g \in G$ and $h \in G_0$, and we denote by $\mathbb{C}[G_0\backslash G/G_0]$ the subalgebra of $\mathbb{C}[G]$ of G_0–bi–invariant functions.

Proposition 2.10.1. *The algebras* $H(G,G_0)$ *and* $\mathbb{C}[G_0\backslash G/G_0]$ *are isomorphic.*

Proof. More generally, consider first an associative algebra A with unit, an idempotent $e \in A$, and the left A–module Ae. It is easy to check that the map $x \mapsto \rho(x) =$ right multiplication by x is an anti–isomorphism from eAe to $\mathrm{End}_A(Ae)$.

Now let $A = \mathbb{C}[G]$; for each $g \in G$, denote by δ_g the characteristic function of $\{g\}$. Set $e = \frac{1}{|G_0|}\sum_{h\in G_0} \delta_h$. Then $Ae = \mathbb{C}[G/G_0]$ and $eAe = \mathbb{C}[G_0\backslash G/G_0]$, so that $H(G,G_0)$

and $\mathbb{C}[G_0\backslash G/G_0]$ are anti-ismorphic. But $\mathbb{C}[G]$ has a canonical anti-isomorphism $\varphi \mapsto \check{\varphi}$, defined by $\check{\varphi}(g) = \varphi(g^{-1})$, which restricts to $\mathbb{C}[G_0\backslash G/G_0]$, so the proposition follows. #

Corollary 2.10.2. *Let* e *be the central idempotent in* $\mathbb{C}[G_0]$ *corresponding to the trivial representation* $G_0 \to GL_1(\mathbb{C})$, *and denote by* $p_1,\cdots,p_m$ *the minimal central idempotents in* $\mathbb{C}[G]$. *Then*

$$H(G,G_0) \approx \oplus\, ep_i\mathbb{C}[G]p_ie$$

where the direct sum is over the i's *with* $ep_i \neq 0$. *The Bratteli diagram for the pair* $\mathbb{C} \subset H(G,G_0)$ *is that part of the Bratteli diagram for the pair* $\mathbb{C}[G_0] \subset \mathbb{C}[G]$ *which lies above the vertex corresponding to* e.

Proof. This follows from Section 2.3. (See Corollary 11.26 of [CR] for a generalization.) #

As a first example, consider the permutation groups $\mathfrak{S}_2 \subset \mathfrak{S}_3$; the diagram for $\mathbb{C}[\mathfrak{S}_2] \subset \mathbb{C}[\mathfrak{S}_3]$ is

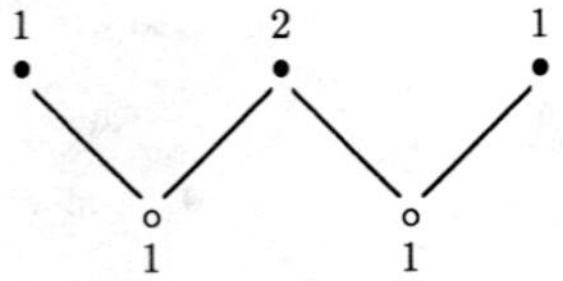

Then $\mathbb{C} \subset H(\mathfrak{S}_3,\mathfrak{S}_2)$ is described by

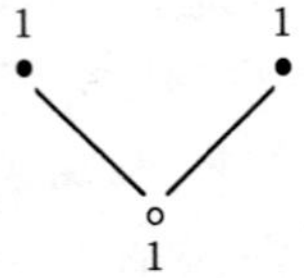

In particular $H(\mathfrak{S}_3,\mathfrak{S}_2) \approx \mathbb{C} \oplus \mathbb{C}$. It is easy to check that there are two double cosets in $\mathfrak{S}_2\backslash\mathfrak{S}_3/\mathfrak{S}_2$. One shows similarly that $H(\mathfrak{S}_{k+1},\mathfrak{S}_k) \approx \mathbb{C} \oplus \mathbb{C}$ for any integer $k \geq 1$.

But the case of main interest here is when q is a prime power, $G = GL_n(q)$ for some $n \geq 2$, and G_0 is the (Borel) subgroup B of upper triangular matrices. (The letter q will no longer denote an idempotent below.) Identifying the double cosets is a special case of the "Bruhat decomposition"

$$GL_n(q) = \coprod_{w \in W} BwB$$

where W is the "Weyl group", namely here the symmetric group $\mathfrak{S}_n$ embedded in $GL_n(q)$ as permutation matrices (see §IV.2 in [BLie]). Thus to each permutation $w \in \mathfrak{S}_n$ there is associated an element a_w of the Hecke algebra $H(G,G_0)$, which is the characteristic function of BwB divided by the order of B. For $i = 1,2,\cdots,n-1$, let s_i be the element of W given by the matrix

$$\begin{bmatrix} 1 & & & & & & \\ & \ddots & & & & & \\ & & 1 & & & & \\ & & & 0 & 1 & & & \\ & & & 1 & 0 & & & \\ & & & & & 1 & & \\ & & & & & & \ddots & \\ & & & & & & & 1 \end{bmatrix}$$

where the first diagonal 0 is the (i,i)th entry, and set $g_i = a_{s_i}$.

Proposition 2.10.3. *With the notation above, one has*

(a) $g_i^2 = (q-1)g_i + q \qquad i = 1,\cdots,n-1$

(b) $g_i g_{i+1} g_i = g_{i+1} g_i g_{i+1} \qquad i = 1,\cdots,n-2$

(c) $g_i g_j = g_j g_i$ *if* $|i-j| \geq 2 \qquad i,j = 1,\cdots,n-1$

Furthermore the elements g_i $(1 \leq i \leq n-1)$ *generate the Hecke algebra* $H(GL_n(q),B)$.

Proof. (see [BLie] as well as Propositions 11.30 and 11.34 in [CR]). For each permutation $w \in \mathfrak{S}_n$, set $C(w) = BwB$. Let $a_w \in \mathbb{C}[B\backslash G/B]$ be the quotient by $|B|$ of the characteristic function of $C(w)$; then $(a_w)_{w\in\mathfrak{S}_n}$ is a $\mathbb{C}$-basis of the Hecke algebra. For $w,w',w'' \in \mathfrak{S}_n$ and for $g \in C(w'')$, one has

$$\begin{aligned}(a_w * a_{w'})(g) &= \sum_{h\in G} a_w(h) a_{w'}(h^{-1}g) \\ &= \frac{1}{|B|^2} |C(w) \cap gC(w')^{-1}| \qquad (*)\end{aligned}$$

If $C(w) \cap gC(w')^{-1}$ is not empty, there exist $b_1,\cdots,b_4 \in B$ with $b_1 w b_2 = g b_3 w'^{-1} b_4$, so that $g \in C(w)C(w')$.

For s in the set $S = \{s_1, \cdots, s_{n-1}\}$ of generators of $\mathfrak{S}_n$, we need to compute $|C(s)|$. Observe more generally that, for any $h \in G$, the map

$$\begin{cases} B/(B \cap hBh^{-1}) \to (BhB)/B \\ \text{class of } b \mapsto \text{class of } bh \end{cases}$$

is well defined (if b,b' are in the same class modulo $B \cap hBh^{-1}$, there exists $b'' \in B$ with $b' = bhb''h^{-1}$, and $b'hB = bhB$) and bijective. Then the number of left classes modulo B in BhB is the index $[B : B \cap hBh^{-1}]$. It follows that

$$|C(s)| = |B|\,[B : B \cap sBs] = |B|q.$$

Let us compute $(a_w * a_{w'})(g)$ when $w = w' = s$. As $C(s)C(s) = B \cup C(s)$ this is zero unless $g \in B \cup C(s)$. For $g \in B$ one has by (∗)

$$(a_s * a_s)(g) = \frac{1}{|B|^2}|C(s)| = qa_1(g).$$

As a_1 is a convolution unit in $\mathbb{C}[B\backslash G/B]$, this implies

$$a_s * a_s = \lambda a_s + q$$

for some $\lambda \in \mathbb{C}$. Introduce the restriction μ to $\mathbb{C}[B\backslash G/B]$ of the augmentation homomorphism $\mathbb{C}[G] \to \mathbb{C}$, mapping φ to $\sum_{g \in G} \varphi(g)$. Then

$$\mu(a_s * a_s) = \frac{1}{|B|^2}|C(s)|^2 = q^2$$

$$\mu(\lambda a_s + q) = \lambda \frac{1}{|B|}|C(s)| + q = (\lambda+1)q$$

and consequently $\lambda = q - 1$. This shows (a).

Introduce the length function $\ell : \mathfrak{S}_n \to \{0,1,2,\cdots\}$ with respect to the generators S. Then

$$a_s * a_w = a_{sw} \quad \text{if } \ell(sw) > \ell(w) \qquad (**)$$

Indeed, if $\ell(sw) > \ell(w)$, then $C(s)C(w) = C(sw)$ by n° IV.2.4 in [BLie], so that $a_s * a_w$ is a scalar multiple of a_{sw} by (∗). Let $g,h \in C(s)$ and $u,v \in C(w)$ with $gu = hv$; then $vu^{-1} = h^{-1}g \in C(s)C(s) = B \cup C(s)$; but $vu^{-1} \in C(s)$ would imply $v \in C(s)C(w) =$

C(sw), which is incompatible with $v \in C(w)$; hence $g \in hB$, and thus any element in C(sw) can be written in exactly $|B|$ ways as a product of one element in C(s) by one in C(w). This shows that $a_s * a_w = a_{sw}$. It follows in particular that $\{a_s\}$ generates $H(GL_n(q),B)$.

Consider finally $s,t \in S$ with $(st)^3 = 1$. Then $\ell(s) = 1$, $\ell(st) = 2$, $\ell(sts) = 3$ and thus $a_s * a_t * a_s = a_{sts}$ by (**). Similarly $a_t * a_s * a_t = a_{tst}$, and (b) holds. Claim (c) follows in the same way. #

Now remember that the symmetric group in n letters has a presentation with generators the transpositions $s_i = (i,i+1)$ for $1 \leq i \leq n-1$ and relations

$$s_i^2 = 1 \quad s_is_{i+1}s_i = s_{i+1}s_is_{i+1} \quad s_is_j = s_js_i \text{ if } |i-j| \geq 2.$$

There is an easy proof of this which shows at the same time that the abstract algebra generated by $n-1$ generators subjected to the relations of 2.10.3 is of dimension at most n!. (See the beginning of §4 in [HKW].) For q a prime power, it follows then that the relations of 2.10.3 give a presentation of the Hecke algebra $H(GL_n(q),B)$. But we shall see that it is important to consider a more general family of algebras, defined for all $q \neq 0$.

2.10.b – The Hecke algebras $H_{q,n}$.

Let $\mathbb{K}$ again be an arbitrary field. Consider an integer $n \geq 1$ and a parameter $q \in \mathbb{K}$. We define $H_{q,n}$ to be the associative $\mathbb{K}$-algebra with unit presented by

generators: $g_1, g_2, \cdots, g_{n-1}$

relations: as in 2.10.3.

Proposition 2.10.4. *One has* $\dim_{\mathbb{K}} H_{q,n} = n!$ *for all* $q \in \mathbb{K}$ *and for all* $n \geq 1$.

Proof. We take for granted the presentation of $\mathfrak{S}_n$ in terms of the transpositions $\{s_i\}$. Each of the n! elements π of $\mathfrak{S}_n$ can be written uniquely as a reduced word w in the $\{s_i\}$ with

(i) minimum length among all words representing π,

(ii) the largest s_i in w appearing only once, and moved as far to the right as possible, and

(iii) all subwords of w reduced according to criteria (i) and (ii).

The corresponding n! words in the generators $\{g_i\}$ of $H_{q,n}$ span $H_{q,n}$ linearly, because the Hecke algebra relations 2.10.3(a)–(c) can be used

(i) to reduce the length of a word in the $\{g_i\}$ (i.e., to write it as a linear combination of shorter words), and

(ii) to reduce the number of occurences of the largest g_i in a word, and to move it to the right,

whenever the corresponding operation can be performed on the corresponding word in the $\{s_i\}$. It follows that $\dim_{\mathbb{K}} H_{q,n}$ is at most $n!$. On the other hand, we will exhibit below a sufficient family of inequivalent irreducible representation of $H_{q,n}$ to obtain the other inequality. See [HKW,§4] for a more explicit proof. #

For convenience we take $\mathbb{K} = \mathbb{C}$ in the following discussion. For q a prime power, $H_{q,n}$ is the same as $H(GL_n(q),B)$ in 2.10.a, and is in particular semi-simple. But we have no reason *a priori* to believe that there is any relationship between these algebras for different values of q. Also, the decomposition of any $H_{q,n}$ as a direct sum of matrix algebras is not obvious, each summand corresponding to some irreducible representation of $GL_n(q)$.

Observe however that, if we put $q = 1$, we recognize $H_{1,n}$ as the algebra $\mathbb{C}[\mathfrak{S}_n]$ of the symmetric group, so $H_{1,n}$ is semi-simple. A necessary and sufficient condition for semi-simplicity of $H_{q,n}$ is the non-degeneracy of the Killing trace $x \longmapsto tr(\lambda(x))$, where tr denotes the trace on $End_{\mathbb{C}}(H_{q,n})$. (For a finite dimensional $\mathbb{C}$-algebra A, the radical $rad(A)$ coincides with $A^{\perp} := \{x \in A:\ tr(\lambda(xy)) = 0 \text{ for all } y \in A\}$. In fact, both $rad(A)$ and $A^{\perp}$ are ideals which contain every nil ideal, and to show equality one shows that each is a nil ideal.) From the proof of Proposition 2.10.4 one obtains a basis $\{g_\sigma : \sigma \in \mathfrak{S}_n\}$ of $H_{q,n}$ and polynomial structure constants $p^{\mu}_{\sigma,\tau}(q)$ such that $g_\sigma g_\tau = \sum_\mu p^{\mu}_{\sigma,\tau}(q)\, g_\mu$. It follows that degeneracy is determined by a polynomial equation in q, so for all but a finite set of $q \in \mathbb{C}$ (n fixed), $H_{q,n}$ is semi-simple of dimension $n!$. Also $H_{q,n-1}$ embeds in $H_{q,n}$ via the obvious identification of the generators g_i for $1 \leq i \leq n-2$.

We now argue intuitively, though extremely plausibly. For the values of q for which $H_{q,n-1}$ and $H_{q,n}$ are semi- simple, the inclusion $H_{q,n-1} \subset H_{q,n}$ is completely described by a vector of integers (for the dimensions of the factors in $H_{q,n}$) and an integer valued matrix (the inclusion matrix). As these should vary continuously with q, they should be independent of q for these values. In particular they can be determined by examining the case $q = 1$. But then they are determined entirely by the dimensions of the different representations of $\mathfrak{S}_{n-1}$ and $\mathfrak{S}_n$ and the restriction rule from $\mathfrak{S}_n$ to $\mathfrak{S}_{n-1}$. For this reason we shall now describe this structure. In 2.10d we will identify a certain singular set

$\Omega \subset \mathbb{K}$ and construct for $q \notin \Omega$ a complete family of irreducible representations of $H_{q,n}$; this will demonstrate that $H_{q,n} \cong \mathbb{K}[\mathfrak{S}_n]$ for all n and for $q \notin \Omega$.

2.10.c – Complex representations of the symmetric group.

The conjugacy classes of the group $\mathfrak{S}_n$ are naturally indexed by partitions of n, two permutations being conjugate if and only if they have the same cycle structure. Thus there are as many irreducible representations of $\mathfrak{S}_n$ (over $\mathbb{C}$) as there are partitions. Although one cannot expect a natural correspondence between representations and partitions on the above grounds, it has long been known how to construct an irreducible representation from a partition. It is convenient to represent partitions by "Young diagrams", as amply illustrated by the following example.

Example 2.10.5. To the partition $\lambda = [\lambda_1, \lambda_2, \lambda_3, \lambda_4, \lambda_5] = [5,3,2,2,1]$ of 13, one associates the Young diagram

The most important rule is the restriction rule: if one restricts the representation of $\mathfrak{S}_n$ corresponding to a Young diagram Y to $\mathfrak{S}_{n-1}$, it is isomorphic to the direct sum of all representations corresponding to all Young diagrams Y′ obtained by removing one box from Y, all occuring with multiplicity one.

Thus the irreducible representations of $\mathfrak{S}_n$ (and hence the Bratteli diagram for $\mathbb{C}\mathfrak{S}_1 \subset \mathbb{C}\mathfrak{S}_2 \subset \mathbb{C}\mathfrak{S}_3 \subset \cdots$) are conveniently pictured by the following important diagram:

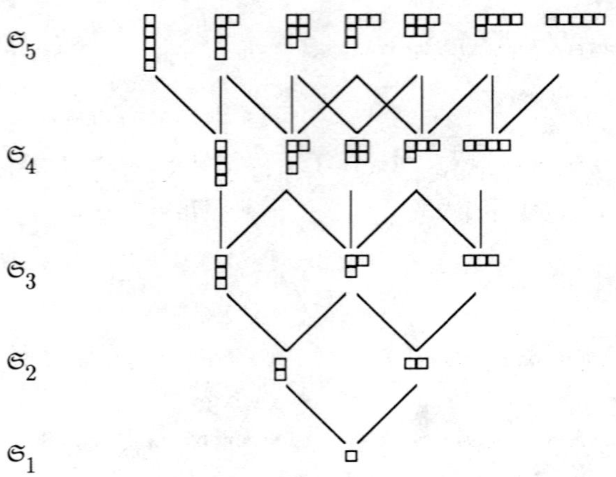

figure 2.10.6

The dimensions of the corresponding representations are given by the number of ascending paths on 2.10.6 beginning with □ and ending at the Young diagram in question. The above facts will actually follow from the construction of irreducible representations for the Hecke algebras $H_{q,n}$, to which we now return.

2.10.d – Irreducible representations of $H_{q,n}$ for $q \notin \Omega$.

The material that follows is taken from Wenzl's thesis [Wen2]. The $\mathbb{K}$-algebra $H_{q,n}$ is that defined at the beginning of 2.10.b; in particular, the field $\mathbb{K}$ is arbitrary.

By our intuitive argument, we expect that figure 2.10.6 should also represent the structure of $H_{q,n}$ for all but a countable number of values of q. While this could be proved in an elegant manner due to Tits (see exercise 26 in §IV.2 of [BLie], or Lemma 85 in [Ste2]), two important pieces of information would be missing: there would be no indication of which values of q are "bad" (and that would be particularly frustrating for $\mathbb{K}$ countable!), and there would be no construction of concrete representations of $H_{q,n}$.

We shall now show how to construct an irreducible representation of $H_{q,n}$ for each partition of n, provided q is not in the set Ω defined below. It is first convenient to dispose of another presentation of $H_{q,n}$ than that of 2.10.b.

Proposition 2.10.7. *Consider a number* $q \in \mathbb{K}^*$ *and an integer* $n \geq 1$. *Assume* $q \neq -1$ *and set*

(a) $e_i = \dfrac{g_i + 1}{q+1}$ $\quad i = 1, \cdots, n-1$

These generate $H_{q,n}$ *and constitute with the relations*

(b) $e_i^2 = e_i$ $\qquad i = 1,\cdots,n-1$

(c) $e_ie_{i+1}e_i - q(q+1)^{-2}e_i = e_{i+1}e_ie_{i+1} - q(q+1)^{-2}e_{i+1}$ $\qquad i = 1,\cdots,n-2$

(d) $e_ie_j = e_je_i$ when $|i-j| \geq 2$ $\qquad i,j = 1,\cdots,n-1$

a presentation of $H_{q,n}$.

Proof. A straightforward computation. #

Naturally this demands comparison with the definition of $\mathcal{A}_{\beta,n}$ in Section 2.8. However we shall postpone further comments on this until the next section.

Define the subset Ω of $\mathbb{K}$ to be the union of $\{0\}$ with those q for which there exists an integer $n \geq 1$ with $\sum_{j=0}^{n} q^j = 0$. Thus $\Omega \setminus \{0\}$ is the set of non-trivial roots of 1 in characteristic 0 and the set of all roots of 1 in finite characteristic. (As already noticed in section 2.8, if $q \notin \Omega$ then $\beta = q^{-1}(q+1)^2$ is generic.) For each $d \in \mathbb{Z} \setminus \{0\}$ and $q \in \mathbb{K} \setminus \Omega$ define

$$a_d(q) = \begin{cases} \dfrac{1-q^{d+1}}{(1+q)\,(1-q^d)} & \text{if } q \neq 1 \\[2ex] \dfrac{d+1}{2d} & \text{if } q = 1 \end{cases}$$

(Remark: When $d > 0$, then $a_d(q) = \dfrac{1+q+\cdots+q^d}{(1+q)(1+q+\cdots+q^{d-1})}$. Note also that $\prod_{1 \leq d \leq k} a_d(q) = Q_k(q)$ where Q_k is as in Proposition 2.8.3.iv.) Suppose given a partition of n, say $\lambda = [\lambda_1,\cdots,\lambda_k]$, where we allow some of the last λ_j's to be zero. We think of λ as a Young diagram. Let V_λ be the free $\mathbb{K}$-vector space on the set of ascending paths p from $\square$ to λ on figure 2.10.6, and denote by $\{v_p\}$ its canonical basis. We define now endomorphisms $f_1,\cdots,f_{n-1}$ of V_λ.

Let $i \in \{1,\cdots,n-1\}$. For each ascending path

$$p = (\lambda^1 = \square, \lambda^2,\cdots,\lambda^n = \lambda)$$

we have to define f_iv_p. The partition λ^{i+1} is obtained from λ^{i-1} in one of three ways

(a) By adding two boxes to the same column of λ^{i-1}. In this case $f_i v_p = v_p$.

(b) By adding two boxes to the same row of λ^{i-1}. In this case $f_i v_p = 0$.

(c) By adding boxes in different rows and columns of λ^{i-1}; more precisely there is pair of integers (r,s) with $r \neq s$ and $\lambda_r^{i-1} \neq \lambda_s^{i-1}$ such that $\lambda_r^i = \lambda_r^{i-1} + 1$ and $\lambda_s^{i+1} = \lambda_s^{i-1} + 1$. In this case there is precisely one ascending path from $\square$ to λ which differs from p in its ith vertex only; we call this path p′. For example:

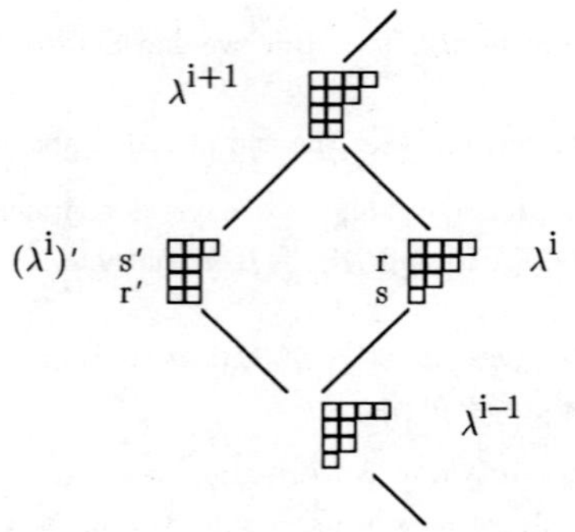

Set

$$d = (\lambda_r^{i+1} - r) - (\lambda_s^{i+1} - s) = (s-r) + (\lambda_r^{i+1} - \lambda_s^{i+1})$$

and observe that $d \neq 0$. Define d' in the same way for the path p′ and note that $d' = -d$. Finally, define

$$f_i v_p = a_d(q) v_p + (1 - a_d(q)) v_{p'}.$$

Observe that f_i leaves invariant the subspace $\mathbb{K}v_p \oplus \mathbb{K}v_{p'}$ of V_λ as well as its canonical complement; on $\mathbb{K}v_p \oplus \mathbb{K}v_{p'}$, it is described by the matrix

$$\begin{bmatrix} a_d(q) & a_d(q) \\ 1-a_d(q) & 1-a_d(q) \end{bmatrix} \qquad (2.10.8).$$

We have taken advantage of the equality $a_d(q) + a_{d'}(q) = 1$, which follows from the definition of a_d and from $d + d' = 0$.

The verification that $f_1, \cdots, f_{n-1}$ satisfy the relations (b) and (d) of Proposition 2.10.7 is trivial. They also satisfy (c), but this is more tedious to check and we refer to [Wen2]. We conclude that, for each partition λ of n there is a representation π_λ of $H_{q,n}$ in V_λ defined by $\pi_\lambda(e_i) = f_i$. A remarkably easy inductive argument shows that the π_λ's are irreducible and mutually inequivalent when λ runs over the set $\mathcal{P}_n$ of all partitions of n (for $q \in \mathbb{K} \setminus \Omega$). Indeed, these representations are absolutely irreducible, because the same argument applies to any extension of $\mathbb{K}$. By theorems of Burnside and Frobenius–Schur, this implies that $H_{q,n}$ has a quotient isomorphic to the multi-matrix algebra $\bigoplus_{\lambda \in \mathcal{P}_n} \mathrm{End}_{\mathbb{K}}(V_\lambda)$, of dimension n!. But we have already reported that the dimension of $H_{q,n}$ is no more than n!. (See the end of 2.10.a above, and §4 in [HKW].) Consequently the dimension is precisely n!, we have a complete set of irreducible representations of $H_{q,n}$ for $q \in \mathbb{K} \setminus \Omega$, and $H_{q,n}$ is isomorphic to $\bigoplus_{\lambda \in \mathcal{P}_n} \mathrm{End}_{\mathbb{K}}(V_\lambda)$. (In particular, setting $q = 1$, this gives for $\mathbb{K} = \mathbb{Q}$ the usual complete set of irreducible representations of the symmetric group $\mathfrak{S}_n$.)

Another trivial consequence of the construction is that the restriction of a representation π_λ of $H_{q,n}$ to $H_{q,n-1}$ is a direct sum $\oplus \pi_{\lambda'}$, where λ' runs over all partitions of n–1 obtained from the partition of λ of n by removing one box from the Young diagram. We reformulate this as follows.

Theorem 2.10.9. *Let $\mathbb{K}$ be a field and let $\Omega \subset \mathbb{K}$ be the union of $\{0\}$, of the non trivial roots of* 1, *and of* 1 *in case* $\mathrm{char}(\mathbb{K}) \neq 0$. *Consider $q \in \mathbb{K} \setminus \Omega$, an integer $n \geq 1$, and the Hecke algebra $H_{q,n}$ generated by $g_1, \cdots, g_{n-1}$ with the relations of* 2.10.b, *or equivalently by $e_1, \cdots, e_{n-1}$ and the relations of Proposition* 2.10.7. *Then*

(a) $H_{q,n}$ *is of dimension* n!.

(b) $H_{q,n}$ *is a multi–matrix algebra.*

(c) *The natural mapping $H_{q,n} \to H_{q,n+1}$ is an imbedding.*

(d) *The structure of the chain $H_{q,1} \subset H_{q,2} \subset \cdots \subset H_{q,n} \subset \cdots$ is given by figure* 2.10.6.

We make one further comment on Wenzl's paper, and for this we assume $\mathbb{K} = \mathbb{C}$. His exposition does not involve the matrix 2.10.8, but rather the related one

$$\begin{bmatrix} a_d & \sqrt{a_d(1-a_d)} \\ \sqrt{a_d(1-a_d)} & 1-a_d \end{bmatrix}$$

with a_d written for $a_d(q)$. It follows that $H_{q,n}$ has a C^*–algebra structure for $q \in \mathbb{R}$ and $q > 0$, for which e_i is an orthogonal projection: In fact, the main interest of [Wen2] is in the values $q = e^{\pm 2\pi i/n} \in \Omega$, for which Wenzl has constructed C^*–algebras which are quotients of the corresponding Hecke algebras.

2.11. The relationship between $\mathcal{A}_{\beta,n}$ and Hecke algebras.

It was first pointed out to V. Jones by R. Steinberg that the defining relations of $\mathcal{A}_{\beta,n}$ (see section 2.8) actually imply the Hecke relations. This is obvious from the definition of $\mathcal{A}_{\beta,n}$ and Proposition 2.10.7, but we would rather state this in terms of the generators g_i again.

Proposition 2.11.1. *Consider* $q \in \mathbb{K} \setminus \{-1,0\}$ *and* $\beta = 2 + q + q^{-1} \in \mathbb{K}^*$, *an integer* $n \geq 1$, *and the algebra* $\mathcal{A}_{\beta,n}$. *Set*

$$\gamma_i = (q+1)\epsilon_i - 1 \quad \textit{so that} \quad \epsilon_i = \frac{\gamma_i + 1}{q+1} \qquad i = 1,\cdots,n-1.$$

These generate $\mathcal{A}_{\beta,n}$ *and constitute with*

$$\gamma_i^2 = (q-1)\gamma_i + q \qquad i = 1,\cdots,n-1$$

$$\gamma_i\gamma_{i+1}\gamma_i = \gamma_{i+1}\gamma_i\gamma_{i+1} \qquad i = 1,\cdots,n-2$$

$$\gamma_i\gamma_j = \gamma_j\gamma_i \quad \textit{when } |i-j| \geq 2 \qquad i,j = 1,\cdots,n-1$$

$$\gamma_i\gamma_{i+1}\gamma_i + \gamma_i\gamma_{i+1} + \gamma_{i+1}\gamma_i + \gamma_i + \gamma_{i+1} + 1 = 0 \qquad i = 1,\cdots,n-2$$

a presentation of $\mathcal{A}_{\beta,n}$.

Proof. A straightforward computation. #

Corollary 2.11.2. *There is a surjective algebra morphism*

$$\psi_n : H_{q,n} \to \mathcal{A}_{\beta,n}$$

defined by $\psi_n(g_i) = \gamma_i$ *for* $i = 1,\cdots,n-1$. *If* $n = 1$ *or* $n = 2$ *it is an isomorphism. If* $n \geq 3$ *its kernel* I_n *is the two-sided ideal of* $H_{q,n}$ *generated by*

$$g_1g_2g_1 + g_1g_2 + g_2g_1 + g_1 + g_2 + 1.$$

Moreover the diagram

$$\begin{array}{ccc} H_{q,n} & \longrightarrow & H_{q,n+1} \\ \Big\downarrow \psi_n & & \Big\downarrow \psi_{n+1} \\ \mathcal{A}_{\beta,n} & \longrightarrow & \mathcal{A}_{\beta,n+1} \end{array}$$

commutes.

Proof. The existence of ψ_n follows from the definition of $H_{q,n}$ and Proposition 2.11.1. It is clear that ψ_1 and ψ_2 are isomorphisms, and that for $n \geq 3$ the kernel of ψ_n is generated by

$$x_i = g_ig_{i+1}g_i + g_ig_{i+1} + g_{i+1}g_i + g_i + g_{i+1} + 1$$

for $i = 1,\cdots,n-2$. As $q \neq 0$, each g_i is invertible with inverse $q^{-1}(g_i+1-q)$. By relations (b) and (c) in Proposition 2.10.3 on has

$$(g_1g_2\cdots g_{n-1})g_k(g_{n-1}^{-1}\cdots g_2^{-1}g_1^{-1}) = g_{k+1} \qquad k = 1,\cdots,n-2$$

and consequently

$$(g_1g_2\cdots g_{n-1})x_i(g_{n-1}^{-1}\cdots g_2^{-1}g_1^{-1}) = x_{i+1} \qquad i = 1,\cdots,n-3$$

Thus I_n is generated by x_1. The last claim is obvious. #

Thus, if q is a value for which $H_{q,n}$ is semi–simple, it must be possible to identify $\mathcal{A}_{\beta,n}$ with a certain ideal of $H_{q,n}$, given by some subset of the set of partitions of n. We shall show that this subset is precisely the set of all partitions with at most two rows; this was also explained to us by R. Steinberg.

We recall that Ω has been defined in 2.10.d, and that the relations which constitute with $g_1,\cdots,g_{n-1}$ a presentation of $H_{q,n}$ are written in Proposition 2.10.3 (see also 2.10.b).

Lemma 2.11.3. *Let* $q \in \mathbb{K} \setminus \Omega$, *let* $n \geq 3$ *be an integer, let* λ *be Young diagram with* n *boxes and with at most two rows, and let*

$$\pi_\lambda : H_{q,n} \to \mathrm{End}(V_\lambda)$$

be as in 2.10.d. *If* $x_i = g_i g_{i+1} g_i + g_i g_{i+1} + g_{i+1} g_i + g_i + g_{i+1} + 1$, *then* $\pi_\lambda(x_i) = 0$ *for* $i = 1, \cdots, n-2$.

Proof. We set first $n = 3$. As there are two partitions of 3 with at most two rows, we split the proof in two steps.

First $\lambda = [3]$, pictured as . Here, according to the definition of $\pi_\lambda(g_1) = \pi_\lambda(g_2) = -1$ and $\pi_\lambda(x_1) = 0$.

Second $\lambda = [2,1]$, pictured as . Here, instead of using π_λ, we may argue with any 2–dimensional irreducible representation of $H_{q,3}$, for example that defined by

$$\pi(g_1) = \begin{bmatrix} q & 0 \\ q & -1 \end{bmatrix} \qquad \pi(g_2) = \begin{bmatrix} -1 & 1 \\ 0 & q \end{bmatrix}$$

(which is irreducible if q is neither 0, nor -1, nor a nontrivial cube root of 1). A routine calculation shows that $\pi(x_1) = 0$.

Assume now $n \geq 4$, and that the lemma holds for $n-1$. By the proof of 2.11.2, it is enough to check that $\pi_\lambda(x_1) = 0$.

We recall from 2.10.d that

$$\pi_\lambda \Big|_{H_{q,n-1}} = \bigoplus_{\lambda'} \pi_{\lambda'}$$

where λ' has one less box than λ. In particular λ' has at most two rows, and $\pi_{\lambda'}(x_1) = 0$ by the induction hypothesis. Consequently $\pi_\lambda(x_1) = 0$. #

Consider $q \in \mathbb{K} \setminus \Omega$ and an integer $n \geq 3$. Let $\mathcal{P}_n$ be the set of partitions of n. We know from section 2.10.d that $H_{q,n}$ is a direct sum $\bigoplus_{\lambda \in \mathcal{P}_n} I_\lambda$ of simple two sided ideals, the notation being such that, for each $\lambda_0 \in \mathcal{P}_n$, the representation π_{λ_0} of $H_{q,n}$ restricts to an isomorphism $I_{\lambda_0} \to \mathrm{End}(V_{\lambda_0})$ and maps I_λ to $\{0\}$ when $\lambda \neq \lambda_0$. We denote by $\mathcal{P}_n^2$ the subset of $\mathcal{P}_n$ of partitions with at most two rows.

Proposition 2.11.4. *Let* $q \in \mathbb{K} \setminus \Omega$, *let* $n \geq 3$ *be an integer, and let*

$$I_n = \mathrm{Ker}(\psi_n : H_{q,n} \to \mathcal{A}_{\beta,n}) = \langle g_1g_2g_1+g_1g_2+g_2g_1+g_1+g_2+1\rangle$$

be as in Corollary 2.11.2. *Then*

$$I_n = \bigoplus_{\lambda \in \mathcal{P}_n \setminus \mathcal{P}_n^2} I_\lambda.$$

Proof. By the previous lemma for $\lambda \in \mathcal{P}_n^2$, one has $\pi_\lambda(I_n) = 0$, so that $I_\lambda \cap I_n = (0)$. Consequently $I_n \subset \bigoplus_{\lambda \in \mathcal{P}_n \setminus \mathcal{P}_n^2} I_\lambda$.

Let $\lambda = [r,s] \in \mathcal{P}_n^2$. From the definition of V_λ (see 2.10.d) one has

$$\begin{aligned}
&\dim V_{[r,s]} = \dim V_{[r,s-1]} && \text{if } r = s,\\
&\dim V_{[r,s]} = \dim V_{[r,s-1]} + \dim V_{[r-1,s]} && \text{if } r > s \geq 1,\\
&\dim V_{[r,0]} = 1.
\end{aligned}$$

By induction on r and s, ones deduces from this

$$\dim V_{[r,s]} = \begin{bmatrix} n \\ s \end{bmatrix} - \begin{bmatrix} n \\ s-1 \end{bmatrix}.$$

Thus

$$\sum_{\lambda \in \mathcal{P}_n^2} (\dim V_\lambda)^2 = \frac{1}{n+1}\begin{bmatrix} 2n \\ n \end{bmatrix},$$

by Lemma 2.8.2.

Suppose that $I_{\lambda_0} \not\subset I_n$ for some $\lambda_0 \in \mathcal{P}_n \setminus \mathcal{P}_n^2$, so that $I_n \subset \ker(\pi_{\lambda_0})$; we shall arrive at a contradiction. For $\lambda \in \mathcal{P}_n^2 \cup \{\lambda_0\}$, the representation π_λ of $H_{q,n}$ defines a representation $\hat{\pi}_\lambda$ of $\mathcal{A}_{\beta,n}$ in V_λ. As the π_λ's are pairwise inequivalent, so are the $\hat{\pi}_\lambda$'s, and

$$\dim \mathcal{A}_{\beta,n} \geq \frac{1}{n+1}\begin{bmatrix} 2n \\ n \end{bmatrix} + (\dim V_{\lambda_0})^2.$$

But this contradicts Proposition 2.8.1. #

Observe that the proof above shows again the equality

$$\dim \mathcal{A}_{\beta,n} = \frac{1}{n+1}\begin{bmatrix} 2n \\ n \end{bmatrix} \text{ for } \beta \text{ generic}$$

of Theorem 2.8.5.

To sum up, we have shown that, for generic β and corresponding q, the algebra $\mathcal{A}_{\beta,n}$ is isomorphic to the quotient of $H_{q,n}$ by the two-sided ideal generated by $g_1g_2g_1 + g_1g_2 + g_2g_1 + g_1 + g_2 + 1$. This ideal corresponds precisely to the direct summands I_λ of $H_{q,n}$ given by all Young diagrams $\lambda \in \mathcal{P}_n$ having at least 3 rows.

CHAPTER 3

Finite von Neumann Algebras with Finite Dimensional Centers

3.1. Introduction.

In this chapter we study pairs of finite von Neumann algebras with finite dimensional centers, and the index of such pairs.

Sections 2 to 4 are purely expository, and may be taken as an encouragement to the reader having essentially no previous experience with von Neumann algebras. Sections 5 to 7 present a generalization to the present setting of some of the ideas of [Jo1] for pairs of factors. Though this chapter cannot be so self-contained as the previous ones, we have tried to minimize the technical background in operator algebras assumed from the reader.

Let us first describe Sections 2 to 4. Let M be a von Neumann algebra which is a factor of type II_1. (The definition is given in Section 3.2.) We denote by $tr : M \longrightarrow \mathbb{C}$ the normalized trace on M. For every Hilbert space H on which M acts, Murray and von Neumann have defined a positive number (possibly ∞) called the coupling constant between M and its commutant; we denote this number by $\dim_M(H)$. Two representations of M by operators on two separable Hilbert spaces H and H′ are equivalent if and only if $\dim_M(H) = \dim_M(H')$. Section 3.2 is an exposition of the definition and the basic properties of these coupling constants. Except for the presentation, all this material comes from the original papers by Murray and von Neumann.

In Section 3.3, we present some *geometric examples* of coupling constants arising in the theory of discrete series representations of Lie groups; they are borrowed from Atiyah–Schmid [AS]. In particular, we show:

Theorem 3.1.1. *Let* G *be a connected real semi–simple, non–compact Lie group without center. Let* Γ *be a lattice in* G, *and let* M *be the von Neumann algebra of the discrete group* Γ. *Then* M *is a* II_1 *factor. If* $\pi : G \longrightarrow U(H)$ *is an irreducible discrete series representation of* G, *then* $\pi|_\Gamma$ *extends to a representation of* M *on* H, *and*

$$\dim_M(H) = \mathrm{covol}(\Gamma)\, d_\pi,$$

where d_π *is the formal dimension of* π.

In Section 3.4, we consider a pair $N \subset M$ of finite factors and we recall some aspects of the original work [Jo1] on this subject. First the index of N in M is now defined to be

$$[M:N] = \dim_N(L^2(M))$$

where $L^2(M)$ is the Hilbert space obtained by completion of M for the scalar product $\langle x|y\rangle = tr(x^*y)$. It was shown in [Jo4] that this definition of index agrees with the purely ring-theoretic definition of Chapter 2.

If $[M:N] < \infty$, the pair $N \subset M$ generates a tower of II_1-factors

$$\mathbb{1} \in M_0 = N \subset M_1 = M \subset \cdots \subset M_{k-1} \subset M_k \subset \cdots$$

by a fundamental construction which is defined as follows. The natural conditional expectation from M_k onto M_{k-1} can be seen as an orthogonal projection $e_k : L^2(M_k) \to L^2(M_{k-1})$, and M_{k+1} is the von Neumann algebra of operators on $L^2(M_k)$ generated by M_k and e_k. This M_{k+1} is again a II_1 factor. It is a particular case of Proposition 3.1.4 below that this way to define the fundamental construction agrees with that of Chapter 2. Moreover the Markov relation holds:

$$[M:N]tr_{k+1}(xe_k) = tr_k(x) \qquad \text{for all } x \in M_k,$$

where tr_k and tr_{k+1} denote the normalized traces on M_k and M_{k+1} respectively.

The sequence $(e_k)_{k\geq 1}$ of projections in $\bigcup_{k\geq 0} M_k$ satisfy the relations

$$[M:N]\, e_i e_j e_i = e_i \qquad \text{if } |i-j| = 1$$
$$e_i e_j = e_j e_i \qquad \text{if } |i-j| \geq 2$$

and provide consequently a representation of the algebras $\mathcal{A}_{\beta,k}$, with $\beta = [M:N]$. (See Section 2.8 and Theorem II.16.) From this follows

Theorem 3.1.2. *If* $N \subset M$ *is a pair of* II_1*-factors, either* $[M:N] = 4\cos^2(\pi/q)$ *for some integer* $q \geq 3$ *or* $[M:N] \in [4,\infty]$.

There is substantial overlap between Sections 3.2 to 3.5 and Sections I to III of Connes' report [Con].

Let us now describe Sections 5 to 7, where we consider a pair $N \subset M$ of *finite von Neumann algebras with finite dimensional centers.* There are projections $p_1,\cdots,p_m$ which are central in M and projections $q_1,\cdots,q_n$ which are central in N such that

$$p_1M, \cdots, p_mM, q_1N, \cdots, q_nN \text{ are finite factors, and}$$
$$N = \bigoplus_{j=1}^{n} q_jN \subset M = \bigoplus_{i=1}^{m} p_iM.$$

If $\dim_{\mathbb{C}}(M) < \infty$, this is the situation of Chapter 2. At this stage, let us assume that each of the factors p_iM, q_jN is of type II_1 (see the comment after 3.5.4).

As in Section 2.3, we define an <u>index matrix</u> $\Lambda_N^M = (\lambda_{i,j}) \in \mathrm{Mat}_{m,n}(\mathbb{R}_+ \cup \{\infty\})$ by

$$\lambda_{i,j} = \begin{cases} 0 & \text{if } p_i q_j = 0 \\ [M_{i,j} : N_{i,j}]^{1/2} & \text{if not} \end{cases}$$

where $N_{i,j} = p_iq_jN$ is a subfactor of the factor $M_{i,j} = p_iq_jMp_iq_j$. We say that N is of <u>finite index</u> in M if Λ_N^M does not have any infinite entry.

For an analysis of traces on M and N (see Section 2.4 when $\dim_{\mathbb{C}}(M)$ is finite), we define also the <u>trace matrix</u> $T_N^M = (c_{i,j}) \in \mathrm{Mat}_{m,n}(\mathbb{R}_+)$ by

$$c_{i,j} = \mathrm{tr}_{p_iM}(p_iq_j),$$

where tr_{p_iM} denotes the normalized trace on the factor p_iM. A trace on M is described by the vector $\overleftarrow{s} \in (\mathbb{R}_+)^m$ with $s_i = \mathrm{tr}(p_i)$, and its restriction to N by the vector $\overleftarrow{s}T_N^M \in (\mathbb{R}_+)^n$. Traces are always assumed to be positive in this chapter, so that $s_i \geq 0$ for $i = 1, \cdots, m$.

If $\dim_{\mathbb{C}}(M) < \infty$ the matrices Λ_N^M and T_N^M are simply related by $c_{i,j} = \lambda_{i,j}\nu_j\mu_i^{-1}$, with $\mu_i^2 = \dim_{\mathbb{C}}(p_iM)$ and $\nu_j^2 = \dim_{\mathbb{C}}(q_jN)$. This relation has no analogue when the p_iM's and the q_jN's are factors of type II_1:

<u>Proposition 3.1.3.</u> *Consider two irredundant matrices*

$$\Lambda = (\lambda_{i,j}) \in \mathrm{Mat}_{m,n}(\{2\cos(\pi/q)\}_{q \geq 2} \cup [2,\infty]) \text{ and } T = (c_{i,j}) \in \mathrm{Mat}_{m,n}(\mathbb{R}_+)$$

satisfying:

$$\lambda_{i,j} = 0 \Leftrightarrow c_{i,j} = 0 \text{ and } \sum_{1 \leq j \leq n} c_{i,j} = 1 \text{ for } i \in \{1, \cdots, m\}.$$

Then there exists a pair $N \subset M$ *as above with* $\Lambda = \Lambda_N^M$ *and* $T = T_N^M$. *Moreover,* M *and* N *may be chosen hyperfinite.*

The Skolem–Noether theorem does not hold for II_1-factors and Proposition 2.3.3 does not carry over to the present setting: the matrices Λ_N^M and T_N^M do not characterize N as a subalgebra of M.

Once a faithful trace is given on M, the fundamental construction gives a new algebra $\langle M,e_N\rangle$, just as described above in the case of factors.

Proposition 3.1.4. *Let* N *be of finite index in* M.

(a) *The algebras* $\langle M,e_N\rangle$ *and* $\mathrm{End}_N^r(M)$ *are isomorphic.*

(b) *The algebra* $\langle M,e_N\rangle$ *is again a finite sum of* II_1 *factors. There is a natural bijection between the minimal central idempotents of* N *and those of* $\langle M,e_N\rangle$.

A convenient isomorphism is described in Corollary 3.6.5, and the bijection of (b) appears in Proposition 3.6.1.iv.

The partial description of $N \subset M$ by Λ_N^M and T_N^M is useful because, if N is of finite index in M and if $L = \langle M,e_N\rangle$, one may compute Λ_M^L and T_M^L from Λ_N^M and T_N^M. Indeed

$$\Lambda_M^L = (\Lambda_N^M)^t \qquad T_M^L = F_N^M \tilde{T}_N^M$$

where $\tilde{T} = \tilde{T}_N^M \in \mathrm{Mat}_{n,m}(\mathbb{R}_+)$ is defined by

$$\tilde{T}_{j,i} = \begin{cases} 0 & \text{if } p_i q_j = 0 \\ \lambda_{i,j}^2 c_{i,j}^{-1} & \text{if not}, \end{cases}$$

and where F_N^M is a diagonal matrix ensuring that $\sum_{1 \le i \le m} (T_M^L)_{j,i} = 1$ for $j = 1,\cdots,n$.

See Propositions 3.6.6 and 3.6.8 for the details.

As in Chapter 2, a trace on M is said to be a Markov trace of modulus β for the pair $N \subset M$ if it extends to a trace tr on $\langle M,e_N\rangle$ for which

$$\beta\, \mathrm{tr}(xe_N) = \mathrm{tr}(x) \qquad x \in M.$$

There exists at most one such extension. As traces are positive in this chapter, β has to be a positive number. The analogues of Theorem 2.1.3 and 2.1.4 hold as follows. Recall that a pair $N \subset M$ is connected if the intersection $Z(M) \cap Z(N)$ of the centers is reduced to $\mathbb{C}1$.

Theorem 3.1.5. *Let* $M = \bigoplus_{i=1}^{m} p_i M$ *and* $N = \bigoplus_{j=1}^{n} q_j N$ *be finite direct sums of* II_1 *factors, let* N *be a subalgebra of* M *of finite index, and write* $T, \tilde{T}$ *for* $T_N^M, \tilde{T}_N^M$.

(a) *Let* $tr : M \to \mathbb{C}$ *be a trace, let* $\overleftarrow{s} \in \mathbb{R}_+^m$ *be defined by* $s_i = tr(p_i)$, *and let* $\beta \in \mathbb{R}_+^*$. *Then* tr *is a Markov trace of modulus* β *for the pair* $N \subset M$ *if and only if*

$$\overleftarrow{s}\, T\tilde{T} = \beta \overleftarrow{s}.$$

(b) *If the conditions of (a) hold, then the Markov extension* $\langle M,e_N\rangle \to \mathbb{C}$ *of* tr *is a Markov trace of modulus* β *for the pair* $M \subset \langle M,e_N\rangle$.

(c) *If* $N \subset M$ *is connected, there exists a unique normalized Markov trace on* $N \subset M$, *and its modulus* β *is the spectral radius of* $T\tilde{T}$.

Comparing Theorems 2.1.4 and 3.1.5, we may define the index of N in M as

$$[M:N] = \rho(T\tilde{T})$$

where ρ denotes spectral radius.

Corollary 3.1.6. *Theorem 3.1.2 holds for finite direct sums of* II_1 *factors.*

We note that the definition of [M:N] given above is not the same as that of Chapter 2. However, P. Jollissaint has shown, in unpublished work, that the two definitions of index coincide.

If $N \subset M$ is a connected pair of finite dimensional multi-matrix algebras with $[M:N] \leq 4$, we have shown in Theorems 2.1.1 and 1.1.2 that the corresponding graph is a Coxeter graph of one of the types A,D,E. The chief result of Section 3.7 is that connected pairs $N \subset M$ of finite direct sums of II_1-factors with $[M:N] \leq 4$ give rise to all possible Coxeter graphs associated with finite and affine groups.

Theorem 3.1.7. *Let* $N \subset M$ *be a connected pair of finite direct sums of* II_1*-factors. Assume that* N *is of finite index in* M *and let* $\Lambda = \Lambda_N^M$ *be the inclusion matrix.*

(a) *If* $[M:N] < 4$, *then* Λ *is the matrix associated (in Theorem 1.1.3) to a bicoloration of one of the following Coxeter graphs:*

$$A_\ell \ (\ell \geq 2),\ B_\ell \ (\ell \geq 2),\ D_\ell \ (\ell \geq 4),\ E_\ell \ (\ell = 6,7,8),$$
$$F_4,\ G_2,\ H_\ell \ (\ell = 3,4),\ I_2(p) \ (p = 5 \ \textit{or} \ p \geq 7).$$

Moreover $[M:N] = \|\Lambda\|^2 = 4\cos^2(\pi/h)$, *where* h *is the Coxeter number. (See tables 1.4.5 and 1.4.7.)*

(*b*) *If* $[M:N] = 4$, then Λ_N^M *corresponds to one of the graphs:*

$$A_\ell^{(1)} \ (\ell \text{ odd}, \ell \geq 1),\ B_\ell^{(1)} \ (\ell \geq 2),\ C_\ell^{(1)} \ (\ell \geq 3),$$
$$D_\ell^{(1)} \ (\ell \geq 4),\ E_\ell^{(1)} \ (\ell = 6,7,8),\ F_4^{(1)},\ G_2^{(1)}.$$

so that $[M:N] = \|\Lambda\|^2$. *(See tables 1.4.6 and 1.4.7.).*

The index range described by Theorem 3.1.2 appears also in the remarkable family of Hecke groups, which are discrete subgroups of $PSL(2,\mathbb{R})$ generated by two parabolic transformations. We have included an Appendix III on these groups. Its purpose is to expose the spectacular comparison with Theorem 3.1.2 as well as to illustrate Section 3.3.

3.2. The coupling constant: definition.

Let H be a (complex) Hilbert space. We denote by $B(H)$ the $*$-algebra of all bounded operators on H, with x^* the adjoint of the operator $x \in B(H)$. Besides the topology associated to the norm

$$\|x\| = \operatorname{Sup}\{\|x\xi\| : \xi \in H \text{ and } \|\xi\| \leq 1\}$$

the algebra has also the ultraweak topology or w-topology which is defined by the semi-norms

$$x \mapsto \left|\sum_{j=1}^{\infty} \langle x\xi_j \mid \eta_j\rangle\right| \qquad \begin{array}{l} \xi_1,\xi_2,\cdots \in H \text{ with } \sum_{j=1}^{\infty} \|\xi_j\|^2 < \infty \\ \eta_1,\eta_2,\cdots \in H \text{ with } \sum_{j=1}^{\infty} \|\eta_j\|^2 < \infty. \end{array}$$

Whenever necessary , we assume H to be separable.

A von Neumann algebra acting on H, or a von Neumann subalgebra of $B(H)$, is a w-closed $*$-subalgebra of $B(H)$ which contains the identity. If M_j is a von Neumann subalgebra of $B(H_j)$ for $j = 1,2$ and if $\varphi : M_1 \to M_2$ is a $*$-isomorphism, it is known that φ is continuous with respect to the w–topology on both M_1 and M_2 (corollary 5.13 in [SZ] or section I.4.3 in [DvN]). A von Neumann algebra is a *–algebra M which is *–isomorphic to a von Neumann subalgebra of B(H) for some H; by the result just recalled, such an algebra has a well–defined w–topology.

A factor is a von Neumann algebra M with center Z_M reduced to the scalar multiples of the identity. Von Neumann algebras are known to be principal in the sense that any w–closed two–sided ideal is generated by a central projection (see section I.3.4 in [DvN]). Thus a von Neumann algebra M is a factor if and only if any two–sided ideal $J \neq 0$ in M is w–dense. There is not any continuity problem for representations of a factor M in the following sense: any *–homomorphism $M \to B(H)$ is w–continuous. (See theorem V.5.1 in [Tak]; the separability of H is crucial here.)

A II_1 factor is an infinite dimensional factor M which admits a normalized finite trace $tr : M \to \mathbb{C}$ such that

(i) $tr(\mathbb{1}) = 1$

(ii) $tr(xy) = tr(yx)$ $x,y \in M$

(iii) $tr(x^*x) \geq 0$ $x \in M$.

It is known that, on a II_1–factor, such a trace is unique in two senses. First, in the usual sense for operator algebras: tr is the unique linear form satisfying (i), (ii) and (iii); see [DvN], n^{os} I.6.4 and III.2.7; moreoever one has $tr(x^*x) > 0$ for $x \neq 0$. But also secondly, in the naive sense: tr is the unique linear form satisfying (i) and (ii), by [FH]. The existence of II_1–factors which may act on separable Hilbert spaces is one of the basic discoveries in the first paper by Murray and von Neumann [MvN I].

A finite factor is a von Neumann algebra which is either a II_1–factor, or isomorphic to B(H) for some H of finite dimension. Such a factor is simple as a complex algebra by [DvN], III.5.2. Here is a characterization of finite factors; for more on this, see [KvN].

Proposition 3.2.1. *Let* M *be a* C^**–algebra with unit and with center reduced to the scalar multiplies of* $\mathbb{1}$. *Let* $tr : M \to \mathbb{C}$ *be a faithful normalized trace (namely a linear form satisfying (i), (ii), (iii) above and* $tr(x^*x) > 0$ *for* $x \neq 0$*). Assume that the unit ball of* M *is complete with respect to the metric* $d(x,y) = \|x-y\|_2$, *where* $\|x\|_2 = (tr(x^*x))^{1/2}$. *Then* M *is a finite factor.*

Proof. Let $H = L^2(M,tr)$ be the Hilbert space obtained by completion of M with respect to the scalar product defined by $\langle x|y\rangle = tr(x^*y)$ for $x,y \in M$. Let $\pi : M \to B(H)$ be the *–representation of M on H, with $\pi(x)$ being the extension to H of the left multiplication by x on M. Then π is injective because tr is faithful. Let $\pi(M)''$ denote the double commutant of M in B(H), which is, by von Neumann's bicommutant theorem, the w–closure of $\pi(M)$ in B(H).

To show that M is a von Neumann algebra, it is enough to show that the inclusion of M in $\pi(M)''$ is surjective. Let $a \in \pi(M)''$ with $\|a\| = 1$. By Kaplansky's density theorem, there is a net (x_α) in M with $\|x_\alpha\| \leq 1$ for all α such that $\pi(x_\alpha)$ converges strongly to a; that is, $\pi(x_\alpha)\xi$ converges to $a\xi$ for all $\xi \in H$. Taking $\xi = 1$, this means that $(\pi(x_\alpha))$ is a Cauchy net for the $\|\cdot\|_2$-distance, so converges with respect to this distance to some element $\pi(x_0)$ by the assumed completeness of the ball of M. One can check that the strong topology and the $\|\cdot\|_2$-topology coincide on the unit ball of $\pi(M)''$, so $a = \pi(x_0) \in \pi(M)$. #

Let M be a finite factor acting on some Hilbert space H. We are going to define the coupling constant $\dim_M(H)$ which is a measure of the size of H as an M-module, the definition being made so that the standard M-module $L^2(M) = L^2(M,tr)$ has size 1. Before comparing other M-modules to that one, we recall the following facts.

Lemma 3.2.2. (a) *Let* $J : L^2(M) \to L^2(M)$ *be the conjugate linear isometry which extends* $\begin{cases} M \to M \\ x \mapsto x^* \end{cases}$. *Then* JMJ *is the commutant* $\mathrm{End}_M(L^2(M))$ *of* M *in* $B(L^2(M))$.

(b) *Let* K *be a Hilbert space and let* M *act on* $L^2(M) \otimes K$ *by the diagonal action* $x(\eta\otimes\theta) = (x\eta) \otimes \theta$. *Then* $JMJ \otimes B(K)$ *is the commutant of* M *in* $B(L^2(M) \otimes K)$.

(c) *Assume that the space* K *of (b) is infinite dimensional. For any* M*-module* H, *there exists an isometry*

$$u : H \to L^2(M) \otimes K$$

which is M*-linear, namely which intertwines the actions of* M.

Proof. (a) Let $x,y,z \in M$. By definition of J

$$JxJy = (xy^*)^* = yx^* = yJx.$$

Applying this twice we get

$$JxJyz = yzJx = yJxJz,$$

and setting $z = \mathbf{1}$,

$$(JxJ)y = y(JxJ).$$

Thus $JMJ \subset M'$ where $M' = \mathrm{End}_M(L^2(M))$.

Let moreover $a \in M'$. By definition of the adjoint

$$\langle y^*x^* | a\rangle = \langle x^* | ya\rangle = \langle x^* | ay\rangle = \langle a^*x^* | y\rangle = \langle x^*a^* | y\rangle = \langle a^* | xy\rangle.$$

Now one has $\langle J\eta | \theta\rangle = \overline{\langle \eta | J\theta\rangle}$ for all $\eta, \theta \in L^2(M)$, and consequently

$$\overline{\langle xy | Ja\rangle} = \langle y^* x^* | a\rangle = \overline{\langle xy | a^*\rangle}$$

so that $Ja = a^*$. Thus the first computation shows also that $JM'J \subset M''$ and, taking adjoints, $M' \subset JM''J$.

By von Neumann's bicommutant theorem, one has $M' = JMJ$.

(b) Let $x \in B(L^2(M) \otimes K)$. Choose an orthonormal basis $(\epsilon_i)_{i \in I}$ of K, and represent x by a matrix $(x_{i,j})_{i,j \in I}$ over $B(L^2(M))$. If x commutes to the action of M, this matrix must have entries in $\mathrm{End}_M(L^2(M))$, and thus $x \in \mathrm{End}_M(L^2(M)) \otimes B(K)$. Conversely any bounded matrix $(x_{i,j})$ with entries in $\mathrm{End}_M(L^2(M))$ commutes with the diagonal action of M.

(c) Consider $H \oplus (L^2(M) \otimes K)$ as an M–module for the diagonal action $x(\zeta \oplus (\eta \otimes \theta)) = x\zeta \oplus (x\,\eta \otimes \theta)$. Then $0 \oplus 1$ is an infinite projection in the commutant of M. By the Murray– von Neumann comparison theory for projections, there exists a partial isometry $\tilde{u}$ in the commutant $\mathrm{End}_M(H \oplus (L^2(M) \otimes K))$ from $1 \oplus 0$ to a subprojection of $0 \oplus 1$. One may view $\tilde{u}$ as an isometry

$$u : H \longrightarrow L^2(M) \otimes K$$

which intertwines the actions. #

As there will be many traces with various normalizations in the sequel, we introduce the following convention. If M is a finite factor, tr_M will denote its normalized trace. So if Tr is any other trace on M, then $\mathrm{Tr} = \mathrm{Tr}(\mathbb{1})\mathrm{tr}_M$, a formula which we will use often. Occasionally, we will have to consider a trace Tr on a factor P which is not finite (for example $B(H)$ or $M \otimes B(H)$, with H of infinite dimension). Let P_+ denote the positive cone of P, consisting of those element of the form z^*z with $z \in P$. Then a trace Tr is a map $P_+ \longrightarrow [0,\infty]$ such that

(i) $\mathrm{Tr}(x+y) = \mathrm{Tr}(x) + \mathrm{Tr}(y) \quad x,y \in P_+$

(ii) $\mathrm{Tr}(\lambda x) = \lambda \mathrm{Tr}(x) \quad \lambda \in \mathbb{R}_+, x \in P_+ \quad (\text{with } 0 \cdot \infty = 0)$

(iii) $\mathrm{Tr}(uxu^*) = \mathrm{Tr}(x) \quad x \in P_+, u$ a unitary in P.

Given a finite factor M acting in a Hilbert space H as in Lemma 3.2.2, we define now the natural trace $\mathrm{Tr}_{M'}$ on its commutant. It is crucial for what follows that $\mathrm{Tr}_{M'}$ is *not* necessarily normalized.

First, if $H = L^2(M)$ as in (a), we define $\mathrm{Tr}_{M'}(JxJ) = \mathrm{tr}_M(x)$ for all $x \in M$; in this case, $\mathrm{Tr}_{M'}$ *is* normalized. Secondly, if $H = L^2(M) \otimes K$ as in (b), consider an orthonormal basis $(\epsilon_i)_{i\in I}$ of K; then any element x in the commutant $\mathrm{End}_M(L^2(M) \otimes K)$ is represented by a matrix $(Jx_{i,j}J)_{i,j\in I}$; when x is moreover positive, then the diagonal elements $x_{i,i}$ are also positive, and we define

$$\mathrm{Tr}_{M'}(x) = \sum_{i\in I} \mathrm{tr}_M(x_{i,i}) \in [0,\infty].$$

For example,

$$\mathrm{Tr}_{M'}(JxJ\otimes p) = \mathrm{tr}_M(x) \dim_{\mathbb{C}}(pK)$$

if $x \in M_+$ and if $p \in B(K)$ is a projection.

Let $\mathcal{F}(K)$ denote the finite-rank operators on K. If $x \in JMJ \otimes \mathcal{F}(K) \subset \mathrm{End}_M(L^2(M) \otimes K)$, that is if all but finitely many of the matrix entries $x_{i,j}$ are zero, but x is not necessarily positive, then $\mathrm{Tr}_{M'}(x)$ is well-defined by the same formula. Furthermore, $x \longmapsto \mathrm{Tr}_{M'}(x)$ is a positive trace on the $*$-algebra $JMJ \otimes \mathcal{F}(K)$.

Third, for H arbitrary and for u as in (c) of Lemma 3.2.2, we define

$$\mathrm{Tr}_{M'}(x) = \mathrm{Tr}_{M'}(uxu^*)$$

for $x \in \mathrm{End}_M(H)_+$, and thus $uxu^* \in \mathrm{End}_M(L^2(M) \otimes K)_+$. If u_1,u_2 are two possible choices for u, then $u_1^*u_1 = u_2^*u_2 = \mathrm{id}_H$ and $u_2xu_2^* = u_2u_1^*u_1xu_2^*$ for $x \in M$; as $\mathrm{Tr}_{M'}$ is a trace,

$$\mathrm{Tr}_{M'}(u_2xu_2^*) = \mathrm{Tr}_{M'}(u_1xu_2^*u_2u_1^*) = \mathrm{Tr}_{M'}(u_1xu_1^*)$$

and $\mathrm{Tr}_{M'}(x)$ does not depend on the choice of u.

The word "natural" is justified by the following property (which again shows the independence just observed).

Lemma 3.2.3. *Let* H_1,H_2 *be two* M*-modules; let* $a : H_1 \to H_2$ *and* $b : H_2 \to H_1$ *be two* M*-linear bounded operators. Denote by* T_j *the natural trace defined on* $\mathrm{End}_M(H_j)$ *as above, for* $j = 1,2$. *Then*

$$T_2(ab) = T_1(ba).$$

Proof. Let $u_j : H_j \to L^2(M) \otimes K$ be an M-linear isometry. Then

$$\begin{aligned} T_2(ab) &= Tr_{M'}(u_2 ab u_2^*) = Tr_{M'}(u_2 a u_1^* u_1 b u_2^*) \\ &= Tr_{M'}(u_1 b u_2^* u_2 a u_1^*) = T_1(ba). \quad \# \end{aligned}$$

Definition. Let M be a finite factor and let H be a M-module. The coupling constant $\dim_M(H)$ is defined to be $Tr_{M'}(id_H)$, where the natural trace $Tr : End_M(H)_+ \to [0,\infty]$ is defined as above. If u is as in 3.2.2.c, one has also $\dim_M(H) = Tr_{M'}(uu^*)$ by 3.2.3.

Proposition 3.2.4. *Let* M *be a finite factor and let* $H, H', H_1, H_2, \cdots$ *be* M*-modules which are separable as Hilbert spaces. Then*

(a) $\dim_M(H) = \dim_M(H')$ *if and only if* H *and* H' *are isomorphic as* M*-modules,*

(b) $\dim_M(\oplus_i H_i) = \sum_i \dim_M(H_i)$,

(c) $\dim_M(L^2(M)) = 1$,

(d) $\dim_M(H) < \infty$ *if and only if the factor* $End_M(H)$ *is finite.*

Proof. Claim (a) follows from the comparison theorem for projections in the factor $End_M(L^2(M) \otimes K)$, claim (b) from the σ-additivity of the trace $Tr_{M'}$ on the same factor, and (c) is obvious.

In all cases, $End_M(H)$ is a semi-finite factor, and thus admits a non-zero trace which is unique up to a multiplicative constant. Claim (d) holds because $End_M(H)$ is finite if and only if it has a finite trace. #

In the next proposition, we continue with properties of $\dim_M$. The deep result is (f). We now describe the main step, the proof of which is in [MvN I] and [MvN IV] (see Theorem X in both papers). Again, let M be a finite factor and let H be a M-module; let tr be the normalized trace on M and let Tr' be the natural trace on $End_M(H)$. Choose $\xi \in H$ with $\xi \neq 0$. Denote by e_ξ the orthogonal projection of H onto the closure of the cyclic module $End_M(H)\xi$, and by e'_ξ that onto $\overline{M\xi}$; observe that $e_\xi \in M$ and $e'_\xi \in End_M(H)$. The basic (and difficult) fact is that the ratio

$$\tilde{c}_M = tr(e_\xi)/Tr'(e'_\xi)$$

is independent of ξ. (When M and H are finite dimensional, this basic fact reduces to Proposition 2.2.7.) Murray and von Neumann define the coupling constant of M and $\mathrm{End}_M(H)$ to be

$$c_M = \mathrm{tr}(e_\xi)/\mathrm{tr}'(e'_\xi) = (\mathrm{Tr}'(\mathbb{1}))\tilde{c}_M \in \mathbb{R}_+^*$$

if $\mathrm{End}_M(H)$ is finite, with tr′ the normalized trace on $\mathrm{End}_M(H)$ and Tr′ the natural trace. In case $\mathrm{End}_M(H)$ is infinite, they define $c_M = +\infty$.

The M–module H gives rise to other modules as follows. Let $e \in B(H)$ be a projection $(e \neq 0)$, with range denoted by eH. If $e \in \mathrm{End}_M(H)$ then eH is naturally a M–module (a submodule of H); if moreover $\mathrm{End}_M(H)$ is finite, the value D(e) of the normalized trace of $\mathrm{End}_M(H)$ on e is called the dimension of e. On the other hand, if $e \in M$, then eH is a eMe–module; the algebra eMe is a finite factor (because it is simple, a fact easy to check) which is called the reduction of M by e. Following common practice, we also write M_e for eMe.

Proposition 3.2.5. *Let* M *be a finite factor and let* H *be a* M*–module. Assume that the factor* $\mathrm{End}_M(H)$ *is finite (namely that* $\dim_M(H) < \infty$*). Then*

(e) $\dim_M(eH) = D(e)\dim_M(H)$ *for any non-zero projection* $e \in \mathrm{End}_M(H)$.

(f) $\dim_M(H) = c_M$, *the coupling constant of Murray and von Neumann.*

(g) $\dim_M(H)\dim_{\mathrm{End}_M(H)}(H) = 1$,

(h) $\dim_{eMe}(eH) = \frac{1}{D(e)}\dim_M(H)$ *for any non-zero projection* $e \in M$, *where* $D(e) = \mathrm{tr}(e)$.

(i) *If* L *is a finite dimensional Hilbert space, then* $\dim_M(H \otimes L) = \dim_M(H)\dim_{\mathbb{C}}(L)$.

Proof. For (e), one may view e as an M–linear isometry from eH to H. Then if $u : H \to L^2(M) \otimes K$ is an M–linear isometry, we have by definition of $\dim_M(\cdot)$ and by Lemma 3.2.3

$$\begin{aligned}\dim_M(eH) &= \mathrm{Tr}_{\mathrm{End}_M(L^2(M)\otimes K)}(ueu^*) \\ &= \mathrm{Tr}_{\mathrm{End}_M(H)}(e)\end{aligned}$$

$$= \mathrm{Tr}_{\mathrm{End}_M(H)}(\mathrm{id}_H)\mathrm{tr}_{M'}(e)$$
$$= \dim_H(H)D(e),$$

where each Tr_* denotes a natural trace.

Next we show how (f) reduces to the result of Murray and von Neumann recalled above. Replacing H by an isomophic submodule of $L^2(M) \otimes K$, we can assume $H \subset L^2(M) \otimes K$. Let $p \in \mathrm{End}_M(L^2(M) \otimes K)$ denote the orthogonal projection from $L^2(M) \otimes K$ onto H. Then by definition

(3.2.5.1) $$\mathrm{Tr}_{\mathrm{End}_M(L^2(M)\otimes K)}(p) = \dim_M(H)$$

Let $\xi \in H$ with $\xi \neq 0$ and let $\eta \in L^2(M) \otimes K$ with $\eta \neq 0$. As earlier, denote by $e_\xi \in M$ and $e'_\xi \in \mathrm{End}_M(H)$ the projections of H onto $\overline{\mathrm{End}_M(H)\xi}$ and $\overline{M\xi}$. Likewise denote by $f_\eta \in M$ and $f'_\eta \in \mathrm{End}_M(L^2(M) \otimes K)$ the projections of $L^2(M) \otimes K$ onto $\overline{\mathrm{End}_M(L^2(M)\otimes K)\eta}$ and $\overline{M\eta}$. With respect to the orthogonal decomposition $L^2(M) \otimes K = H \oplus H^\perp$, the algebra M acts by operators of the form $\begin{bmatrix} * & 0 \\ 0 & * \end{bmatrix}$, the algebra $\mathrm{End}_M(L^2(M) \otimes K)$ is of the form $\begin{bmatrix} \mathrm{End}_M(H) & * \\ * & * \end{bmatrix}$ and the space $\overline{\mathrm{End}_M(L^2(M) \otimes K)\xi}$ is of the form $\begin{bmatrix} \overline{\mathrm{End}_M(H)\xi} \\ * \end{bmatrix}$. It follows that $pf_\xi = e_\xi p$, or in matrix form that $f_\xi = \begin{bmatrix} e_\xi & 0 \\ 0 & * \end{bmatrix}$, so that it is the same element in M which acts as f_ξ on $L^2(M) \otimes K$ and as e_ξ on H. Consequently

(3.2.5.2) $$\mathrm{tr}_M(f_\xi) = \mathrm{tr}_M(e_\xi).$$

Observe also that, more simply

(3.2.5.3) $$pf'_\xi p = f'_\xi p = e'_\xi$$

because $\overline{M\xi} \subset H$.

To compute $\tilde{c}_2 = \mathrm{tr}_M(f_\eta) \div \mathrm{Tr}_{\mathrm{End}_M(L^2(M)\otimes K)}(f'_\eta)$, we may choose $\eta = \mathbf{1} \otimes \kappa$ with $\mathbf{1} \in M \subset L^2(M)$ and $\kappa \neq 0$ in K. Then f_η is the identity on $L^2(M) \otimes K$ and f'_η is the projection onto $L^2(M) \otimes \mathbb{C}\kappa$. Consequently

$$\mathrm{tr}_M(f_\eta) = 1 = \mathrm{Tr}_{\mathrm{End}_M(L^2(M)\otimes K)}(f'_\eta)$$

and $\tilde{c}_2 = 1$. But $\tilde{c}_2$ can also be computed using $\xi \in H$, so one has

(3.2.5.4) $$\mathrm{tr}_M(f_\xi) = \mathrm{Tr}_{\mathrm{End}_M(L^2(M)\otimes K)}(f'_\xi).$$

The coupling constant of Murray and von Neumann for M and $\mathrm{End}_M(H)$ is

$$c_M = \mathrm{tr}_M(e_\xi)/\mathrm{tr}_{\mathrm{End}_M(H)}(e'_\xi),$$

since we are assuming that $\mathrm{End}_M(H)$ is finite. By uniqueness of the normalized trace on $\mathrm{End}_M(H)$, one has

(3.2.5.5) $$\mathrm{tr}_{\mathrm{End}_M(H)}(pxp) = \mathrm{Tr}_{\mathrm{End}_M(L^2(M)\otimes K)}(x) \div \mathrm{Tr}_{\mathrm{End}_M(L^2(M)\otimes K)}(p)$$

for any $x \in \mathrm{End}_M(L^2(M)\otimes K)$. Putting together (3.2.5.1) to (3.2.5.5) one obtains

$$\begin{aligned} c_M &= \mathrm{tr}_M(e_\xi)/\mathrm{tr}_{\mathrm{End}_M(H)}(e'_\xi) \\ &= \{\mathrm{tr}_M(e_\xi)/\mathrm{Tr}_{\mathrm{End}_M(L^2(M)\otimes K)}(f'_\xi)\}\ \mathrm{Tr}_{\mathrm{End}_M(L^2(M)\otimes K)}(p) \\ &= \{\mathrm{tr}_M(f_\xi)/\mathrm{Tr}_{\mathrm{End}_M(L^2(M)\otimes K)}(f'_\xi)\}\ \dim_M(H) \\ &= \dim_M(H), \end{aligned}$$

and claim (f) is proved.

Claim (g) now follows trivially from (f). As for (h), using (e) and (g) as well as $\mathrm{End}_{eMe}(eH) = e(\mathrm{End}_M(H))e$, we have

$$\begin{aligned} \{\dim_{eMe}(eH)\}^{-1} &= \dim_{\mathrm{End}_{eMe}(eH)}(eH) = D(e)\dim_{\mathrm{End}_M(H)}(H) \\ &= D(e)\{\dim_M(H)\}^{-1}. \end{aligned}$$

Point (i) follows easily from the definition of $\dim_M(\cdot)$.

This ends the proof of Proposition 3.2.5. #

If $M = Mat_\mu(\mathbb{C})$ for some integer $\mu \geq 1$, then $\dim_M(H) = \mu^{-2} \dim_{\mathbb{C}}(H)$ is of the form $\frac{d}{\mu}$ with d an integer as in Proposition 2.2.7. This follows for example from claims (b) and (c) of Proposition 3.2.4. The object of the next section is to describe examples involving factors of type II_1.

3.3. The coupling constant: examples.

The situation for which the coupling constant is computed in this section is of the following kind: G is a non-compact semi-simple connected real Lie group which has the same rank as its maximal compact subgroups, $\pi : G \to U(H)$ is an irreducible representation of G in the discrete series, and $M = W^*(\Gamma)$ is the von Neumann algebra of an appropriate discrete subgroup Γ of G. Then H is naturally an M-module. Theorem 3.3.2 below is a computation of $\dim_M(H)$, due to Atiyah-Schmidt [AS,(3.3)]. First we discuss some background; the knowledgeable reader should jump to Theorem 3.3.2.

3.3.a. Discrete series.

Let G be a locally compact group. We assume that G is unimodular, we choose a Haar measure dg on G, and we denote by $\lambda_G : G \to U(L^2(G,dg))$ the left regular representation of G.

For an irreducible unitary representation $\pi : G \to U(H)$ of G, the following properties are equivalent:

(i) π is a subrepresentation of λ_G; more precisely, there exists a projection p in the commutant of $\lambda_G(G)$ such that the restriction of λ_G to the range of p is equivalent to π;

(ii) There exist $\xi,\eta \in H - \{0\}$ such that $g \mapsto <\pi(g)\xi | \eta>$ is in $L^2(G,dg)$;

(iii) For all $\xi,\eta \in H$ the function $g \mapsto <\pi(g)\xi | \eta>$ is in $L^2(G,dg)$.

If these hold, π is said to belong to the (unitary) discrete series. On may then attach to π a real number $d_\pi > 0$, called its formal dimension, such that Schur's orthogonality relations formally hold. In particular, for any $\pi : G \to U(H)$ in the discrete series

$$\int_G \langle \pi(g)\xi | \eta \rangle \langle \eta' | \pi(g)\xi' \rangle dg = d_\pi^{-1} \langle \xi | \xi' \rangle \langle \eta' | \eta \rangle$$

for $\xi,\xi',\eta,\eta' \in H$.

The formal dimension d_π depends on π *and* on the choice of the Haar measure for G; if $d'g = kdg$ for some constant $k > 0$, the two corresponding formal dimensions of π are related by $d'_\pi = k^{-1}d_\pi$. If G is compact and if $\int_G dg = 1$, then d_π is the dimension of H in the naive sense. For all this, see section 16 in [Rbt] or Chapter 14 in [DC*].

Given an arbitrary (unimodular) group G, its discrete series may be empty. This happens for G infinite abelian, or infinite discrete, or $G = SL(2,\mathbb{C})$, or $G = SL(n,\mathbb{R})$ with $n \geq 3$, to quote but a few examples. When G is a semi-simple connected real Lie group with maximal compact subgroup K, then G has discrete series representations if and only if G and K have the same rank. In particular $SL(2,\mathbb{R})$ has a discrete series, as well as $SO(n,1)^0$ for n even.

3.3.b. Factors defined by icc groups.

On a discrete group Γ, we consider always the counting measure; the space of square summable functions from Γ to $\mathbb{C}$ is denoted by $\ell^2(\Gamma)$. The von Neumann algebra $W^*(\Gamma)$ of Γ is the (ultra)weak closure of the linear span of $\lambda_\Gamma(\Gamma)$ in $B(\ell^2(\Gamma))$; by von Neumann's theorem, it is also the bicommutant of $\lambda_\Gamma(\Gamma)$ in $B(\ell^2(\Gamma))$, and $W^*(\Gamma)$ is thus also denoted by $\lambda_\Gamma(\Gamma)''$.

Let $\delta_e \in \ell^2(\Gamma)$ be the function which takes the value 1 at the identity e of Γ and 0 elsewhere. It is easy to check that $x \longmapsto x(\delta_e)$ is a linear injection of $W^*(\Gamma)$ in $\ell^2(\Gamma)$, and that the map $tr(x) = \langle x(\delta_e) | \delta_e \rangle$ is a normalized finite faithful trace on $W^*(\Gamma)$; see the end of 4.2 in [Sak]. It follows that the von Neumann algebra $W^*(\Gamma)$ is finite, and that the Hilbert space $L^2(W^*(\Gamma),tr)$ defined before Lemma 3.2.2 is canonically isomorphic to $\ell^2(\Gamma)$.

Moreover $W^*(\Gamma)$ is a factor (and thus a factor of type II_1) if and only if Γ is an infinite conjugacy class group, or for short an icc group (Lemma 4.2.18 in [Sak]). The following lemma exhibits a rich class of icc groups. Before this, we recall that the quotient G/Γ of a unimodular locally compact group G by a discrete subgroup Γ has always a G-invariant measure, which is unique up to a scalar factor; by definition, Γ is a lattice in G if the meaure of G/Γ is finite.

Lemma 3.3.1. *A lattice* Γ *in a connected semi-simple real Lie group* G *without center and without a compact factor is an icc group, and* $W^*(\Gamma)$ *is consequently a* II_1*-factor.*

Proof. The main point is Borel's density theorem, which we quote without proof (see [Bor] or [Zim]): Γ is Zariski–dense in G.

Consider $h \in \Gamma$ and its conjugacy class C_h in Γ. The map $\begin{cases} \Gamma \to C_h \\ \gamma \mapsto \gamma h \gamma^{-1} \end{cases}$ extends by continuity to the Zariski closure $\begin{cases} G \to \overline{C_h} \\ g \mapsto ghg^{-1} \end{cases}$. If C_h is finite, then $\overline{C_h} = C_h$ and $\{g \in G \mid gh = hg\}$ is a closed subgroup of finite index in G. But the algebraic group corresponding to G is Zariski– connected, and it follows that $\{g \in G \mid gh = hg\} = G$. Thus h is central in G, so that $h = e$. This shows that Γ is an icc group. #

A final remark about this: let $\Gamma_1 \subset G_1$ and $\Gamma_2 \subset G_2$ be two examples of the situation in the previous lemma. Assume moreover that G_1 and G_2 have real rank at least two. It is a conjecture, due to A. Connes and "beyond Mostow and Margulis", that $W^*(\Gamma_1)$ is isomorphic to $W^*(\Gamma_2)$ if and only if Γ_1 and Γ_2 are isomorphic.

3.3.c. $W^*(\Gamma)$–modules associated to subrepresentations of λ_G.

Let G be a unimodular Lie group with Haar measure dg and let Γ be a discrete subgroup of G. In the present context, it is convenient to define a fundamental domain for Γ in G to be a subset D of G which is measurable and satisfies

$\gamma_1 D \cap \gamma_2 D$ has null measure for $\gamma_1, \gamma_2 \in \Gamma$ with $\gamma_1 \neq \gamma_2$ and

$G \setminus \bigcup_{\gamma \in \Gamma} \gamma D$ has null measure.

Such a D always exists. Indeed, as $G \to \Gamma \backslash G$ is a topological covering, it has a Borel section, and the image of such a Borel section is a convenient D. The measure of D does not depend on D itself and is called the covolume of Γ. (If dg is defined via a differential form Ω of maximal degree on G, there is a unique form ω on $\Gamma\backslash G$ which pulls back to Ω, and the covolume of Γ is $\int_{\Gamma\backslash G} \omega$.)

Of course, $\mathrm{covol}(\Gamma)$ does depend on the choice of the Haar measure on G. If $d'g = k dg$ for some constant $k > 0$, the two corresponding covolumes of Γ are related by $\mathrm{covol}'(\Gamma) = k\, \mathrm{covol}(\Gamma)$.

Given $\Gamma \subset G$ and D as above, there is an isomorphism from $L^2(G,dg)$ onto $\ell^2(\Gamma) \otimes L^2(D,dg)$ which maps φ to $\sum_{\gamma \in \Gamma} \delta_\gamma \otimes \varphi_\gamma$, where $\delta_\gamma \in \ell^2(\Gamma)$ is the characteristic function of $\{\gamma\}$ in Γ, and where $\varphi_\gamma(g) = \varphi(\gamma g)$ for $\gamma \in \Gamma$, $g \in D$. It follows from the definitions of λ_G and λ_Γ that the restriction $\lambda_G|_\Gamma$ to Γ of the left regular representation of G is the tensor product of λ_Γ with the trivial representation of Γ on

$L^2(D,dg)$. Hence the von Neumann algebra $\lambda_G(\Gamma)''$ is isomorphic to $W^*(\Gamma) \otimes \mathbb{C} \cong W^*(\Gamma)$.

More generally, let $p \in B(L^2(G,dg))$ be a projection which commutes with $\lambda_G(\Gamma)$. Denote by H_p the range of p, by $\pi_p : \Gamma \to U(H_p)$ the corresponding subrepresentation of $\lambda_G|_\Gamma$, and by $\pi_p(\Gamma)''$ the von Neumann algebra generated by $\pi_p(\Gamma)$ in $B(H_p)$. Then the W^*–morphism $\begin{cases} \lambda_G(\Gamma)'' \to \pi_p(\Gamma)'' \\ x \mapsto px \end{cases}$ is obviously surjective. If Γ is moreover an icc group, then $\lambda_G(\Gamma)'' \cong W^*(\Gamma)$ is a factor of type II_1 and is in particular a simple ring, so that the map $\lambda_G(\Gamma)'' \to \pi_p(\Gamma)''$ is an isomorphism.

We shall particularize below to the case in which the projection p commutes with all of $\lambda_G(G)$, and defines an irreducible representation of G in the discrete series.

3.3.d The formula $\dim_M(H) = \mathrm{covol}(\Gamma)d_\pi$.

Now the relevant background has been established, and we demonstrate the main result of this section.

Theorem 3.3.2. *Let* G *be a connected semi–simple real Lie group with Haar measure* dg, *let* Γ *be a discrete subgroup in* Γ, *let* M *denote* $W^*(\Gamma)$ *and let* $\pi : G \to U(H)$ *be an irreducible representation in the discrete series. Assume that* Γ *is an icc group. Then* $\dim_M(H) = \mathrm{covol}(\Gamma)d_\pi$.

Observations. (1) Lemma 3.3.1 says that Γ is automatically an icc group in case it is a lattice in a connected simple noncompact Lie group without center.

(2) Both $\mathrm{covol}(\Gamma)$ and d_π depend on dg, but these dependences cancel out in the product.

Proof. From the discussion in 3.3.c, we may assume that H is included as an M–module in $L^2(G,dg)$. This inclusion, say u, satisfies $u^*u = id_H$ and $uu^* = p$, where p is the orthogonal projection from $L^2(G,dg)$ onto H. Also, $L^2(G,dg)$ may be identified with $L^2(M) \otimes K$, where $L^2(M)$ is the canonical M–module, and where K is the trivial M–module $L^2(D,dg)$ associated to some fundamental domain D of Γ in G. Thus we have

$$\dim_M(H) = \mathrm{Tr}_{M'}(p);$$

in this proof, M' denotes the commutant of M in $L^2(G,dg)$ or in $L^2(M) \otimes K$, and $\mathrm{Tr}_{M'}$ is the natural trace on M'.

By Lemma 3.2.2.b, this commutant M' is generated by finite sums of the form $x = \sum_{\gamma\in\Gamma} \rho_\gamma \otimes a_\gamma$. For each $\gamma \in \Gamma$, the symbol ρ_γ stands for $J\lambda_\Gamma(\gamma)J \in \mathrm{End}_M(L^2(M))$ and a_γ is a finite rank operator in $B(K)$. Let $(\epsilon_n)_{n\in\mathbb{N}}$ be an orthonormal basis of K. Let $\overline{\epsilon_m} \otimes \epsilon_n$ denote the operator $\xi \mapsto \langle \epsilon_m | \xi \rangle \epsilon_n$ on K. One may write $a_\gamma = \sum_{m,n\in\mathbb{N}} a_{\gamma,m,n} \overline{\epsilon_m} \otimes \epsilon_n$, where the $a_{\gamma,m,n}$ are complex numbers. By definition of $\mathrm{Tr}_{M'}$ one has

$$\mathrm{Tr}_{M'}(\rho_\gamma \otimes a_\gamma) = \mathrm{tr}_M(\lambda_\gamma) \sum_{m\in\mathbb{N}} a_{\gamma,m,m} = \begin{cases} 0 & \text{if } \gamma \neq e \\ T_K(a_e) & \text{if } \gamma = e \end{cases}$$

where tr_M is the normalized trace on M and where T_K is the trace on $B(K)$ normalized by $T_K(\overline{\epsilon_m} \otimes \epsilon_m) = 1$ for all $m \in \mathbb{N}$. With x as above, one has consequently

$$\mathrm{Tr}_{M'}(x) = T_K(a_e).$$

Let $q : L^2(G,dg) \to K$ be the orthogonal projection given by restricting functions from G to D, and let T denote the trace on $B(L^2(G,dg))$ taking value 1 on projections of rank one. Then $T_K(y) = T(qyq)$ for $y \in B(K)_+$ or $y \in \mathcal{F}(K)$. In particular, for x of the form $x = \sum_{\gamma\in\Gamma} \rho_\gamma \otimes a_\gamma$ we have

$$\mathrm{Tr}_{M'}(x) = T_K(a_e) = T(qa_eq) = T(qxq). \tag{3.3.2.1}$$

Finally any $x \in M'_+$ is the strong limit of an increasing net of operators of the form $\sum_{\gamma\in\Gamma} \rho_\gamma \otimes a_\gamma$; as the traces are normal, the formula (3.3.2.1) holds for all $x \in M'_+$, and in particular

$$\dim_M(H) = \mathrm{Tr}_{M'}(p) = T(qpq).$$

The right–hand term is explicitly given by

$$T(qpq) = \sum_{n\in\mathbb{N}} \langle qpq\epsilon_n | \epsilon_n \rangle = \sum_{n\in\mathbb{N}} \|p\epsilon_n\|^2.$$

Recall that δ_γ denotes the characteristic function of $\{\gamma\}$ in Γ, and that ϵ_n, which is a function on D, is also naturally a function on G (vanishing outside D). Thus the orthonormal basis $\{\delta_\gamma \otimes \epsilon_n\}_{\gamma \in \Gamma, n \in \mathbb{N}}$ of $\ell^2(\Gamma) \otimes K$ is more conveniently viewed as the basis $\{\lambda_G(\gamma)\epsilon_n\}_{\gamma \in \Gamma, n \in \mathbb{N}}$ of $L^2(G,dg)$. Let η be a unit vector in $L^2(G,dg)$; assume that $\eta \in H$, namely that $p\eta = \eta$. For any $g \in G$ one has (writing λ instead of λ_G)

$$1 = \|\lambda(g)\eta\|^2 = \sum_{\gamma \in \Gamma} \sum_{n \in \mathbb{N}} |\langle \lambda(g)\eta | \lambda(\gamma)\epsilon_n \rangle|^2.$$

Consequently, as p commutes with $\lambda(G)$:

$$\begin{aligned} \operatorname{covol}(\Gamma) &= \int_D dg = \sum_{n \in \mathbb{N}} \sum_{\gamma \in \Gamma} \int_D |\langle \lambda(\gamma^{-1}g)p\eta | \epsilon_n \rangle|^2 dg \\ &= \sum_{n \in \mathbb{N}} \int_G |\langle p\lambda(g)\eta | \epsilon_n \rangle|^2 dg \\ &= \sum_{n \in \mathbb{N}} \int_G |\langle \lambda(g)\eta | p\epsilon_n \rangle|^2 dg. \end{aligned}$$

By Schur's relations

$$\operatorname{covol}(\Gamma) = \sum_{n \in \mathbb{N}} \frac{1}{d_\pi} \|\eta\|^2 \|p\epsilon_n\|^2 = \frac{1}{d_\pi} \dim_M(H)$$

and the proof is complete. #

Corollary 3.3.3. *In the situation of the previous theorem,* Γ *is a lattice if and only if the commutant of* M *in* $B(H)$ *is a finite factor.*

Proof. The last condition holds if and only if $\operatorname{covol}(\Gamma)$ is finite. #

We now particularize G to the group $PSL(2,\mathbb{R})$. For each integer $k \geq 2$, let H_k be the space of holomorphic functions on the Poincaré half-plane $\mathcal{P}$ which are square-summable for the measure $y^{k-2}dxdy$. (The open unit disc Δ in the complex plane with the corresponding measure is equally good). As G acts on $\mathcal{P}$ by fractional linear transformations, there is a natural unitary representation π_k of G in H_k. It is a standard result that H_k is an infinite dimensional Hilbert space and that π_k is an irreducible discrete series representation. (These π_k consitute the holomorphic discrete series, and the full discrete series contains a second "half", the anti-holomorphic part.)

Define the Haar measure dg on G as follows: let $\mathbb{T} = SO(2)/\{\pm 1\}$ be the maximal compact subgroup of G, such that $\begin{cases} G \to \mathcal{P} \\ g \mapsto g(i) \end{cases}$ induces a diffeomorphism $G/\mathbb{T} \cong \mathcal{P}$; then

$d\mu(z) = y^{-2}dxdy$ is a G–invariant measure on $\mathcal{P}$; if φ is a continuous function $G \to \mathbb{C}$ with compact support, set

$$\int_G \varphi(g)dg = \int_{\mathcal{P}} d\mu(z) \int_{\mathbb{T}} dt\varphi(gt) \qquad z = g(i)$$

where dt is the Haar measure on $\mathbb{T}$ of total measure 1.

Then the virtual dimension of π_k is known to be given by $d_k = \frac{k-1}{4\pi}$; see theorem 17.8 in [Rbt]. (Warning: under the Cayley transform $\begin{cases} \mathcal{P} \to \Delta \\ z=x+iy \mapsto \frac{z-i}{z+i} = u+iv \end{cases}$ the measure $\frac{dxdy}{y^2}$ corresponds to $\frac{4dudv}{(1-u^2-v^2)^2}$, which is 4 times that in [Rbt]! The measure chosen in the present section is that which is defined by the Riemannian structure for which $\mathcal{P}$ has constant curvature –1; the computation may be found, for example, in Section 5.10 of [Car].)

Now consider an integer $q \geq 3$, set $\lambda = 2\cos(\pi/q)$ and let Γ_λ be the Hecke subgroup of $PSL(2,\mathbb{R})$ generated by the classes modulo ± 1 of the matrices

$$\begin{bmatrix} 1 & \lambda \\ 0 & 1 \end{bmatrix} \text{ and } \begin{bmatrix} 0 & 1 \\ -1 & 0 \end{bmatrix}.$$

Then $covol(\Gamma_\lambda) = \pi(1 - \frac{2}{q})$ by the Gauss–Bonet formula, because Γ_λ has a triangular fundamental domain with angles $0, \frac{\pi}{q}, \frac{\pi}{q}$ (see Appendix III).

Altogether, we have shown:

<u>Example 3.3.4</u>. Given integers $q \geq 3$ and $k \geq 2$, consider the II_1–factor $M = W^*(\Gamma_\lambda)$ defined by the Hecke group Γ_λ with $\lambda = 2\cos(\pi/q)$ and the Hilbert space H_k of the holomorphic discrete series of $PSL(2,\mathbb{R})$. Then H_k is a M–module of coupling constant

$$\dim_M(H_k) = \frac{k-1}{4}(1 - \frac{2}{q}).$$

<u>3.3.e. A digression on the Peterson inner product.</u>

We particularize further, and set $q = 3$ in example 3.3.4. That is, we consider $\Gamma = PSL(2,\mathbb{Z})$ as a discrete subgroup of $PSL(2,\mathbb{R})$.

Given an integer $p \geq 1$, recall that a <u>cusp form of weight</u> p is (in this situation) a holomorphic function $f : \mathcal{P} \to \mathbb{C}$ on the Poincaré half–plane satisfying two conditions. The first one is an invariance:

$$f(z) = (cz+d)^{-p} f\left[\frac{az+b}{cz+d}\right], \quad z \in \mathcal{P}, \quad \begin{bmatrix} a & b \\ c & d \end{bmatrix} \in SL(2,\mathbb{Z}).$$

The second one is a growth condition: observe that $f(z) = f(z+1)$, so that $\tilde{f}$ can be defined on the punctured unit disc Δ^* by $\tilde{f}(e^{2i\pi z}) = f(z)$; the second defining condition is that the Laurent expansion of $\tilde{f}$ in Δ^* is of the form $\tilde{f}(w) = \sum_{n\geq 1} a_n w^n$ for $w \in \Delta^*$.

It is a result of Hecke that a cusp form f of weight p satisfies $|f(x+iy)| \leq By^{-p/2}$ for all $x + iy \in \mathcal{P}$ and for some constant B; see page I.24 in [Ogg].

Let $M = W^*(\Gamma)$. Consider an integer $k \geq 2$ and the M–module H_k of example 3.3.4. Given a cusp form f of weight p, the growth condition implies that f induces a multiplication operator $A_f : H_k \to H_{k+p}$, defined by $(A_f\varphi)(z) = f(z)\varphi(z)$, which is bounded (in fact $\|A_f\| \leq B$ with B as above). The invariance condition implies that A_f is M–linear. Consequently, given two cusp forms f,g of weight p, the operator $A_f^*A_g : H_k \to H_k$ is in the commutant $End_M(H_k)$. Let T_k denote the natural trace on $End_M(H_k)$. Then the space of cusp forms of weight p has a natural hermitian form

$$\langle f|g\rangle_k = T_k(A_f^*A_g).$$

A computation in the same spirit as that presented in the proof of Theorem 3.3.2 shows that

$$\langle f|g\rangle_k = \frac{k-1}{\pi}\int_D \overline{f(z)}g(z)y^{p-2}dxdy$$

with D a fundamental domain for Γ in $\mathcal{P}$. Up to a constant factor $\frac{k-1}{\pi}$, this is known as the Peterson scalar product for cusp forms.

This suggests a natural project, which could be interesting for the study of cusp forms: evaluate L^q–norms defined by

$$\|f\|_{k,q} = \{T_k((A_f^*A_f)^{q/2})\}^{1/q}.$$

The equality $T_k(A_f^*A_g) = T_{k+p}(A_gA_f^*)$ should be useful.

3.4. Index for subfactors of II_1 factors.

There were two main motivations for the introduction in [Jo1] of the concept of index for subfactors. The first was that, if $\Gamma_1 < \Gamma_2$ are two icc discrete groups, the II_1 factor

$N = \lambda(\Gamma_1)''$ acts in an obvious way on $\ell^2(\Gamma_2)$ and $\dim_N(\ell^2(\Gamma_2)) = [\Gamma_2:\Gamma_1]$. Furthermore $\ell^2(\Gamma_2)$ is the same as $L^2(M)$ where M is $\lambda(\Gamma_2)''$. This suggested the following definition:

Definition 3.4.1. The index of a subfactor N of a finite factor M is

$$[M:N] = \dim_N(L^2(M)).$$

This was the original definition of index; it was shown in [Jo4] that this definition agrees with the ring–theoretic one which we have given in Chapter 2, when M and N are finite factors. The index can also be computed as $[M:N] = \dim_N(H)/\dim_M(H)$, where H is any M–module of finite dimension over M; see Proposition 3.4.6.

The second motivation was a result of M. Goldman [Gol], who showed that, if $N \subset M$ are II_1 factors (always with the same identity $\mathbb{1}$) then, if $\dim_N(L^2(M)) = 2$, there is a crossed product decomposition $M = N \rtimes \mathbb{Z}/2\mathbb{Z}$. Consequently if one defines $[M:N]$ as above, Goldman's result is seen to be a beautiful analogue of the fact that a subgroup of index 2 of a group is normal.

It would also have been nice to have been able to call a subfactor $N \subset M$, normal when its (unitary) normalizer generates M. But unfortunately standard terminology reserves "normal" for subfactors N such that $(N' \cap M)' \cap M = N$, and the term regular is used for subfactors with the normalizer property described above. We take this opportunity to introduce some more terminology.

Definition 3.4.2. If $N \subset M$ are factors we say that N is irreducible if $N' \cap M = \mathbb{C}$.

It is not hard to see that a regular irreducible subfactor has integer index (or ∞ which we shall treat as an integer) — see [Jo7]. A more refined analysis based on [Jo6] shows that all regular subfactors have integer index. On the other hand $\dim_M(H)$ can be any positive real number so the question naturally arose:

(a) What are the possible values of $[M:N]$?
(b) What are the possible values of $[M:N]$ for an irreducible pair $N \subset M$?

Question (a) was settled completely in [Jol] for $M = R$, the hyperfinite II_1 factor. Question (b) remains open even for $M = R$, and question (a) is open for arbitrary II_1 factors M. We summarize the most important known results as follows:

Theorem 3.4.3. *Let* N *be a subfactor of a* II_1 *factor* M.

(i) *Either* $[M:N] = 4\cos^2\pi/q$ *for some integer* $q \geq 3$, *or* $[M:N] \geq 4$.

(ii) *If* $[M:N] < 4$, *then* N *is automatically irreducible in* M.

(iii) *There exist subfactors of the hyperfinite* II_1 *factor* R *with any of the index values allowed by* (i).

(iv) *There are examples of factors* M *for which the set of all possible values* [M:N] *is countable.*

Remarks: Statements (i) to (iii) are from [Jo1]. We prove (i) below. A generalization to finite direct sums of II_1 factors is shown in Corollary 3.7.6. A second proof of (i) occurs in Corollary 4.6.6, as a byproduct of the analysis of "derived towers".

Statement (ii) is proved as Corollary 3.6.2(c).

We will verify (iii) by giving several constructions of subfactors of R. The first construction, in this section, works for all allowed index values. Another construction, valid for the index values $4\cos^2\pi/q$ is given in Theorem 4.4.2. A third construction, in Section 4.5, produces irreducible pairs; the index values $4\cos^2\pi/q$ are obtained once more, as well as sporadic values greater than 4. In Section 4.7.d, we give examples of non-conjugate irreducible subfactors of R of index 4. We would also like to mention the work of Wenzl [Wen2], in which a family of irreducible subfactors of R of index greater than 4 is produced by a construction involving the Hecke algebras $H_\infty(q)$ for q a primitive root of unity.

Statement (iv) is from [PP2], and will not be proved here.

For arbitrary II_1 factors, the question of existence of subfactors of index $4\cos^2\pi/q$ remains open, more precisely we know of no example of a full II_1 factor M having a subfactor of index $4\cos^2\pi/q$, $q \neq 3,4,6$. (A II_1 factor is called "full" if the group of inner automorphisms is closed in the topology of pointwise strong convergence in the whole automorphism group — an example of such a factor is $\lambda(PSL(2,\mathbb{Z}))''$.)

Proof of 3.4.3 (i). As for finite dimensional algebras (2.6.2) there is always a (unique) faithful trace-preserving conditional expectation from M onto N, which, viewed as an operator on $L^2(M)$ is the orthogonal projection e_N onto $L^2(N)$. The fundamental construction again yields a II_1-factor

$$\mathrm{End}_N^r(L^2(M,tr)) = \mathrm{End}_N^r(M) = \langle M,e_N\rangle.$$

(See Theorem 3.4.6 below for the first equality.)

We claim that the normalized trace of $\langle M, e_N\rangle$ has the Markov property

$$[M:N]\mathrm{tr}(e_N x) = \mathrm{tr}(x) \qquad \text{for all } x \in M.$$

Indeed, the linear form defined on N by $x \mapsto \mathrm{tr}(e_N x)$ is a trace (3.6.1.iii). As $1 = \mathrm{tr}(e_N)\,[M:N]$ by Proposition 3.2.5.e applied to the N–module $L^2(M)$, the property is valid for $x \in N$, by uniqueness of the normalized trace on N. But then for $x \in M$, we have $[M:N]\mathrm{tr}(e_N x) = [M:N]\mathrm{tr}(e_N x e_N) = [M:N]\mathrm{tr}(e_N E_N(x)) = \mathrm{tr}(E_N(x)) = \mathrm{tr}(x)$, using 3.6.1.i .

Now the tower construction of Chapter 2 works and yields an increasing sequence of II_1–factors

$$M_0 = N \subset M_1 = M \subset \cdots \subset M_k \subset M_{k+1} \subset \cdots,$$

and a sequence of self–adjoint projections $(e_i)_{i\geq 1}$ satisfying

(3.4.3.1)
$$\beta\, e_i e_{i\pm 1} e_i = e_i$$
$$e_i e_j = e_j e_i \quad \text{if } |i-j| \geq 2,$$

with $\beta = [M:N]$. Claim (i) now follows from Theorem II.16.

An alternative proof using the trace goes as follows: The trace tr on $\bigcup_k M_k$ has the Markov property

(3.4.3.2)
$$\beta\, \mathrm{tr}(w e_j) = \mathrm{tr}(w) \qquad \text{for } j \geq 1 \text{ and } w \in \mathrm{alg}\,\{\mathbb{1}, e_1, \cdots, e_{j-1}\}.$$

where $\beta = [M:N]$. Now suppose that $\beta < 4$ but $\beta \notin \{4\cos^2\pi/q : q \geq 3\}$. Using 2.8.5 and 2.8.7 (note that the number β is generic) as well as the relations 3.4.3.1 and 3.4.3.2, we obtain for each $k \geq 1$ a trace preserving isomorphism of the algebra $B_{\beta,k}$ of Section 2.8 onto the algebra $C_k = \{\mathbb{1}, e_1, \cdots, e_{k-1}\}''$. By 2.8.4(vii), for each k the trace of the minimal central projection Q_0^k (necessarily a self–adjoint projection in C_k) is $P_k(\beta^{-1})$. But by 2.8.3(iii), if $4\cos^2(\pi/k) < \beta < 4\cos^2(\pi/k+1)$, then $P_k(\beta^{-1}) < 0$, contradicting the positivity of the trace. It follows that if $\beta < 4$, then $\beta \in \{4\cos^2\pi/q : q \geq 3\}$. #

Proof of 3.4.3(iii). Fix $\beta \in \mathbb{R}$ with $\beta = 4\cos^2\pi/q$ for some integer $q \geq 3$, or $\beta \geq 4$. Consider a sequence of self–adjoint projections $(e_i)_{i\geq 1}$ on a Hilbert space, together with a faithful normal tracial state tr on $R = \{\mathbb{1}, e_1, e_2, \cdots\}''$ satisfying the relations 3.4.3.1 as well as the Markov property 3.4.3.2.

First we must recall how such a sequence of projections and such a trace can be constructed. In 2.8.4 (in case $\beta \geq 4$) and in Section 2.9 (in case $\beta = 4\cos^2\pi/q$ for some

q) we have constructed an increasing sequence of finite dimensional C^*–algebras $(B_{\beta,k})_{k\geq1}$, with $B_{\beta,k}$ generated by its identity and self-adjoint projections $e_1,\cdots e_{k-1}$ satisfying the relations 3.4.3.1, and a positive faithful normalized trace tr on $B_{\beta,k}$ satisfying the relation 3.4.3.2 for $1 \leq j \leq k$. Since tr is faithful, the trace representation π_{tr} is faithful as well, and we can take R to be $\pi_{tr}(\cup B_{\beta,k})''$.

A simpler procedure is available when β is the square of the norm of a non-negative integer valued matrix (i.e. $\sqrt{\beta} \in \mathcal{N}(\mathbb{N})$). In this case there is a connected pair of finite dimensional C^*–algebras $B \subset A$ with $[A:B] = \beta$, and the tower construction for this pair yields a sequence of projections $(e_i)_{i\geq1}$ satisfying 3.4.3.1, and a positive faithful trace on alg $\{\mathbb{1},e_1,\cdots\}$ satisfying 3.4.3.2. Cf. 2.7.5 and the discussion at the end of Appendix IIa.

Lemma 3.4.4. [Jo1] *With the notation above,* R *is the hyperfinite* II_1 *factor.*

Proof. It is clear that R is a finite, hyperfinite von Neumann algebra. We claim that if z is in the center of R, then

$$\mathrm{tr}(zx) = \mathrm{tr}(z)\mathrm{tr}(x) \quad \text{for all } x \in R.$$

It will follow from this and the faithfulness of tr that $z = \mathrm{tr}(z)\mathbb{1}$, so R is a factor.

For each k, let $C_k = \mathrm{alg}\,\{\mathbb{1},e_1,\cdots e_{k-1}\}$. By 2.9.6(e) (in case $\beta < 4$) or by 2.8.7(a) and 2.8.5(b) (in case $\beta \geq 4$), the map $\epsilon_j \mapsto e_j$ (on the generators $\{\epsilon_j\}$ of $\mathcal{A}_{\beta,k}$) induces a trace preserving isomorphism of $B_{\beta,k}$ onto C_k. It then follows from 2.9.6(g) (for $\beta < 4$) or from 2.8.5(f) (for $\beta \geq 4$) that $e_j \mapsto e_{k-j}$ extends to an inner automorphism of C_k, and hence to an inner automorphism α_k of R.

Note that tr has the multiplicative property $\mathrm{tr}(y_1y_2) = \mathrm{tr}(y_1)\mathrm{tr}(y_2)$ whenever $y_1 \in C_s$ and $y_2 \in \mathrm{alg}\,\{\mathbb{1},e_s,\cdots e_{s+m}\}$. (One can verify this directly or use the isomorphism $C_m \cong B_{\beta,m}$ together with 2.8.5(e) or 2.9.6(f).)

It will suffice to verify the relation $\mathrm{tr}(zx) = \mathrm{tr}(z)\mathrm{tr}(x)$ when $x \in C_k$ for some k. Let $\epsilon > 0$, and choose $y \in C_\ell$ for some ℓ, such that $\|z-y\|_2 < \epsilon$. Then $\mathrm{tr}(y\alpha_{k+\ell}(x)) = \mathrm{tr}(y)\mathrm{tr}(x)$, since $\alpha_{k+\ell}(x) \in \mathrm{alg}\,\{\mathbb{1},e_{\ell+1},\cdots e_{\ell+k-1}\}$. Consequently,

$$\begin{aligned} &|\mathrm{tr}(zx) - \mathrm{tr}(z)\mathrm{tr}(x)| \\ &= |\mathrm{tr}(z\alpha_{k+\ell}(x)) - \mathrm{tr}(z)\mathrm{tr}(x)| \qquad \text{(since } \alpha_{k+\ell} \text{ is inner)} \\ &\leq |\mathrm{tr}((z-y)\alpha_{k+\ell}(x))| + |\mathrm{tr}(y\alpha_{k+\ell}(x)) - \mathrm{tr}(z)\mathrm{tr}(x)| \end{aligned}$$

$$= |\mathrm{tr}((z-y)\alpha_{k+\ell}(x))| + |(\mathrm{tr}(y)-\mathrm{tr}(z))\mathrm{tr}(x)|$$
$$\leq 2\,\epsilon\,\|x\|_2.$$

Since ϵ is arbitrary, this finishes the proof. #

Lemma 3.4.5. [Jo1] Set $R_\beta = \{\mathbb{1}, e_2, e_3, \cdots\}''$. Then $[R:R_\beta] = \beta$.

Proof. We know (by 2.8.5 and 2.8.7 or by 2.9.6) that for each $k \geq 2$, the relation $\beta\,\mathrm{tr}(e_1 x) = \mathrm{tr}(x)$ holds when $x \in \mathrm{alg}\,\{\mathbb{1}, e_2, \cdots e_k\}$, and, taking limits, we have the same relation also for $x \in R_\beta$. Therefore $E_{R_\beta}(e_1) = \beta^{-1}\mathbb{1}$.

Similarly $E_N(e_2) = \beta^{-1}\mathbb{1}$, where $N = \{\mathbb{1}, e_3, e_4, \cdots\}''$. For $k \geq 3$, any $x \in \mathrm{alg}\,\{\mathbb{1}, e_2, \cdots e_k\}$ is of the form $x = a + \sum_i b_i e_2 c_i$, with $a, b_i, c_i \in \mathrm{alg}\,\{\mathbb{1}, e_3, \cdots e_k\}$. Consequently, $E_N(x) = a + \beta^{-1} \sum_i b_i c_i$ and $e_1 x e_1 = E_N(x) e_1$. Taking limits again, we have

$$(*) \qquad e_1 x e_1 = E_N(x) e_1 \ \text{ for all } x \in R_\beta.$$

One next verifies that $x e_1 = \beta\, E_{R_\beta}(x e_1) e_1$ for all $x \in R$, by first checking this for $x \in \mathrm{alg}\,\{\mathbb{1}, e_1, \cdots e_k\}$ (that is, for x of the form $x = a + \sum_i b_i e_1 c_i$, with $a, b_i, c_i \in \mathrm{alg}\,\{\mathbb{1}, e_2, \cdots e_k\}$) and then by taking limits. Consequently $Re_1 = R_\beta e_1$, and $Re_1 R = R_\beta e_1 R_\beta$. Observe also that $R = Re_1 R$, because finite factors are algebraically simple ([DvN], Cor. III.5.3).

Let e be the orthogonal projection of $L^2(R)$ onto $L^2(R_\beta)$. One has $exe = E_{R_\beta}(x) e$ for all $x \in R$, by 3.6.1.i. below, so that in particular, $e e_1 e = \beta^{-1} e$. We claim that also $e_1 e e_1 = \beta^{-1} e_1$. Since $R = R_\beta e_1 R_\beta$, it suffices to check this equality on vectors $x e_1 y \Omega$, where $x, y \in R_\beta$ and Ω is the trace vector for R. But

$$\begin{aligned}
e_1 e e_1 (x e_1 y \Omega) &= e_1 e e_1 E_N(x) y \Omega && \text{(by } (*)\text{)}\\
&= e_1 E_{R_\beta}(e_1 E_N(x) y) \Omega && \text{(by definitions of } e \text{ and } E_{R_\beta})\\
&= e_1 E_{R_\beta}(e_1) E_N(x) y \Omega && \text{(by } R_\beta\text{-linearity of } E_{R_\beta})\\
&= \beta^{-1}\, e_1 E_N(x) y \Omega &&\\
&= \beta^{-1}\, e_1 (x e_1 y \Omega) && \text{(by } (*)\text{)}.
\end{aligned}$$

It follows from the relations $ee_1e = \beta^{-1}e$ and $e_1ee_1 = \beta^{-1}e_1$ that e and e_1 are equivalent projections in $\langle R,e\rangle$. Since e is finite in $\langle R,e\rangle$ by 3.6.1(v), the projection e_1 is finite in $\langle R,e\rangle$. But $\mathbb{1}$ is the sum of finitely many projections each equivalent in R to a subprojection of e_1, so $\langle R,e\rangle$ is finite. Hence $[R:R_\beta] = tr(e)^{-1} = tr(e_1)^{-1} = \beta$.

This completes the proof of the lemma, and also of 3.4.3(iii). #

It is tempting to guess that the pair $R \supset R_\beta$ is irreducible, also for $\beta > 4$, since on a purely algebraic level it is easy to see that there is no element of the algebra generated by $\{e_1, e_2, \cdots\}$ which commutes with $\{e_2, e_3, \cdots\}$. V. Jones confesses to spending considerable effort to prove this, but it turned out that R_β has non-trivial relative commutant in R when $\beta > 4$. A laborious proof of this non-obvious fact was given in [Jo1] and a simpler proof in [PP1]; we will give a proof due to Popa in 4.7.5. The difficulty is that one cannot write down an explicit form for an element in $R_\beta' \cap R$ without invoking a beautiful representation of $\{e_1, e_2, \cdots\}''$ discovered by Pimsner and Popa.

We have seen that one way to obtain a sequence of projections $(e_i)_{i\ge1}$ satisfying the relations 3.4.3.1 is to form the tower from an indecomposable pair $B \subset A$ of finite dimensional C^*-algebras. Then, as we have observed in Chapter 2, the restrictions on index are related to restrictions on the type of inclusions $B \subset A$ which yield a modulus $\beta < 4$. This is where the Coxeter graphs of types A, D, and E enter the picture. But to create the sequence $(e_i)_{i\ge1}$ one can also use a pair $N \subset M$ of finite direct sums of II_1-factors. In the following sections we will see how, if one allows this extra freedom, the remaining Coxeter graphs appear!

We finish this section by recording one useful fact on index of subfactors from [Jo1].

Proposition 3.4.6. *Let* $N \subset M$ *be finite factors and let* H *be any* M*-module such that* $\dim_M(H)$ *is finite. Then* $[M:N] = \dfrac{\dim_N(H)}{\dim_M(H)}$. *(In particular,* $\dim_N(H) \ge \dim_M(H)$.*)*

Proof. If H_1 and H_2 are any two M-modules such that $\dim_M(H_i)$ is finite for $i = 1,2$, then there is a finite dimensional Hilbert space K and an M-invariant projection q such that $H_1 \cong q(H_2 \otimes K)$ as M-modules. Then $\dim_N(H_1) \le \dim_N(H_2 \otimes K) = \dim_N(H_2)\dim_{\mathbb{C}}(K)$, by 3.2.5(i), so $\dim_N(H_1)$ is finite if and only if $\dim_N(H_2)$ is. In particular, $[M:N]$ is finite if and only if $\dim_N(H)$ is.

Assuming that [M:N] is finite and choosing an M–module isomorphism $H \cong q(L^2(M) \otimes K)$, as above, we have

$$\begin{aligned}\dim_N(H) &= \dim_N(q(L^2(M) \otimes K)) \\ &= tr_{N'}(q) \dim_N(L^2(M) \otimes K) \qquad \text{(by 3.2.5(e))} \\ &= tr_{M'}(q) \dim_{\mathbb{C}}(K) \dim_N(L^2(M)) \quad \text{(by 3.2.5(i))},\end{aligned}$$

while $\dim_M(H) = tr_{M'}(q) \dim_{\mathbb{C}}(K)$. #

3.5. Inclusions of finite von Neumann algebras with finite dimensional centers.

We saw in Chapter 2 that a unital inclusion $B \subset A$ of finite dimensional C^*–algebras can be specified by the inclusion matrix $\Lambda \in \mathrm{Mat}_{fin}(\mathbb{N})$ and a vector $\vec{\nu} \in \mathbb{N}^n$ for some n, specifying the algebra B up to isomorphism. It is impossible to specify an inclusion so precisely in the II_1–case since, for example, it is possible to find infinitely many non–conjugate subfactors of index 4 in R, even irreducible ones, as we shall see in Chapter 4. What we will do is specify enough information to be able to calculate all the needed coupling constants, which will enable us to find the Markov traces as in Section 2.7.

The situation will differ in two ways from the finite dimensional case. The first is that there are no minimal projections around, so integers do not appear in this way. The second is that the subfactors can have indices different from squares of integers. This extra freedom allows the appearance of new Coxeter graphs.

First some notation. Let $M = \bigoplus_{i=1}^{m} M_i$ be a direct sum of finite factors with corresponding minimal central projections $p_1, \cdots, p_m$. Since the trace on a finite factor is unique up to a scalar multiple, a trace on M is completely specified by a row vector $\overleftarrow{s} = (s_1, \cdots, s_m)$, with $s_i = tr(p_i)$. (*Warning*: This is *not* the same vector which was used in Chapter 2 to specify a trace on a direct sum of finite dimensional factors; there we used the vector whose i^{th} component is the trace of a minimal projection in M_i.) A trace is positive (i.e., trace $(a^*a) \geq 0$) if and only $\overleftarrow{s}$ has non–negative components. *We adopt the convention that "trace" means "positive trace".* A trace is faithful (i.e., trace $(a^*a) = 0$ implies $a = 0$) if none of the components of $\overleftarrow{s}$ are zero, and normalized if $\sum_{i=1}^{m} s_i = 1$. A trace is automatically normal; i.e., if $\{f_i\}$ is a family of mutually orthogonal projections, then trace $(\bigvee_{i=1}^{\infty} f_i) = \sum_{i=1}^{\infty} \mathrm{trace}(f_i)$.

Recall that if P is a finite factor, tr_P denotes its unique normalized trace, and if Tr is any other trace on P, then $\mathrm{Tr} = \mathrm{Tr}(\mathbb{1})\mathrm{tr}_P$.

Let $N = \bigoplus_{j=1}^{n} N_j$ be another direct sum of finite factors, contained in M and having the same identity. Let $q_1, \cdots, q_n$ be the minimal central projections of N.

Definition 3.5.1. If $N \subset M$ are as above, we define the m–by–n matrix $T_N^M = (c_{i,j})$ by

$$c_{i,j} = \mathrm{tr}_{p_iM}(p_iq_j).$$

Proposition 3.5.2.

(i) *The matrix* T_N^M *is row–stochastic; i.e.,* $c_{i,j} \geq 0$ *and* $\sum_j c_{i,j} = 1$ *for all* i.

(ii) *If* $\overleftarrow{s}$ *specifies a trace on* M, *then* $\overleftarrow{s}\, T_N^M$ *specifies its restriction to* N.

(iii) *If* $N \subset M \subset L$ *are finite direct sums of finite factors, then* $T_N^L = T_M^L T_N^M$.

Proof. (i) $\sum_j c_{i,j} = \sum_j \mathrm{tr}_{p_iM}(p_iq_j) = \mathrm{tr}_{p_iM}(p_i) = 1$, since $\sum_j q_j = \mathbb{1}$.

(ii) As $\sum_i p_i = \mathbb{1}$,

$$\mathrm{trace}(q_j) = \sum_i \mathrm{trace}(p_iq_j) = \sum_i \mathrm{trace}(p_i)\mathrm{tr}_{p_iM}(p_iq_j) = \sum_i s_ic_{i,j}.$$

(iii) Let $\{r_k\}$ denote the minimal central projections of L, so that T_N^L is the matrix whose (k,j) entry is $\mathrm{tr}_{r_kL}(r_kq_j)$. Since $q_j = \sum_i p_iq_j$, one has

$$(T_N^L)_{k,j} = \sum_i \mathrm{tr}_{r_kL}(r_kp_iq_j).$$

But in the finite factor $A = r_kL$, if $e \leq f$ are two projections, then $\mathrm{tr}_A(e) = \mathrm{tr}_A(f)\mathrm{tr}_{fAf}(e)$. Thus

$$(T_N^L)_{k,j} = \sum_i \mathrm{tr}_{r_kL}(r_kp_i)\mathrm{tr}_{r_kp_iLr_kp_i}(r_kp_iq_j).$$

If $r_k p_i \neq 0$, then $x \mapsto tr_{r_k p_i L r_k p_i}(r_k x)$ is a trace on $p_i M$ whose value at p_i is 1, so in fact $tr_{r_k p_i L r_k p_i}(r_k x) = tr_{p_i M}(x)$. Hence

$$(T_N^L)_{k,j} = \sum_i tr_{r_k L}(r_k p_i) tr_{p_i M}(p_i q_j)$$

$$= \sum_i (T_M^L)_{k,i} (T_N^M)_{i,j},$$

as desired. #

A second piece of data needed is the matrix of indices of the "partial embeddings". Note that $N_{i,j} = Np_i q_j = \{p_i q_j x : x \in N\}$ is a finite factor, a subfactor of $M_{i,j} = p_i q_j M p_i q_j$.

Definition 3.5.3. (i) With notation as above, define an m–by–n matrix Λ_N^M with entries

$$\lambda_{i,j} = [M_{i,j} : N_{i,j}]^{1/2}.$$

(We note that this expression is the same as in the finite dimensional case. Observe that in the finite dimensional case Λ_N^M determines T_N^M, namely

$$(T_N^M)_{i,j} = \lambda_{i,j} \nu_j / \mu_i$$

where $p_i M \cong Mat_{\mu_i}(\mathbb{C})$, and $q_j N \cong Mat_{\nu_j}(\mathbb{C})$.)

(ii) The inclusion $N \subset M$ is called connected if $Z(M) \cap Z(N) = \mathbb{C}1$. This is true if and only if Λ_N^M is indecomposable.

(iii) A representation π of M on a Hilbert space H is called a finite representation of the pair $N \subset M$ if $\pi(N)'$ is a finite von Neumann algebra.

(iv) We say that N is of finite index in M if $N \subset M$ admits a finite faithful representation.

(Note that parts (ii), (iii), and (iv) make sense for arbitrary pairs of finite von Neumann algebras – not necessarily with finite dimensional centers.)

Lemma 3.5.4. *Suppose* $N \subset M$ *are finite direct sums of finite factors. The following are equivalent:*

(i) N *is of finite index in* M.

(ii) *The matrix* Λ_N^M *has only finite entries.*

(iii) *For any faithful trace* tr *on* M, *the regular representation of* M *on* $L^2(M,tr)$ *is a finite representation of the pair* $N \subset M$.

(iv) *For any faithful representation* $\{\pi, \mathfrak{H}\}$ *of* M *such that* $\pi(M)'$ *is finite, the algebra* $\pi(N)'$ *is also finite.*

Proof. (iv) $\Rightarrow$ (iii) $\Rightarrow$ (i) is evident.

(i) $\Rightarrow$ (ii). If π is a faithful finite representation of the pair $N \subset M$ on H, then the commutant of $\pi(N_{i,j})$ on $\pi(p_iq_j)H$ is $\pi(p_iq_j)\pi(N)'\pi(p_iq_j)$, which is finite. It follows that $\dim_{N_{i,j}}(\pi(p_iq_j)H) < \infty$ (Proposition 3.2.4.d), and

$$[M_{i,j} : N_{i,j}] = \dim_{N_{i,j}}(L^2(M_{i,j})) = \frac{\dim_{N_{i,j}}(\pi(p_iq_j)H)}{\dim_{M_{i,j}}(\pi(p_iq_j)H)}$$

(by 3.4.6.), which is finite.

(ii) $\Rightarrow$ (iv). Consider a faithful M–module H for which M' is finite. Since $\mathbb{1} = \sum_{i,j} p_iq_j$, to show that N' is finite, it suffices to show that each p_iq_j is a finite projection in N' (because a sum of finite projections is finite.) If $p_iq_j \neq 0$, then $p_iq_jN'p_iq_j$ is the commutant of $N_{i,j}$ on p_iq_jH. By 3.4.6 and 3.2.5.h,

$$\begin{aligned}\dim_{N_{i,j}}(p_iq_jH) &= \lambda_{i,j}^2 \dim_{M_{i,j}}(p_iq_jH) \\ &= \lambda_{i,j}^2 \mathrm{tr}_{p_iM}(p_iq_j)^{-1} \dim_{p_iM}(p_iH).\end{aligned}$$

Since M' is finite on H, so is $p_iM' = (p_iM)'$ on p_iH, so by 3.2.4.d, $\dim_{p_iM}(p_iH) < \infty$. Hence also $\dim_{N_{i,j}}(p_iq_jH) < \infty$, and by 3.2.4.d again, $(N_{i,j})'$ is finite. #

Observe that the analogue for Λ of Proposition 3.5.3.iii does not hold. For example, let R be the hyperfinite II_1 factor, let p be a non–trivial projection in R, let φ be an isomorphism from R_p to $R_{\mathbb{1}-p}$, and set

$$N = \{y \in R\colon\ y = x + \varphi(x) \text{ for some } x \in R_p\}, \text{ and}$$

$$M = R_p \oplus R_{\mathbb{1}-p}.$$

Then

$$\Lambda_M^R \Lambda_N^M = (1\ 1)\begin{bmatrix}1\\1\end{bmatrix} = 2,$$

and

$$\Lambda_N^R = [R:N]^{1/2} = \{\frac{1}{tr(p)} + \frac{1}{tr(\mathbf{1}-p)}\}^{1/2}$$

by Corollary 2.2.5 of [Jo1] or 4.7.2. These are not equal, unless $tr(p) = 1/2$.

Of course, if $N \subset M \subset L$ is a triple of finite factors, then $[L:N] = [L:M]\ [M:N]$ by Proposition 3.4.6.

If N and M are as in 3.5.4, and the inclusion $N \subset M$ is connected, then all factors of N and M are of type II_1, or $\dim_{\mathbb{C}}(M) < \infty$. It is also known that all factors of N and M share (or do not share) the property of being hyperfinite (Lemma 2.1.8 in [Jo1]) or the property T (see [Ana] and [PP2]).

If π is a finite faithful representation of the pair $N \subset M$ on H, then the centers of $\pi(M)'$ and $\pi(N)'$ are the same as those of M and N respectively, and the rows and columns of $\Lambda_{\pi(M)'}^{\pi(N)'}$ are naturally indexed by the columns and rows of Λ_N^M. The generalization of Proposition 2.3.5 to this setting is the following.

Lemma 3.5.5. *Let* $N \subset M$ *be a pair of finite direct sums of finite factors, as above, as suppose* π *is a faithful finite representation of the pair. Then*

$$\Lambda_{\pi(M)'}^{\pi(N)'} = (\Lambda_N^M)^t.$$

Proof. If M and N are factors, the equality holds because $[\pi(N)':\pi(M)'] = [M:N]$ by Propositions 3.4.6 and 3.2.5.g. To extend the equality to the general case, one proceeds exactly as in the finite dimensional case (Proposition 2.3.5), with Proposition 2.2.5b being replaced by [DvN], Proposition 1 of §I.2, which says: if Q is a von Neumann algebra on H and p is a projection in Q or in Q', then $End_{pQp}(pH))$ equals $pEnd_Q(H)p$. #

Also note that $\pi(M)'$ is of finite index in $\pi(N)'$ by Lemmas 3.5.4. and 3.5.5.

Proposition 3.5.6. *Given an irredundant* m*–by–*n *matrix* Λ *over* $\{0\} \cup \{2\cos\pi/q : q \geq 3\} \cup [2,\infty]$, *and an* m*–by–*n *row stochastic matrix* T *having the same pattern of zero entries as* Λ, *there exists a pair* $N \subset M$ *(both hyperfinite) with* $\Lambda_N^M = \Lambda$ *and* $T_N^M = T$.

Proof. Take M to be the direct sum of m copies of R, the unique hyperfinite II_1 factor, denoted R_i. In each R_i, choose a partition of unity $\{q_{i,j} : 1 \leq j \leq n\}$ with $tr(q_{i,j}) = (T)_{i,j}$. If $(T)_{i,j}$ is non–zero choose a II_1 subfactor $P_{i,j}$ of $R_{i,j} = q_{i,j}R_iq_{i,j}$

with $[R_{i,j} : P_{i,j}]^{1/2} = (\Lambda)_{i,j}$ (possible by [Jo1], Theorem 4.3.2). For each i and j such that $(T)_{i,j} \neq 0$, choose an isomorphism $\theta_{i,j} : R \to P_{i,j}$ (possible since all the factors are II_1 and hyperfinite). Set $q_j = \sum_i q_{i,j}$, put $N_j = \{\sum_i \theta_{i,j}(x) : x \in R\}$, and $N = \bigoplus_{j=1}^{n} N_j$. Then $q_j N = N_j$, and N is the required subalgebra. #

3.6. The fundamental construction.

The discussion of the fundamental construction in Chapter 2 was purely ring theoretic. In the von Neumann algebra framework, where the preferred modules are Hilbert spaces, it is natural to make a construction which, apparently, depends on the choice of a trace on M. We begin by showing that in fact the ring theoretic construction is exactly the same.

First we recall some notions from [Jo1] which work for arbitrary finite von Neumann algebras exactly as for factors. Let $N \subset M$ be finite von Neumann algebras with the same identity. Given a faithful normalized trace on M, there is a unique faithful normal conditional expectation $E_N : M \to N$ determined by $tr(xy) = tr(E_N(x)y)$ for $x \in M$ and $y \in N$. In fact E_N is the restriction to M of the orthogonal projection $e_N : L^2(M,tr) \to L^2(N,tr)$. We denote by $\langle M,e_N \rangle$ the von Neumann algebra on $L^2(M,tr)$ generated by M and e_N.

We let J denote the conjugate linear isometry of $L^2(M,tr)$ extending the map $x \mapsto x^*$ on M.

Proposition 3.6.1.

(i) $e_N x e_N = E_N(x) e_N$ *for* $x \in M$

(ii) $J e_N J = e_N$

(iii) *For* $x \in M$, x *commutes with* e_N *if and only if* $x \in N$.

(iv) $\langle M,e_N \rangle = JN'J$

(v) *The map* $\psi \begin{cases} N \longrightarrow \langle M, e_N \rangle \\ y \longmapsto y e_N \end{cases}$ *is an injective morphism onto* $e_N \langle M,e_N \rangle e_N$.

(vi) *The central support of* e_N in $\langle M,e_N \rangle$ is $\mathbf{1}$.

(vii) *The space* Me_NM, *which denotes the linear span of* $\{x' e_N x'' : x', x'' \in M\}$, *is a strongly dense *–subalgebra of* $\langle M,e_N \rangle$.

Proof. (cf. [Jo1]).

(i) It suffices to check that $E_N(xE_N(y)) = E_N(x)E_N(y)$, but this follows from the N–linearity of E_N.

(ii) Follows from $E_N(x^*) = E_N(x)^*$.

(iii) Note that x commutes with e_N if and only if left multiplication by x commutes with E_N. This is clearly so for $x \in N$. On the other hand, if $x \in M$ and x commutes with E_N, then $x = xE_N(\mathbb{1}) = E_N(x) \in N$.

(iv) By (iii) $N = M \cap \{e_N\}'$, so $N' = (M' \cup \{e_N\})'' = \langle M', e_N \rangle$. But $JM'J = M$ and $Je_NJ = e_N$, so $JN'J = \langle M, e_N \rangle$.

(v) By (i), the indicated map is an epimorphism. Let Ω denote the canonical trace vector in $L^2(M,tr)$. If $ye_N = 0$, then $ye_N\Omega = y\Omega = 0$ and $y = 0$ because Ω is separating, so ψ is an isomorphism.

(vi) Let z be the central support of e_N in N'. Then $z \in N \cap N'$ and $\psi(z-\mathbb{1}) = ze_N - e_N = 0$, by definition of a central support, so $z = \mathbb{1}$ by (v). Now (vi) follows from (iv) and (ii).

(vii) First note that by (i), the set

$$X = \{x_0 + \sum_{i=1}^{n} x_i e_N y_i : n \in \mathbb{N}, x_i, y_i \in M\}$$

is a $*$-subalgebra of $\langle M, e_N \rangle$ containing M and e_N, so the strong closure of X is $\langle M, e_N \rangle$. If

$$Y = \{\sum x_i e_N y_i : x_i, y_i \in M\},$$

then Y is a two sided ideal in X, so by the Kaplansky density theorem and the joint strong continuity of muliplication on the unit ball, the strong closure $\overline{Y}$ of Y is a two sided ideal in $\langle M, e_N \rangle$. But $\overline{Y}$ contains the central support of e_N, which is $\mathbb{1}$ by point (vi), so $\overline{Y} = \langle M, e_N \rangle$. #

We now specialize to the case where N and M are direct sums of finitely many II_1 factors with minimal central projections $\{q_j; j = 1, \cdots, n\}$ and $\{p_i; i = 1, \cdots, m\}$ respectively. By the equality (iv) above, $\langle M, e_N \rangle$ is also a finite direct sum of II_1 factors, with minimal central projections $\{Jq_jJ: j=1, \cdots, n\}$.

Lemma 3.6.2.

(a) *If* $N \subset M$ *are type* II_1 *von Neumann algebras with finite dimensional centers and* N *is of finite index in* M, *then* $\dim_{\mathbb{C}}(N' \cap M) < \infty$.

(b) *If* $N \subset M$ *are* II_1 *factors, then* $\dim_{\mathbb{C}}(N' \cap M) \leq [M:N]$.

(c) *If* $N \subset M$ *are* II_1 *factors with* $[M:N] < 4$, *then* $N' \cap M = \mathbb{C}\mathbb{1}$.

Proof. We first consider the case that N and M are factors. Let $H = L^2(M)$ and write $\mathrm{Tr}_{N'}$ for the natural trace on $\mathrm{End}_N(H)$. If f is a projection in $N' \cap M$, then

$$\begin{aligned} \mathrm{Tr}_{N'}(f) &= \dim_N(fH) && \text{(by definition of } \dim_N) \\ &\geq \dim_{fMf}(fH) && \text{(by 3.4.6)} \\ &= \mathrm{tr}_M(f)^{-1} && \text{(by 3.2.5(h))} \\ &\geq 1 \end{aligned}$$

Suppose $N' \cap M$ contains k mutually orthogonal projections $f_1, \cdots, f_k$ with $\sum f_i = \mathbb{1}$. Then

$$\begin{aligned} [M:N] &= \mathrm{Tr}_{N'}(\mathbb{1}) = \sum_i \mathrm{Tr}_{N'}(f_i) \\ &\geq \sum_i \mathrm{tr}_M(f_i)^{-1} \geq k^2. \end{aligned}$$

In particular, if $N' \cap M \neq \mathbb{C}\mathbb{1}$. then $[M:N] \geq 4$, and if $N' \cap M$ is infinite dimensional, then $[M:N] = \infty$. Suppose $[M:N] < \infty$, and let $f_1, \cdots, f_k$ be a *maximal* family of mutually orthogonal projections in $N' \cap M$; then $[M:N] \geq k^2 \geq \dim_{\mathbb{C}}(N' \cap M)$. This proves all the assertions in the case of factors.

Now return to the situation where N and M are finite direct sums of finite factors. The projections $p_i q_j$ are central projections in $N' \cap M$ and $p_i q_j (N' \cap M) = N'_{p_i q_j} \cap M_{p_i q_j}$. So if $\dim_{\mathbb{C}}(N' \cap M) = \infty$ there must be a pair (i,j) for which $\dim_{\mathbb{C}}(N'_{p_i q_j} \cap M_{p_i q_j}) = \infty$. But this contradicts the observation just made for the case of factors, and completes the proof of (a). #

The next results (3.6.3–3.6.5) depend on ideas of Pimsner and Popa [PP1].

Lemma 3.6.3. *Let* $N \subset M$ *be finite direct sums of type* II_1 *factors with* N *of finite index in* M, *and let* tr *be a faithful trace on* M. *If* $x \in \langle M, e_N \rangle$, *there is a unique* $y \in M$ *for which* $x e_N = y e_N$.

Proof. Regard $N \subset M$ represented on $L^2(M)$.

Let us first check uniqueness. Suppose $y,y' \in M$ with $xe_N = ye_N = y'e_N$. If Ω is the trace vector in $L^2(M)$, then

$$(y-y')\Omega = (y-y')e_N\Omega = 0,$$

so $y' = y$ because Ω is separating.

To prove existence, we have to show that $\langle M,e_N\rangle e_N = Me_N$ and we proceed as follows.

As N' is finite, $\langle M,e_N\rangle$ is finite by 3.6.1.iv, and there exists a faithful normal conditional expectation F from $\langle M,e_N\rangle$ onto M (see Proposition II.5 for the proof of this latter fact). We claim that $F(e_N)$ is invertible in M. Since F is an M–M–bimodule map, $F(e_N)$ belongs to $N' \cap M$, which is finite dimensional by Lemma 3.6.2. Consequently, to show that the self-adjoint element $F(e_N)$ is invertible, it is enough to check that $xF(e_N)x \neq 0$ for any positive element $x \neq 0$ in $N' \cap M$. But if

$$0 = xF(e_N)x = F(xe_Nx),$$

then $xe_Nx = 0$, since F is faithful. And $xe_Nx = (e_Nx)^*(e_Nx)$, so $e_Nx = 0$. Hence

$$0 = e_Nxe_N = E_N(x)e_N,$$

which implies $x = 0$ by 3.6.1.v and the faithfulness of E_N. This proves the claim that $F(e_N)$ is invertible.

Now we may obtain a formula for xe_N. Suppose first that x is in Me_NM, namely that x is a finite sum $\sum a_je_Nb_j$ with $a_j,b_j \in M$. Then $F(xe_N) = \sum a_jE_N(b_j)F(e_N)$ and

$$F(xe_N)F(e_N)^{-1}e_N = xe_N.$$

This formula holds for any $x \in \langle M,e_N\rangle$ because both sides are strongly continuous in x and because Me_NM is strongly dense in $\langle M,e_N\rangle$ by Proposition 3.6.1.vii. Thus $xe_N \in Me_N$ for any $x \in \langle M,e_N\rangle$. #

Theorem 3.6.4. *Let* $N \subset M$ *be type* II_1 *von Neumann algebras with finite dimensional centers and let* tr *be a faithful normal trace on* M *for which* N' *is finite on* $L^2(M,tr)$.

Then

(i) *As a right module over* N, *the algebra* M *is projective of finite type.*

(ii) *The conditional expectation* $E_N : M \to N$ *is very faithful (in the sense of Section 2.6).*

(iii) $\langle M,e_N\rangle = Me_NM := \{\sum_{j=1}^{n} a_je_Nb_j : n \geq 1, a_j,b_j \in M\}$.

(iv) *If* $\alpha : M \to M$ *is a right* N*-module map, then* α *extends uniquely to an element of* $\langle M,e_N\rangle = JN'J$ *on* $L^2(M,tr)$.

(v) *If* $x \in JN'J$ *then* $x(M) \subset M$, *where* M *is viewed as a dense subspace of* $L^2(M,tr)$.

Proof. (i) Any strongly closed right ideal in N is projective of finite type, and in fact of the form pN with p a projection in N. (See [Tak], II.3.12.) We are going to show that M is isomorphic, as a right N-module, to a finite direct sum of such ideals. In the course of doing so we exhibit a basis $\{v_i : 1 \leq i \leq n\}$ of M over N with the following properties:

(a) $E_N(v_i^*v_j) = 0$ if $i \neq j$.

(b) $f_i := E_N(v_i^*v_i)$ is a projection in N, $v_if_i = v_i$, and $E_N(v_i^*x) = f_iE_N(v_i^*x)$, for $1 \leq i \leq n$ and $x \in M$.

(c) Every x in M has a unique expansion

$$x = \sum_{i=1}^{n} v_iy_i, \text{ with } y_i \in N.$$

In fact $v_iy_i = v_iE_N(v_i^*x)$.

Since the central support of e_N in $\langle M,e_N\rangle$ is $\mathbf{1}$ and since $\langle M,e_N\rangle$ is finite with finite dimensional center by 3.6.1(iv), there exists a finite set $w_1,\cdots,w_n$ of partial isometries in $\langle M,e_N\rangle$ with $w_j^*w_j \leq e_N$ and $\sum w_jw_j^* = \mathbf{1}$; in particular the w_j have mutually orthogonal range projections. (See [Tak], V.1.34.) As $w_je_N = w_j$, there are, by 3.6.3, elements $v_1,\cdots,v_n \in M$ with $w_j = v_je_N$ for all j. We verify that the v_i have the properties (a)–(c). For $i \neq j$

$$0 = w_i^*w_j = e_Nv_i^*v_je_N = E_N(v_i^*v_j)e_N,$$

so $E_N(v_i^*v_j) = 0$ by 3.6.1(v). Similarly, since $w_i^*w_i$ is a projection in $\langle M,e_N\rangle$ and $w_i^*w_i = E_N(v_i^*v_i)e_N$, 3.6.1(v) implies that $f_i := E_N(v_i^*v_i)$ is a projection in N.

Furthermore

$$\begin{aligned} v_i f_i e_N &= v_i e_N v_i^* v_i e_N = w_i w_i^* w_i \\ &= w_i = v_i e_N, \end{aligned}$$

so that $v_i f_i = v_i$, by the uniqueness statement of 3.6.3. Therefore, since $f_i \in N$,

$$f_i E_N(v_i^* x) = E_N(f_i v_i^* x) = E_N(v_i^* x) \quad \text{for } x \in M.$$

For any $x \in M$,

$$\begin{aligned} x e_N &= \sum_j w_j w_j^* x e_N = \sum_j v_j e_N v_j^* x e_N \\ &= \sum_j v_j E_N(v_j^* x) e_N, \end{aligned}$$

and hence $x = \sum_j v_j E_N(v_j^* x)$, by 3.6.3. To show uniqueness of the expansion, suppose that $x = \sum v_i y_i$ with $y_i \in N$. Then

$$\begin{aligned} v_i E_N(v_i^* x) &= v_i E_N(v_i^* \sum_j v_j y_j) \\ &= v_i E_N(v_i^* v_i) y_i \\ &= v_i f_i y_i = v_i y_i, \end{aligned}$$

using N–linearity of E_N and properties (a) and (b) of $\{v_i\}$. We will refer to a family $\{v_i\}$ having properties (a)–(c) as a Pimsner–Popa basis of M over N; see [PP1].

Now consider the N–linear map

$$\Psi \begin{cases} M \to \bigoplus_{1 \le j \le n} f_j N \\ x \mapsto (E_N(v_j^* x)). \end{cases}$$

It follows from the expansion $x = \sum v_j E_N(v_j^* x)$ that Ψ is injective. On the other hand, if $(y_j) \in \bigoplus_j f_j N$ and $x = \sum v_j y_j$ then by the uniqueness of the expansion, $v_j y_j = v_j E_N(v_j^* x)$ for all j. Multiplying both sides on the left by v_j^* and applying E_N gives $f_j y_j = f_j E_N(v_j^* x)$; since both y_j and $E_N(v_j^* x)$ are in $f_j N$, that is $y_j = E_N(v_j^* x)$. Thus $(y_j) = \Psi(x)$ and Ψ is surjective.

(ii) Let $\alpha : M \to N$ be a right N–linear map and set $a = \sum_j \alpha(v_j)v_j^*$. Recall from Section 2.6 that $E_N^\flat(a) : M \to N$ is defined by $E_N^\flat(a)(x) = E_N(ax)$ for $x \in M$. We have

$$\begin{aligned}\alpha(x) &= \alpha(\sum v_j E_N(v_j^* x)) \\ &= \sum \alpha(v_j) E_N(v_j^* x) && \text{by N–linearity of } \alpha \\ &= E_N(\sum \alpha(v_j) v_j^* x) && \text{by N–linearity of } E_N \\ &= E_N^\flat(a)(x),\end{aligned}$$

so that $\alpha = E_N^\flat(a)$.

(iii) It follows from 3.6.3 that Me_NM is a two–sided ideal in $\langle M,e_N\rangle$. But Me_NM contains $\sum_j v_j e_N v_j^* = \sum_j w_j w_j^* = \mathbb{1}$, so $Me_NM = \langle M,e_N\rangle$.

(iv) If $\alpha : M \to M$ is right N–linear, then for $x \in M$, $\alpha(x) = \alpha \sum_j v_j E_N(v_j^* x) = \sum_j \alpha(v_j) E_N(v_j^* x)$; thus $\alpha = \sum_j \lambda(\alpha(v_j)) \circ E_N \circ \lambda(v_j^*)$, where $\lambda(y)$ denotes left multiplication by y. The unique $\|\cdot\|_2$–continuous extension of α to $L^2(M,tr)$ is $\sum_j \alpha(v_j) e_N v_j^* \in \langle M,e_N\rangle$.

(v) Any $x \in \langle M,e_N\rangle$ is of the form $\sum_j a_j e_N b_j$ by claim (iii). If $y \in M$ then

$$x(y) = \sum a_j E_N(b_j y) \in M. \qquad \#$$

<u>Corollary 3.6.5</u>. *Let* $N \subset M$ *be a pair of von Neumann algebras of type* II_1 *having finite dimensional centers, and suppose that* N *is of finite index in* M. *Let* tr *be any faithful normal trace on* M *and define* e_N *and* E_N *via* tr. *Then*

$$M \otimes_N M \cong \langle M,e_N\rangle \qquad \textit{as N–bimodules, and}$$
$$\mathrm{End}_N^r(M) \cong \langle M,e_N\rangle \qquad \textit{as } \mathbb{C}\textit{–algebras.}$$

<u>Proof</u>. Since $N \subset M$ has finite index, $\langle M,e_N\rangle$ is finite. The isomorphism $\mathrm{End}_N^r(M) \cong \langle M,e_N\rangle$ follows from 3.6.4(iv) or (v); the correspondence is defined by $\sum \lambda(a_j) E_N \lambda(b_j) \mapsto \sum a_j e_N b_j$.

The isomorphism $M \otimes_N M \cong \mathrm{End}_N^r(M)$ extending the map $a \otimes_N b \mapsto \lambda(a) E_N \lambda(b)$ on elementary tensors follows from 3.6.4(i) and (ii) and 2.6.3. One can also verify directly the isomorphism $M \otimes_N M \cong \langle M, e_N \rangle$ by using a Pimsner–Popa basis. #

The next proposition determines one part of the spatial data for the inclusion $M \subset \langle M, e_N \rangle$.

Proposition 3.6.6. *Let* $N \subset M$ *be finite direct sums of finite factors such that* N *is of finite index in* M, *and let* tr *be any faithful trace on* M. *Then* $\Lambda_M^{\langle M, e_N \rangle} = (\Lambda_N^M)^t$.

Proof. This follows from 3.5.4, 3.5.5, and the formulas $JN'J = \langle M, e_N \rangle$, $JM'J = M$. #

To describe $M \subset \langle M, e_N \rangle$ more precisely, we also have to compute the matrix of traces $T_M^{\langle M, e_N \rangle}$. This is the part of the theory which differs most from the finite dimensional case presented in Chapter 2.

Before proceeding, we summarize our notation: $N \subset M$ is a pair of finite von Neuman algebras with finite dimensional centers, with N of finite index in M; the minimal central projections in M and N are respectively $\{p_i : 1 \le i \le m\}$ and $\{q_j : 1 \le j \le n\}$. A trace tr on M is specified by the row vector $\overleftarrow{s}$, $s_i = \mathrm{tr}(p_i)$. Let $H = L^2(M, \mathrm{tr})$. Set

$$\begin{aligned} N_{i,j} &= p_i q_j N p_i q_j, & M_{i,j} &= p_i q_i M p_i q_j, \\ N'_{i,j} &= p_i q_j N' p_i q_j & &= \mathrm{End}_{N_{i,j}}(p_i q_j H), \\ M'_{i,j} &= p_i q_j M' p_i q_j & &= \mathrm{End}_{M_{i,j}}(p_i q_j H), \end{aligned}$$

when $p_i q_j \ne 0$. We have the trace matrix T_N^M with entries $c_{i,j} = \mathrm{tr}_{p_i M}(p_i q_j)$, and the index matrix Λ_N^M with entries

$$\begin{aligned} \lambda_{i,j} &= 0 & & p_i q_j = 0, \\ \lambda_{i,j} &= [M_{i,j} : N_{i,j}]^{1/2} & & \text{if } p_i q_j \ne 0. \end{aligned}$$

Our present goal is to compute the entries of $T_M^{\langle M, e_N \rangle} = T_{N'}^{M'}$, namely

$$d_{j,i} = \mathrm{tr}_{q_j N'}(q_j p_i).$$

Lemma 3.6.7. *If* $p_i q_j \neq 0$, *then*

(i) $\dim_{N'_{i,j}}(p_i q_j H) = \frac{c_{i,j}}{\lambda^2_{i,j}}$, *and*

(ii) $d_{j,i} \dim_{N'_{i,j}}(p_i q_j H) = \dim_{q_j N'}(q_j H)$.

Proof. By 3.4.6,

$$\lambda^2_{i,j} = \frac{\dim_{N_{i,j}}(p_i q_j H)}{\dim_{M_{i,j}}(p_i q_j H)},$$

and by 3.2.5(h),

$$\dim_{M_{i,j}}(p_i q_j H) = \operatorname{tr}_{p_i M}(q_j p_i)^{-1} \dim_{p_i M}(p_i H).$$

But since M is in standard form on H, so is $p_i M$ on $p_i H$, and $\dim_{p_i M}(p_i H) = 1$. Combining these observations,

$$\begin{aligned}
\lambda^2_{i,j} &= \dim_{N_{i,j}}(p_i q_j H) \operatorname{tr}_{p_i M}(p_i q_j) \\
&= \dim_{N_{i,j}}(p_i q_j H)\, c_{i,j} \\
&= \dim_{N'_{i,j}}(p_i q_j H)^{-1} c_{i,j},
\end{aligned}$$

by 3.2.5.g. Hence (i).

(ii) This reads

$$\dim_{N'_{i,j}}(p_i q_j H) = [\operatorname{tr}_{q_j N'}(p_i q_j)]^{-1} \dim_{q_j N'}(q_j H),$$

which follows from 3.2.5(h). #

Notation: For each j, let

$$\varphi_j = \left[\sum_{i=1}^{m}{}' \frac{\lambda^2_{i,j}}{c_{i,j}} \right]^{-1},$$

the sum being over those i such that $p_i q_j \neq 0$, and let F be the diagonal matrix $F = \operatorname{diag}(\varphi_1, \cdots, \varphi_n)$. Furthermore, let $\tilde{T}$ be the n–by–m matrix

$$(\tilde{T})_{j,i} = 0 \quad \text{if } p_i q_j = 0.$$

$$(\tilde{T})_{j,i} = \frac{\lambda_{i,j}^2}{c_{i,j}} \quad \text{if } p_i q_j \neq 0.$$

<u>Proposition 3.6.8.</u> $T_M^{\langle M,e_N\rangle} = F\tilde{T}$.

<u>Proof.</u> Combining 3.6.7(i) and (ii) we get

$$\text{(3.6.8.a)} \qquad d_{j,i} = \frac{\dim_{q_jN'}(q_jH)}{\dim_{N'_{i,j}}(q_jp_iH)} = \frac{\lambda_{i,j}^2}{c_{i,j}} \dim_{q_jN'}(q_jH),$$

if $p_iq_j \neq 0$, and $d_{j,i} = 0$ otherwise. To eliminate $\dim_{q_jN'}(q_jH)$ we use the fact that $T_M^{\langle M,e_N\rangle}$ is row stochastic,

$$1 = \sum_{i=1}^{m} d_{j,i} = \left[\sum_{i=1}^{m}{}' \frac{\lambda_{i,j}^2}{c_{i,j}}\right] \dim_{q_jN'}(q_jH),$$

so

$$\text{(3.6.8.b)} \qquad \dim_{q_jN'}(q_jH) = \varphi_j.$$

Putting this back in (3.6.8.a) gives $d_{j,i} = \varphi_j \frac{\lambda_{i,j}^2}{c_{i,j}}$ if $p_iq_j \neq 0$ and $d_{j,i} = 0$ otherwise, as desired. #

<u>Remark.</u> Let us check what that formula $T_M^{\langle M,e_N\rangle} = F\tilde{T}$ means for finite dimensional algebras. Suppose that $p_iM \cong \mathrm{Mat}_{\mu_i}(\mathbb{C})$ and $q_jN \cong \mathrm{Mat}_{\nu_j}(\mathbb{C})$. As noted before, the inclusion matrix $\Lambda = \Lambda_N^M$ determines the trace matrix $T = T_N^M$ via

$$c_{i,j} = \lambda_{i,j} \frac{\nu_j}{\mu_i},$$

since q_jp_i is the sum of $\lambda_{i,j}\nu_j$ orthogonal minimal projections in p_iM. Setting $\tilde{\mu} = \mathrm{diag}(\mu_1,\cdots,\mu_m)$ and $\tilde{\nu} = \mathrm{diag}(\nu_1,\cdots,\nu_n)$, this can be written

$$T = \tilde{\mu}^{-1}\Lambda\tilde{\nu}.$$

When $p_i q_j \neq 0$, we have $(\tilde{T})_{j,i} = \frac{\lambda_{i,j}^2}{c_{i,j}} = \lambda_{i,j}\frac{\mu_i}{\nu_j}$; and when $p_i q_j = 0$, $(\tilde{T})_{j,i} = 0 = \lambda_{i,j}\frac{\mu_i}{\nu_j}$.

Thus

$$\tilde{T} = \tilde{\nu}^{-1}\Lambda^t\tilde{\mu}.$$

Set $L = \langle M,e_N\rangle$, $L = \bigoplus_{j=1}^n L_j$; then $L_j \cong \mathrm{Mat}_{\kappa_j}(\mathbb{C})$ where $\kappa_j = (\Lambda^t\vec{\mu})_j = \sum_i \mu_i\lambda_{i,j}$. Note that

$$\varphi_j = (\sum_i \tilde{T}_{j,i})^{-1} = (\frac{1}{\nu_j}\sum_i \lambda_{i,j}\mu_i)^{-1} = \frac{\nu_j}{\kappa_j}.$$

Thus

$$(T_M^{\langle M,e_N\rangle})_{j,i} = \varphi_j(\tilde{T})_{j,i} = \frac{\nu_j}{\kappa_j}\frac{\mu_i}{\nu_j}\lambda_{i,j} = \lambda_{i,j}\frac{\mu_i}{\kappa_j},$$

which is in accord with the relation observed above between the inclusion matrix and the index matrix.

We now return to the analysis of the general case.

As the minimal central projection in $\langle M,e_N\rangle = JN'J$ are precisely $\{Jq_jJ : 1 \leq j \leq n\}$, any trace Tr on $\langle M,e_N\rangle$ is specified by a row vector $\overleftarrow{r}$, with $r_j = \mathrm{Tr}(Jq_jJ)$. It will turn out to be useful to calculate the quantities $\mathrm{Tr}(e_NJq_jJ)$. Recall that $Je_N = e_NJ$. Also observe that

$$e_NJq_jJ = e_Nq_j. \tag{3.6.9}$$

In fact, let Ω denote the trace vector in $H = L^2(M,\mathrm{tr})$, i.e. the identity $\mathbf{1}$ of M reqarded as an element of H. The linear space $\{x\Omega : x \in M\}$ is dense in H and we have

$$\begin{aligned} e_NJq_jJ(x\Omega) &= e_NJq_jx^*\Omega = e_Nxq_j\Omega \\ &= E_N(xq_j)\Omega = E_N(x)q_j\Omega \\ &= q_jE_N(x)\Omega = E_N(q_jx)\Omega \\ &= e_Nq_j(x\Omega). \end{aligned}$$

Lemma 3.6.10. *Let* Tr *be any trace on* $\langle M,e_N\rangle$ *and let* $r_j = \mathrm{Tr}(Jq_jJ)$. *Then*

(i) $\quad \mathrm{tr}_{q_jN'}(q_je_N) = \varphi_j = \left[\sum_{i=1}^{m}{}' \frac{\lambda_{i,j}^2}{c_{i,j}}\right]^{-1}.$

(ii) $\quad \mathrm{Tr}(e_Nq_j) = \mathrm{Tr}(e_NJq_jJ) = r_j\varphi_j.$

Proof. (i) Since N is in standard form on e_NH, so is q_jN and its commutant $q_je_NN'q_je_N$ on q_je_NH; hence

$$\begin{aligned}1 &= \dim_{q_je_NN'q_je_N}(q_je_NH)\\ &= [\mathrm{tr}_{q_jN'}(q_je_N)]^{-1}\dim_{q_jN'}(q_jH) \qquad \text{(by 3.2.5(h).)}\\ &= [\mathrm{tr}_{q_jN'}(q_je_N)]^{-1}\varphi_j \qquad \text{(by 3.6.8(b).)}\end{aligned}$$

(ii) Since the map $x \mapsto \mathrm{Tr}(Jx^*J)$ is a trace on the factor q_jN' we have $\mathrm{Tr}(Jx^*J) = \mathrm{Tr}(Jq_jJ)\mathrm{tr}_{q_jN'}(x)$, and in particular, using 3.6.9,

$$\mathrm{Tr}(e_NJq_jJ) = \mathrm{Tr}(Je_Nq_jJ) = \mathrm{Tr}(Jq_jJ)\mathrm{tr}_{q_jN'}(e_Nq_j) = r_j\varphi_j,$$

by part (i). #

3.7. Markov traces on $\mathrm{End}_N(M)$, a generalization of index.

Definition 3.7.1. Let $N \subset M$ be finite von Neumann algebras with N of finite index in M. We say that a faithful trace tr on M is a Markov trace of modulus β for the pair $N \subset M$ if it extends to a trace, also called tr, on $\langle M,e_N\rangle$ for which

(3.7.2) $$\beta\,\mathrm{tr}(xe_N) = \mathrm{tr}(x) \quad \text{for } x \in M.$$

The extension of tr to $\langle M,e_N\rangle$ is uniquely determined by (3.7.2). Also it suffices for (3.7.2) to hold for $x \in N$, since then for $x \in M$

$$\begin{aligned}\mathrm{tr}(xe_N) &= \mathrm{tr}(e_Nxe_N) = \mathrm{tr}(E_N(x)e_N)\\ &= \tfrac{1}{\beta}\mathrm{tr}(E_N(x)) = \tfrac{1}{\beta}\mathrm{tr}(x).\end{aligned}$$

Cf. Lemma 2.7.1.

We restrict our attention to pairs of finite direct sums of finite factors and continue to use the notation of the previous section.

Theorem 3.7.3. *A trace on* M *specified by the vector* $\overleftarrow{s}$, $s_i = tr(p_i)$ *is a Markov trace of modulus* β *if and only if*

$$\overleftarrow{s}\,T_N^M\tilde{T} = \beta\,\overleftarrow{s}.$$

Proof. ($\Rightarrow$). Suppose Tr is a trace on $\langle M,e_N\rangle$ extending the given trace on M and satisfying the Markov property (3.7.2). Let $\overleftarrow{r}$ be the row vector, $r_j = Tr(Jq_jJ)$. By the Markov property we have

$$\beta\,Tr(e_Nq_j) = tr(q_j) = t_j,$$

where $\overleftarrow{t} = \overleftarrow{s}\,T_N^M$ is the vector specifying $tr\big|_N$. Putting this together with 3.6.10(ii) gives

$$t_j = \beta\,Tr(e_Nq_j) = \beta\,r_j\varphi_j, \text{ or}$$

(3.7.3.1) $$\overleftarrow{t} = \beta\,\overleftarrow{r}\,F.$$

Hence

$$\begin{aligned}\beta\,\overleftarrow{s} &= \beta\,\overleftarrow{r}\,T_M^{\langle M,e_N\rangle} = \beta\,\overleftarrow{r}\,F\tilde{T} && \text{(by 3.6.8)}\\ &= \overleftarrow{t}\,\tilde{T} && \text{(by 3.7.3.1)}\\ &= \overleftarrow{s}\,T_N^M\tilde{T}.\end{aligned}$$

($\Leftarrow$) Given a trace tr on M satisfying $\overleftarrow{s}\,T_N^M\tilde{T} = \beta\,\overleftarrow{s}$, define $\overleftarrow{r} = \beta^{-1}\overleftarrow{s}\,T_N^MF^{-1}$ (motivated by 3.7.3.1), and define a trace Tr on $\langle M,e_N\rangle$ by $Tr(Jq_jJ) = r_j$. Then

$$\overleftarrow{r}\,T_M^{\langle M,e_N\rangle} = (\beta^{-1}\overleftarrow{s}\,T_N^MF^{-1})(F\tilde{T}) = \overleftarrow{s},$$

so Tr extends tr on M (3.5.2(ii)).

It remains to show the Markov property, $Tr(xe_N) = \beta^{-1}tr(x)$ for $x \in N$, and by linearity it is enough to check this for $x \in Nq_j$. Now $x \mapsto Tr(xe_N)$ is a trace on the factor Nq_j, so $Tr(xe_N) = Tr(q_je_N)tr_{Nq_j}(x)$; hence it suffices to show that $Tr(q_je_N) = \beta^{-1}tr(q_j) = \beta^{-1}t_j$. But by 3.6.10(ii)

$$\begin{aligned}\mathrm{Tr}(q_j e_N) &= r_j \varphi_j = (\overleftarrow{r}F)_j \\ &= (\beta^{-1}\overleftarrow{s}\, T_N^M F^{-1} F)_j \\ &= \beta^{-1} t_j,\end{aligned}$$

as desired. #

Corollary 3.7.4. *Suppose* $N \subset M$ *are finite direct sums of finite factors, with* N *of finite index in* M. *Set* $T = T_N^M$.

(i) *If* $N \subset M$ *is a connected inclusion, then there is a unique normalized Markov trace on* $N \subset M$; *it is faithful and has modulus equal to the spectral radius of* $T\tilde{T}$.

(ii) *If* tr *is a Markov trace of modulus* β *on* $N \subset M$, *then the unique extension of the trace to* $\langle M, e_N \rangle$ *satisfying (3.7.2) is a Markov trace of modulus* β *(for* $M \subset \langle M, e_N \rangle$*).*

Proof. (i) Since $N \subset M$ is connected, T is indecomposable and $T\tilde{T}$ is irreducible by a straightforward generalization of Lemma 1.3.2.b. Therefore by Perron– Frobenius theory, $T\tilde{T}$ has a unique non–negative eigenvector $\overleftarrow{s}$ with $\sum_i s_i = 1$. Furthermore $s_i > 0$ and the corresponding eigenvalue is the spectral radius of $T\tilde{T}$

(ii) If $\overleftarrow{s}$ is the vector specifying the Markov trace on M, then the extension of the trace to $\langle M, e_N \rangle$ satisfying the Markov condition (3.7.2) is specified by the vector $\overleftarrow{r} = \beta^{-1}\overleftarrow{s}\, TF^{-1}$. Let R denote the matrix $T_M^{\langle M, e_N \rangle} = F\tilde{T}$, with entries

$$R_{j,i} = \begin{cases} \dfrac{\lambda_{i,j}^2}{c_{i,j}} \varphi_j & \text{if } p_i q_j \neq 0 \\ 0 & \text{if } p_i q_j = 0. \end{cases}$$

Since $\Lambda_M^{\langle M, e_N \rangle} = (\Lambda_N^M)^t$, the matrix $\tilde{R}$ (which is to R as $\tilde{T}$ is to T) has entries

$$\tilde{R}_{i,j} = \begin{cases} \lambda_{i,j}^2 / R_{j,i} = c_{i,j} \varphi_j^{-1} & \text{if } p_i q_j \neq 0 \\ 0 & \text{if } p_i q_j = 0. \end{cases}$$

That is $\tilde{R} = TF^{-1}$. But then

$$\begin{aligned}\overleftarrow{r}\, R\tilde{R} &= (\beta^{-1}\overleftarrow{s}\, TF^{-1})(F\tilde{T})(TF^{-1}) \\ &= \beta^{-1}\overleftarrow{s}\, T\tilde{T}TF^{-1} \\ &= \overleftarrow{s}\, TF^{-1} \qquad \text{(by 3.7.3)} \\ &= \beta\, \overleftarrow{r}.\end{aligned}$$

Hence $\overleftarrow{r}$ defines a Markov trace on $\langle M, e_N \rangle$ by Theorem 3.7.3. #

Remark. Before going on, let us see how the analysis above agrees with that in Chapter 2 for finite dimensional algebras. Assume that $Mp_i \cong \mathrm{Mat}_{\mu_i}(\mathbb{C})$ and $Nq_j \cong \mathrm{Mat}_{\nu_j}(\mathbb{C})$. We noted in the remark following 3.6.8 that

$$T = \tilde{\mu}^{-1}\Lambda\tilde{\nu} \text{ and } \tilde{T} = \tilde{\nu}^{-1}\Lambda^t\tilde{\mu},$$

where $\tilde{\mu} = \mathrm{diag}(\mu_1,\cdots,\mu_m)$ and $\tilde{\nu} = \mathrm{diag}(\nu_1,\cdots,\nu_n)$. Thus

$$T\tilde{T} = \tilde{\mu}^{-1}\Lambda\Lambda^t\tilde{\mu}.$$

In this chapter we have been specifying a trace tr on M by the vector $\overleftarrow{s}$ with $s_i = \mathrm{tr}(p_i)$, while in Chapter 2 we specified the trace by $\overleftarrow{s}'$, where s_i' is the trace of a minimal projection in Mp_i. The vector $\overleftarrow{s}$ and $\overleftarrow{s}'$ are related by $\overleftarrow{s} = \overleftarrow{s}'\tilde{\mu}$. The condition given in Chapter 2 for tr to be a Markov trace of modulus β is $\overleftarrow{s}'\Lambda\Lambda^t = \beta\overleftarrow{s}'$. But this is equivalent to

$$\begin{aligned}\overleftarrow{s}(T\tilde{T}) &= (\overleftarrow{s}'\tilde{\mu})(\tilde{\mu}^{-1}\Lambda\Lambda^t\tilde{\mu}) \\ &= \overleftarrow{s}'\Lambda\Lambda^t\tilde{\mu} = \beta\overleftarrow{s}'\tilde{\mu} = \beta\overleftarrow{s}. \qquad \#\end{aligned}$$

Definition 3.7.5. Let $N \subset M$ be finite sums of II_1-factors with the same identity and with N of finite index in M. Let $\Lambda = \Lambda_N^M = (\lambda_{i,j})$ be the matrix of indices and $T = T_N^M = (c_{i,j})$ be the row stochastic matrix of traces as above. Form $\tilde{T} = \tilde{T}(\Lambda,T)$, the matrix whose (j,i) entry is 0 if $c_{i,j} = 0$ and $\dfrac{\lambda_{i,j}^2}{c_{i,j}}$ otherwise. Then the <u>index of N in M</u>, [M:N], is the largest eigenvalue of the matrix $T\tilde{T}$.

Remark. It is easy to see that this definition agrees with that of Section 3.4 when N and M are factors. We mention again that P. Jolissaint has recently shown that this definition always coincides with the ring theoretic definition given in Section 2.1 and [Jo4].

Corollary 3.7.6. *If* $N \subset M$ *are as above and* $[M:N] < 4$, *then* $[M:N] \in \{4\cos^2\pi/q : q \geq 3\}$.

Proof. The index is the largest of the numbers [Mz:Nz], where z is a minimal projection in $Z(M) \cap Z(N)$, so we can assume that $M \supset N$ is connected.

By 3.7.4(i), there is a Markov trace tr on M of modulus [M:N].

Then 3.7.4(ii) allows us to iterate the fundamental construction in the usual way to obtain a tower

$$M_0 = N \subset M_1 = M \subset \cdots \subset M_k \subset M_{k+1} \subset \cdots,$$

a sequence of self-adjoint projections $(e_k)_{k\geq 1}$ with $M_{k+1} = \langle M_k, e_k\rangle$ for all k, and a trace tr on $\bigcup_k M_k$ satisfying the Markov property

$$[M:N]\mathrm{tr}(e_k x) = \mathrm{tr}(x) \quad \text{for} \quad x \in M_k.$$

The projections e_k then satisfy the usual relations and therefore the restriction on [M:N] follows from [Jo1]; see the argument given in Section 3.4. #

Next we provide some examples. Note that by 3.5.6, to construct examples it suffices to give the matrices Λ and T.

Example 3.7.7. The simplest new example is where M is a II_1-factor, p is a projection in M and $N = pMp + (1-p)M(\mathbb{1}-p)$. Here the matrix Λ_N^M is (1 1), and T_N^M is (t 1−t), where $t = \mathrm{tr}_M(p)$. Thus $\tilde{T} = \begin{bmatrix} 1/t \\ 1/1-t \end{bmatrix}$ and $T\tilde{T} = 2$. So $[M:N] = 2$, independent of t! #

Example 3.7.8. Consider an inclusion $N \subset M$ with $\Lambda = \Lambda_N^M = \begin{bmatrix} 1 & 1 \\ 0 & 1 \end{bmatrix}$ and $T = T_N^M = \begin{bmatrix} t & 1-t \\ 0 & 1 \end{bmatrix}$. Then $\tilde{T} = \begin{bmatrix} 1/t & 0 \\ 1/1-t & 1 \end{bmatrix}$ and $T\tilde{T} = \begin{bmatrix} 2 & 1-t \\ 1/1-t & 1 \end{bmatrix}$. The characteristic equation is $\lambda^2 - 3y + 1 = 0$, so $[M:N] = 4\cos^2 \pi/5$, independent of t. #

Example 3.7.9. Take $\Lambda_N^M = \begin{bmatrix} 1 & 1 \\ 1 & 1 \end{bmatrix}$, $T_N^M = \begin{bmatrix} a & 1-a \\ b & 1-b \end{bmatrix}$, with $0 < a,b < 1$. Then $\tilde{T} = \begin{bmatrix} 1/a & 1/b \\ 1/1-a & 1/1-b \end{bmatrix}$ and $T\tilde{T} = \begin{bmatrix} 2 & \frac{a}{b} + \frac{1-a}{1-b} \\ \frac{b}{a} + \frac{1-b}{1-a} & 2 \end{bmatrix}$. The characteristic polynomial is

$$(\lambda - 2 - \sqrt{2+\alpha+\alpha^{-1}})(\lambda - 2 + \sqrt{2+\alpha+\alpha^{-1}}),$$

with $\alpha = \frac{a}{b}\frac{(1-b)}{(1-a)}$. So $[M:N] = 2 + \sqrt{2+\alpha+\alpha^{-1}}$, which can be any real number greater than or equal to 4. #

Example 3.7.10. Let M be a II_1-factor, p a projection of trace t in M and $N = pMp + Q$, where Q is a subfactor of index λ in $(1-p)M(1-p)$. Then $\Lambda_N^M = (1\ \lambda^{1/2})$ and $T_N^M = (t\ 1-t)$. So $\tilde{T} = \begin{bmatrix} 1/t \\ \lambda/1-t \end{bmatrix}$ and $T\tilde{T} = 1 + \lambda$. This is ≤ 4 when $\lambda = 1, 2, 4\cos^2\pi/5$, or 3.

Remark. The index matrices in example 3.7.10 correspond to A_3, B_3, H_3, and $G_2^{(1)}$, respectively, under the corespondence of Theorem 1.1.3, when $\lambda = 1, 2, 4\cos^2\pi/5$, and 3. This is no accident, as we will see.

Proposition 3.7.11. *Let* $\Lambda = (\lambda_{i,j})$ *be an irredundant matrix over* $\{2\cos(\pi/q) : q \geq 2\}$ *and* $T = (c_{i,j})$ *is a row stochastic matrix with the same pattern of zero entries as* Λ. *Let* $\tilde{T} = \tilde{T}(\Lambda,T)$ *be the matrix whose* (j,i)*-entry is zero if* $c_{i,j} = 0$ *and equal to* $\frac{\lambda_{i,j}^2}{c_{i,j}}$ *otherwise.*

If the spectral radius of $T\tilde{T}$ *is less than 4, then it equals* $4\cos^2\pi/q$ *for some* $q \in \{3,4,5\cdots\}$.

Proof. We can suppose that Λ and T are indecomposable. By 3.5.6, there is a connected inclusion $M \supset N$ of finite direct sums of II_1-factors with $\Lambda = \Lambda_N^M$ and $T = T_N^M$. Thus the result is a corollary of 3.7.6. #

Remark. It would be interesting to find a proof of 3.7.11 within usual matrix theory; hopefully this might give information on the spectral radius of $T\tilde{T}$ even when it is larger than 4.

Proposition 3.7.12.

(a) *Let* Λ *be an irredundant* m-by-n *matrix with non-negative real values. Then there is a row stochastic* m-by-n *matrix* T *with the same pattern of zeros such that*

$$\rho(T\tilde{T}) = \|\Lambda\|^2,$$

where ρ *denotes spectral radius and* $\tilde{T} = \tilde{T}(\Lambda,T)$ *is as above.*

(b) *If* Λ *is irredundant with values in* $\{2\cos(\pi/q) : q \geq 2\}$, *then there is a pair* $N \subset M$ *of finite direct sums of* II_1*-factors with* $\Lambda = \Lambda_N^M$ *and* $[M:N] = \|\Lambda\|^2$.

(c) *If* Λ *is any non-zero matrix over* $\{2\cos(\pi/q) : q \geq 2\}$ *and* $\|\Lambda\| < 2$, *then*

$$\|\Lambda\| \in \{2\cos(\pi/q) : q \geq 3\}.$$

Proof. (a) As Λ is irredundant, we can define a row stochastic matrix $T = (c_{i,j})$ by

$c_{i,j} = \left[\sum_j \lambda_{i,j}\right]^{-1} \lambda_{i,j}$, or $T = X\Lambda$, where X is the m-by-m diagonal matrix whose (i,i)-entry is $\left[\sum_j \lambda_{i,j}\right]^{-1}$. Then the (j,i) entry of $\tilde{T}$ is $\lambda_{i,j}\left[\sum_j \lambda_{i,j}\right]$, i.e., $\tilde{T} = \Lambda^t X^{-1}$. Thus $T\tilde{T} = X\Lambda\Lambda^t X^{-1}$, which has the same spectrum as $\Lambda\Lambda^t$.

(b) By 3.5.6 there is a pair $N \subset M$ with $\Lambda = \Lambda_N^M$ and $T = T_N^M$. Then $[M:N] = \rho(T\tilde{T}) = \|\Lambda\|^2$.

(c) It suffices to consider Λ irredundant, so the result follows from (b) and 3.7.6. #

Of course, 3.7.12(c) was already known as a consequence of Theorem 1.1.3. Theorem 1.1.3 suggests (but does not immediately imply) the following, which is the main result of this section.

Theorem 3.7.13. *Let* $N \subset M$ *be a connected inclusion of finite direct sums of* II_1*-factors.*

(a) *If* $[M:N] < 4$, *then* Λ_N^M *is the matrix associated (in Theorem 1.1.3) to a bicoloration of one of the following Coxeter graphs:*

$$A_\ell\ (\ell \geq 2),\ B_\ell\ (\ell \geq 3),\ D_\ell\ (\ell \geq 4),\ E_\ell\ (\ell = 6,7,8),$$
$$F_4,\ G_2,\ H_\ell\ (\ell = 3,4),\ I_2(p)\ (p = 5 \text{ or } p \geq 7).$$

Moreover $[M:N] = \|\Lambda_N^M\|^2 = 4\cos^2 \pi/h$, *where* h *is the Coxeter number. (See tables 1.4.5, 1.4.6, and 1.4.7.)*

(b) *If* $[M:N] = 4$, *then* Λ_N^M *corresponds to one of*

$$A_\ell^{(1)}\ (\ell \text{ odd}, \ell \geq 1),\ B_\ell^{(1)}\ (\ell \geq 2),\ C_\ell^{(1)}\ (\ell \geq 3),$$
$$D_\ell^{(1)}\ (\ell \geq 4),\ E_\ell^{(1)}\ (\ell = 6,7,8),\ F_4^{(1)},\ G_2^{(1)}.$$

Lemma 3.7.14. (Schwenck, [Sch2]) *Let* $A = (a_{i,j})$ *be a non-negative* n*-by-*n *matrix and let* $G = (g_{i,j})$ *be the matrix with entries* $g_{i,j} = (a_{i,j}a_{j,i})^{1/2}$. *Then* $\|G\| = \rho(G) \leq \rho(A)$, *where* ρ *denotes spectral radius.*

Proof. For any k+1-tuple $\alpha = (i_1, i_2, \cdots, i_{k+1})$ with $1 \leq i_j \leq n$, let α^{-1} denote the reversed tuple $\alpha^{-1} = (i_{k+1}, \cdots, i_2, i_1)$, and let

$$a_\alpha = a_{i_1,i_2} a_{i_2,i_3} \cdots a_{i_k,i_{k+1}}.$$

Let W_k be the set of k+1–tuples with $i_1 = i_{k+1}$; thus $\alpha \mapsto \alpha^{-1}$ is bijection of W_k and

$$\mathrm{tr}(A^k) = \sum_{\alpha \in W_k} a_\alpha = \tfrac{1}{2}\sum_{\alpha \in W_k} (a_\alpha + a_{\alpha^{-1}}).$$

Thus

$$\mathrm{tr}(A^k) \geq \sum_{\alpha \in W_k} (a_\alpha a_{\alpha^{-1}})^{1/2} = \sum_{\alpha \in W_k} g_\alpha = \mathrm{tr}(G^k),$$

for all $k \in \mathbb{N}$. When k is even, we have

$$n\rho(A)^k = n\rho(A^k) \geq \mathrm{tr}(A^k) \geq \mathrm{tr}(G^k) \geq \rho(G)^k,$$

where the first equality and last inequality result from considering canonical forms for A and G, noting that the eigenvalues of G^k are positive. Taking k^{th} roots and then the limit as $k \to \infty$ gives the result. #

Lemma 3.7.15. *Let* $\Lambda = (\lambda_{i,j})$ *be an* m–*by*–n *irredundant matrix over* $\{r \in \mathbb{R} : r = 0 \text{ or } r \geq 1\}$. *Let* $T = (c_{i,j})$ *be a row stochastic matrix with the same pattern of zero entries as* Λ. *Let* $\tilde{T}$ *be the* n–*by*–m *matrix whose* (j,i)–*entry is* 0 *if* $c_{i,j} = 0$ *and* $\frac{\lambda_{i,j}^2}{c_{i,j}}$ *otherwise.*

If $\rho(T\tilde{T}) \leq 4$ *then* $\|\Lambda\|^2 \leq \rho(T\tilde{T})$.

Proof. We may assume without loss of generality that Λ is indecomposable. Suppose that there exist indices i_1, i_2, j_1, j_2 such that the four entries λ_{i_μ, j_ν} for $\mu,\nu \in \{1,2\}$ are all non–zero; that is, the graph $\tilde{\Gamma}(\Lambda)$ contains a subgraph of the form

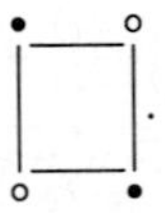

Rearranging the rows and columns, we can suppose $i_1 = j_1 = 1$ and $i_2 = j_2 = 2$. Denote the (i,j)–entry of $T\tilde{T}$ by $\gamma_{i,j}$. Then $\gamma_{i,j} = {\sum_k}' \frac{c_{i,k}\lambda_{j,k}^2}{c_{j,k}}$, the sum being over those k for which $c_{j,k} \neq 0$. In particular $\gamma_{i,j} \geq {\sum_k}' \frac{c_{i,k}}{c_{j,k}}$ with equality everywhere if and only if all the non–zero $\lambda_{j,k}$ are equal to one.

By monotonicity of the Perron–Frobenius eigenvalue, we have

$$\rho(T\tilde{T}) \geq \rho\left(\begin{bmatrix} \gamma_{1,1} & \gamma_{1,2} \\ \gamma_{2,1} & \gamma_{2,2} \end{bmatrix}\right),$$

with equality if and only if Λ is 2–by–n. This in turn is no smaller than

$$\rho\left(\left[\sum_k{}' \frac{c_{i,k}}{c_{j,k}}\right]_{1\leq i,j\leq 2}\right),$$

by the observation above, with equality if and only if all the non–zero $\lambda_{j,k}$ with $j \leq 2$ are equal to one. Truncating the sums defining the entries of the last matrix we see that the spectral radius is at least

$$\rho\left(\begin{bmatrix} 2 & \frac{c_{1,1}}{c_{2,1}} + \frac{c_{1,2}}{c_{2,2}} \\ \frac{c_{2,1}}{c_{1,1}} + \frac{c_{2,2}}{c_{1,2}} & 2 \end{bmatrix}\right),$$

with equality if and only if $\lambda_{j,k} = 0$ for $j = 1,2$ and $k > 2$. If we replace the off–diagonal entries by their geometric mean, we do not alter the spectrum, so the last quantity is equal to

$$\rho\left(\begin{bmatrix} 2 & (2+\alpha+\alpha^{-1})^{1/2} \\ (2+\alpha+\alpha^{-1})^{1/2} & 2 \end{bmatrix}\right),$$

where $\alpha = \dfrac{c_{1,1}c_{2,2}}{c_{2,1}c_{1,2}}$. Finally, this is at least

$$\rho\left(\begin{bmatrix} 2 & 2 \\ 2 & 2 \end{bmatrix}\right) = 4,$$

with equality if and only if $\alpha = 1$. But since $\rho(T\tilde{T}) \leq 4$ by hypothesis, we must have equality at every step: Λ and T are 2–by–2 with $\Lambda = \begin{bmatrix} 1 & 1 \\ 1 & 1 \end{bmatrix}$ and $\alpha = 1$. Since T is row–stochastic this implies $T = \begin{bmatrix} \frac{1}{2} & \frac{1}{2} \\ \frac{1}{2} & \frac{1}{2} \end{bmatrix}$, and $\|\Lambda\|^2 = \rho(T\tilde{T}) = 4$.

If on the other hand $\tilde{\Gamma}(\Lambda)$ contains no subgraph of the form 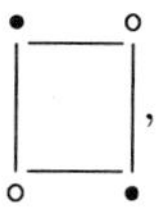, then for $i \neq j$ there is at most one non–zero term in the sum defining $\gamma_{i,j}$,

$$\gamma_{i,j} = \begin{cases} 0 & \text{if the } i^{\text{th}} \text{ and } j^{\text{th}} \text{ rows of } \Lambda \text{ are orthogonal} \\ \dfrac{c_{i,k_0}}{c_{j,k_0}} \lambda_{j,k_0}^2 & \text{for some } k_0, \text{ otherwise.} \end{cases}$$

Observe that k_0 depends on (i,j), and $k_0(i,j) = k_0(j,i)$. On the other hand, $\gamma_{i,i} = \sum_k \lambda_{i,k}^2$. Note that for all pairs (i,j), the (i,j) entry of $\Lambda\Lambda^t$ is the geometric mean of the (i,j) and (j,i)–entries of $T\tilde{T}$; i.e.,

$$(\Lambda\Lambda^t)_{i,j} = \begin{cases} \sum_k \lambda_{i,k}^2 & \text{if } i = j \\ 0 & \text{if the } i^{\text{th}} \text{ and } j^{\text{th}} \text{ rows of } \Lambda \text{ are orthogonal} \\ \lambda_{i,k_0}\lambda_{j,k_0} & \text{otherwise.} \end{cases}$$

Hence by Lemma 3.7.14

$$\|\Lambda\|^2 = \rho(\Lambda\Lambda^t) \leq \rho(T\tilde{T}). \quad \#$$

<u>Proof of Theorem 3.7.13.</u> (a) Let $T = T_N^M$ and $\tilde{T} = \tilde{T}(\Lambda,T)$. By hypothesis $[M:N] = \rho(T\tilde{T}) < 4$, so by 3.7.15 we have $\|\Lambda\|^2 \leq \rho(T\tilde{T})$. Let $\mathcal{S}$ be the (convex) set of all row– stochastic matrices of the same dimension and zero–pattern as Λ and T. For each $S \in \mathcal{S}$ define $\tilde{S} = \tilde{T}(\Lambda,S)$, the matrix whose (j,i)–entry is 0 if $\lambda_{i,j} = 0$ and $\lambda_{i,j}^2/(S)_{i,j}$ otherwise, and $\varphi(S) = \rho(S\tilde{S})$. Then φ is a continuous function of S by elementary perturbation theory, and φ assumes the value $[M:N] = \rho(T\tilde{T})$ as well as the value $\|\Lambda\|^2$, by 3.7.12(a). But by 3.7.11, the set of values of φ less than 4 is discrete, so by convexity of $\mathcal{S}$, φ is constant and

$$\|\Lambda\|^2 = \rho(T\tilde{T}) = [M:N].$$

The classification of Λ_N^M then follows from Theorem 1.1.3.

(b) We have $\|\Lambda\|^2 \leq \rho(T\tilde{T}) = 4$, by 3.7.15. If $\|\Lambda\|^2 < 4$, then the connectedness argument of part (a) would imply that $\|\Lambda\|^2 = \rho(T\tilde{T}) < 4$, a contradiction. Thus $\|\Lambda\|^2 = 4$, and the classification follows from 1.1.3 again. #

CHAPTER 4

Commuting sqares, subfactors, and the derived tower

4.1. Introduction.

There are two main themes in this chapter. The first is the approximation of a pair $N \subset M$ of hyperfinite II_1 factors by pairs $C_n \subset B_n$ of finite dimensional von Neumann algebras, with

$$\begin{array}{ccc} N & \subset & M \\ \cup & & \cup \\ C_{n+1} & \subset & B_{n+1} \\ \cup & & \cup \\ C_n & \subset & B_n \end{array}.$$

and $M = (\cup B_n)''$, $N = (\cup C_n)''$. In order for the approximating "ladder" of finite dimensional algebras to behave well with respect to the fundamental construction and the index, it should behave well with respect to the conditional expectations: $E_N|B_n$ should be the conditional expectation of B_n onto C_n. We are thus led to the following definition which was first introduced by Popa (Lemma 1.2.2 in [Pop1]; see also [Pop2]):

Definition 4.1.1. A diagram

$$\begin{array}{ccc} C_1 & \subset & B_1 \\ \cup & & \cup \\ C_0 & \subset & B_0 \end{array}$$

of finite von Neumann algebras with a finite faithful normal trace tr on B_1 is a commuting square if the diagram

$$\begin{array}{ccc} C_1 & \xleftarrow{E_{C_1}} & B_1 \\ \cup & & \cup \\ C_0 & \xleftarrow[E_{C_0}]{} & B_0 \end{array}$$

commutes.

In Section 4.2, we study the notion of commuting squares and give a number of examples of constructions which produce commuting squares. In particular we consider the behavior of commuting squares under the fundamental construction.

<u>Proposition 4.1.2.</u> *Consider a commuting square* $\begin{matrix} C_1 \subset B_1 \\ \cup \quad \cup \\ C_0 \subset B_0 \end{matrix}$ *with respect to a trace* tr *which is a Markov trace for the pair* $B_0 \subset B_1$ *of finite von Neumann algebras with finite dimensional centers. Let* $B_2 = \langle B_1, e_1 \rangle$ *be the von Neumann algebra obtained via the fundamental construction for* $B_0 \subset B_1$, *and let* $C_2 = \{C_1, e_1\}''$. *Then*

$$\begin{matrix} C_2 \subset B_2 \\ \cup \quad \cup \\ C_1 \subset B_1 \end{matrix}$$

is also a commuting square.

Therefore iterating the fundamental construction will produce an infinite ladder of commuting squares. Now suppose that we have a pair $N \subset M$ of finite von Neumann algebras with a Markov trace tr of modulus β and a ladder of commuting squares

$$\begin{matrix} N & \subset & M \\ \cup & & \cup \\ C_{n+1} & \subset & B_{n+1} \\ \cup & & \cup \\ C_n & \subset & B_n \end{matrix}$$

with $N = (\cup C_n)''$ and $M = (\cup B_n)''$. Let $\langle M, e\rangle$ be the result of the fundamental construction for $N \subset M$ and set $A_n = \langle B_n, e\rangle$ for each $n \geq 1$. We show that the algebras $(A_n)_{n\geq 0}$ generate $\langle M, e\rangle$, and $\begin{matrix} M & \subset & \langle M, e\rangle \\ \cup & & \cup \\ B_n & \subset & A_n \end{matrix}$ is a commuting square with respect to the Markov extension of the trace to $\langle M, e\rangle$.

In Section 4.3 we prove a theorem of H. Wenzl on pairs $N \subset M$ generated by a ladder of commuting squares satisfying a periodicity assumption. (See Section 4.3, Hypothesis (B), for the exact assumption.)

Theorem 4.1.3. *Suppose* $N \subset M$ *is a hyperfinite pair (with a finite faithful trace* tr *on* M) *generated by a ladder of commuting squares*

$$\begin{matrix} C_{n+1} & \subset & B_{n+1} \\ \cup & & \cup \\ C_n & \subset & B_n \end{matrix}$$

of finite dimensional von Neumann algebras. Suppose that the inclusion data for the ladder is periodic, in the sense of Hypothesis B of Section 4.3. Then

(i) N and M *are factors and* $[M:N] < \infty$.

Let e *and* $(A_n)_{n\geq 0}$ *be as above and let* z_n *be the central support of* e *in* A_n.

(ii) *For large* n, $z_n = \mathbb{1}$. *Equivalently,* A_n *is isomorphic to the result of the fundamental construction for* $C_n \subset B_n$.

(iii) *For large* n, $[M:N] = [B_n:C_n] = \|\vec{t}^{(n)}\|^2/\|\vec{s}^{(n)}\|^2$, *where* $\vec{t}^{(n)}$ *and* $\vec{s}^{(n)}$ *are the vectors of the trace on* C_n and B_n *respectively.*

Section 4.4 contains a contruction of (necessarily irreducible) pairs of hyperfinite factors with index less than 4, as follows: Start with a connected pair $C_0 \subset B_0$ of finite dimensional von Neumann algebras with index $\beta < 4$, and let $B_1 = \langle B_0, e_{C_0} \rangle$ be the result of the fundamental construction for $C_0 \subset B_0$, with respect to the Markov trace of modulus β on B_0. Define $q \in \mathbb{T}$ by $\beta = 2 + q + q^{-1}$, set $g = qe_{C_0} - (\mathbb{1}-e_{C_0})$, a unitary element in B_1, and set $C_1 = gB_0g^{-1}$. Then $\begin{matrix} C_1 & \subset & B_1 \\ \cup & & \cup \\ C_0 & \subset & B_0 \end{matrix}$ is a commuting square, with respect to the Markov extension of the trace to B_1. Let $(B_n)_{n\geq 0}$ be the tower obtained by iterating the fundamental construction, with $B_{n+1} = \langle B_n, e_n \rangle$ and set $C_{n+1} = \mathrm{alg}\{C_n, e_n\}$ for $n \geq 1$. Then $\begin{matrix} C_{n+1} & \subset & B_{n+1} \\ \cup & & \cup \\ C_n & \subset & B_n \end{matrix}$ is a ladder of commuting squares, with respect to the Markov trace on $\cup B_n$. It turns out that the inclusion data is periodic and that $B = (\cup B_n)'' \supset C = (\cup C_n)''$ is a pair of factors with index β.

Let $(M_k)_{k\geq 0}$ be the tower obtained from a pair $N \subset M$ of finite von Neumann algebras with finite dimensional centers, with index $\beta < 4$ and let $(e_k)_{k\geq 1}$ be the usual sequence of projections in the tower construction. We already know another construction of an irreducible pair with index β, namely $\{e_1,e_2,\cdots\}'' \supset \{e_2,e_3,\cdots\}''$ (Theorem 3.4.3). An argument due to C. Skau shows that $\{e_1,e_2,\cdots\}' \cap (\cup M_k)'' = N$ (Theorem 4.4.3).

In Section 4.5 we present a construction which yields *irreducible* pairs of hyperfinite II_1 factors, starting with a Coxeter graph Γ of type A, D, or E and a choice of a distinguished vertex w_1 on Γ. In particular for $\Gamma = E_6$ and w_1 an end vertex on one of the long arms of Γ, we obtain the index value $3 + \sqrt{3}$, which is at present the smallest known value larger than 4 of the index of an irreducible pair. The construction goes as follows.

Give Γ the bicoloration with w_1 white and with r white vertices altogether. Let M_0 denote the abelian von Neumann algebra $\mathbb{C}^r$ and M_1 the finite dimensional von Neumann algebra containing M_0 such that Γ is the Bratteli diagram of the inclusion $M_0 \subset M_1$. Form the tower $(M_j)_{j\geq 0}$, starting with the pair $M_0 \subset M_1$ and the Markov trace tr on M_1, and let $(e_j)_{j\geq 1}$ be the usual sequence of projections. Let M be the factor $(\bigcup_{j\geq 0} M_j)''$ and let N be the subfactor of M generated by $\mathbb{1}$ and the e_j's. Set $\beta = [M_1:M_0] = \|\Gamma\|^2$; since $\beta < 4$, Skau's Lemma 4.4.3 applies and $M \cap N' = M_0$. Let p be the minimal projection of M_0 corresponding to the vertex w_1 and set $C = pN$ and $B = pMp$. Then $C \subset B$ is a pair of factors with $C' \cap B = p(N' \cap M)p = pM_0p = \mathbb{C}p$; that is $C \subset B$ is irreducible. The index of this pair can be computed as follows:

Let Γ also denote the matrix of the bicolored graph Γ, and let $\overleftarrow{\xi}$ denote the unique Perron–Frobenius row vector for $\Gamma^t\Gamma$, normalized so that its first co-ordinate (corresponding to the distinguished white vertex w_1) is 1. Let Λ be the Coxeter graph of type A with the same Coxeter number as Γ, and with a bicoloration having at least one white end vertex, which is labelled as the first white vertex. Denote also by Λ the matrix of the graph Λ, and let $\overleftarrow{\eta}$ be the Perron–Frobenius row vector of $\Lambda^t\Lambda$, normalized so that its first co-ordinate (corresponding to the chosen white end vertex) is 1. then $[B:C] = \|\overleftarrow{\eta}\|^2/\|\overleftarrow{\xi}\|^2$. The proof uses Wenzl's index formula from Section 4.3.

The second main topic of Chapter 4, presented in Sections 4.6 and 4.7, is the derived tower and principal graph of a pair of finite factors $N \subset M$ of finite index. The derived tower is the chain of relative commutants $(N' \cap M_k)_{k\geq 0}$, where $(M_k)_{k\geq 0}$ is the tower for the pair $N \subset M$. It follows from 3.6.2 that $\dim (N' \cap M_k) \leq [M:N]^k$ for $k \geq 0$.

Let $(e_k)_{k\geq 0}$ be the projections in the tower construction, let Y_k denote $N' \cap M_k$ and Λ_k the inclusion matrix for $Y_k \subset Y_{k+1}$. The following summarizes the structure of the derived tower.

Theorem 4.1.4.

(i) *The inclusion* $Y_k \subset Y_{k+1}$ *is connected.*

(ii) $Y_k e_k Y_k$ *is an ideal in* Y_{k+1}, *and if* $z_{k+1} = z(e_k)$ *is the corresponding central projection in* Y_{k+1} *then the homomorphism* $\begin{cases} Y_k \longrightarrow Y_k e_k Y_k \\ x \longmapsto x z_{k+1} \end{cases}$ *has inclusion matrix* Λ^t_{k-1}.

(iii) *For all* k, $\|\Lambda_k\|^2 \le [M:N]$.

(iv) *For* $k \ge 2$, *if* $x \in Y_{k+1}$ *and* $x(Y_k e_k Y_k) = 0$, *then* $x(Y_{k-1} e_{k-1} Y_{k-1}) = 0$.

(v) *For all* $k \ge 1$, *the following are equivalent:*

(a) $Y_k e_k Y_k = Y_{k+1}$.

(b) $\overleftarrow{s}^{(k-1)} \Lambda^t_{k-1} \Lambda_{k-1} = [M:N]\, \overleftarrow{s}^{(k-1)}$, *where* $\overleftarrow{s}^{(k-1)}$ *is the vector of the trace on* Y_{k-1}.

(c) $\Lambda_k = \Lambda^t_{k-1}$.

(d) $\|\Lambda_{k-1}\|^2 = [M:N]$.

(vi) *If the equivalent conditions of (v) hold for* k, *then they also hold for* k+1.

We call the ideal $Y_k e_k Y_k$ "the old stuff", since it is determined by $Y_{k-1} \subset Y_k$; the complementary ideal is called "the new stuff". Then (iv) says that "the new stuff comes only from the old new stuff", or $(xz_k)(1 - z_{k+1}) = 0$. The principal graph of the pair $N \subset M$ is obtained as follows: on the Bratteli diagram of the derived tower, delete on each level the vertices corresponding to the old stuff, and the edges emanating from them; the result is a connected bipartite graph with a distinguished vertex $*$, the unique vertex on level 0. The Bratteli diagram of the derived tower can be reconstructed from the principal graph. The pair $N \subset M$ is said to be of finite depth if the principal graph is finite; the depth is the maximum distance from any vertex to $*$.

This analysis, together with the work of Chapter 1, yields a new proof of the restriction on index values:

Corollary. (i) *Suppose* $N \subset M$ *is a pair of* II_1 *factors with* $[M:N] < 4$. *Then*

(a) $[M:N] = 4\cos^2 \pi/h$ *for some integer* $h \ge 3$.

(b) *The depth of* $N \subset M$ *is no greater than* h–2.

(c) *The principal graph of* $N \subset M$ *is a Coxeter graph of type* A, D, *or* E, *whose norm is* $[M:N]^{1/2}$.

(ii) *Suppose* $N \subset M$ *is a pair of* II_1 *factors with* $[M:N] = 4$.

(a) *If* $N \subset M$ *is of finite depth, then the principal graph* Γ *is a completed Coxeter graph of type* A, D, *or* E, *i.e., one of the graphs in Table 1.4.6.*

(b) *If* N ⊂ M *is of infinite depth, then* Γ *is one of the following:*

A_∞ 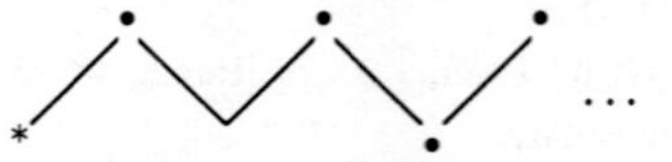

$A_{n,\infty}$ 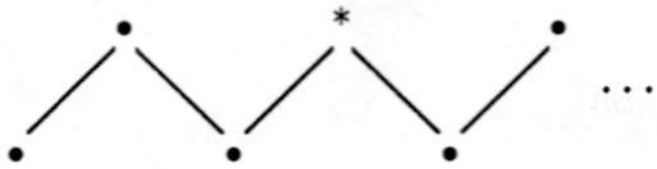 *(end vertex at distance* n *from* $*$*)*

$A_{\infty,\infty} = \mathbb{T}$ 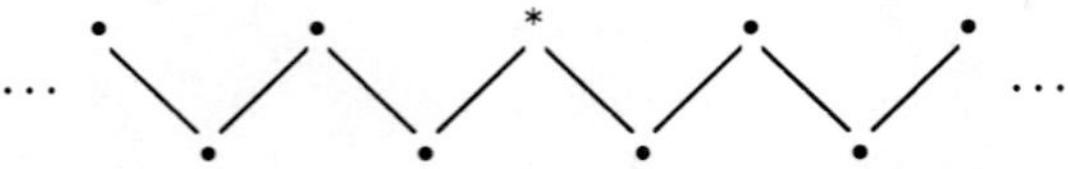

(doubly infinite linear graph)

D_∞ 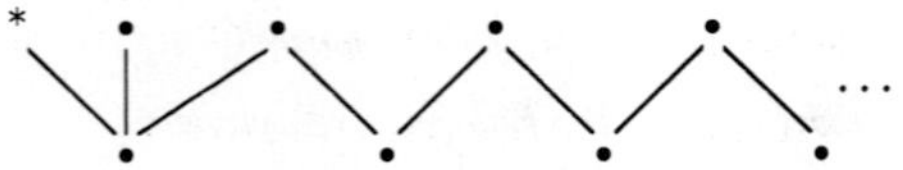

$D_{n,\infty}$ 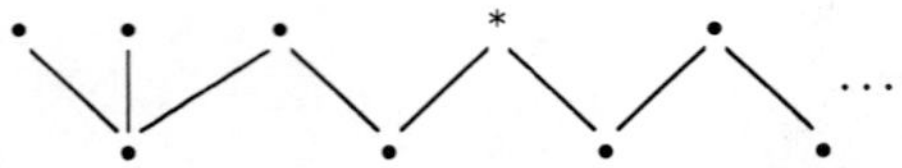

(end vertex at distance n *from* $*$)

Section 4.7 is devoted to computing the derived tower for a number of examples: Crossed–products and fixed point algebras for outer actions of finite groups give examples with depth 2. The pairs $R_\beta \subset R$ (of Proposition 3.4.4) when $\beta < 4$ have principal graphs of type A_n; for $\beta = 4$ the principal graph is A_∞. In 4.7.c we give a general method which allows the computation of the derived tower in many examples coming from group actions. In 4.7.d we use this method to obtain the derived towers for the index 4 subfactors $R^G \subset (R \otimes \mathrm{Mat}_2(\mathbb{C}))^G$, where the hyperfinite II_1 factor R is realized as the weak closure of the CAR algebra $\overset{\infty}{\otimes} \mathrm{Mat}_2(\mathbb{C})$ in the trace representation, and G is a closed subgroup of SU(2) acting by the infinite tensor product of its action by conjugation on $\mathrm{Mat}_2(\mathbb{C})$. In this way one obtains as principal graphs all the affine Coxeter graphs of type A, D, and E, as well as the infinite graphs A_∞, $A_{\infty,\infty}$, and D_∞ listed above. Finally we compute the derived tower for the pair $R_\beta \subset R$ when $\beta > 4$. This is the most difficult result of the chapter, involving a representation of the sequence $(e_i)_{i\geq 1}$ in the CAR algebra due to Pimsner and Popa and a theorem of Popa on the tunnel construction (a mirror image of the tower construction). Ultimately one identifies the pair $R_\beta \subset R$ with the pair $N^{\mathbb{T}} \subset (N \otimes \mathrm{Mat}_2(\mathbb{C}))^{\mathbb{T}}$ where N is the completion of the CAR algebra with respect to a certain Powers state. The principal graph is therefore $A_{\infty,\infty}$.

4.2. Commuting squares.

We begin with a proposition, inspired by Lemma 2.1 of [Pop2], which gives a number of equivalent conditions for a commuting square.

<u>Proposition 4.2.1.</u> *Consider a diagram*

$$\begin{array}{ccc} C_1 & \subset & B_1 \\ \cup & & \cup \\ C_0 & \subset & B_0 \end{array}$$

of finite von Neumann algebras and a finite faithful normal trace tr *on* B_1. *All conditional expectations being with respect to* tr, *the following are equivalent.*

(i) $E_{C_1}(B_0) \subset C_0$.

(ii) $E_{C_1}E_{B_0} = E_{C_0}$.

(iii) $E_{C_1}E_{B_0} = E_{C_0}E_{B_0}$.

(iv) $E_{C_1}E_{B_0} = E_{B_0}E_{C_1}$ *and* $B_0 \cap C_1 = C_0$.

(v) *The diagram*
$$\begin{array}{ccc} C_1 & \xleftarrow{E_{C_1}} & B_1 \\ \cup & & \cup \\ C_0 & \xleftarrow[E_{C_0}]{} & B_0 \end{array}$$
commutes.

(vi) $E_{C_0}(b_0c_1) = E_{C_0}(b_0)E_{C_0}(c_1)$ *for* $b_0 \in B_0$ *and* $c_1 \in C_1$.

(vii) $E_{C_0}(b_0c_1) = 0$ *for* $b_0 \in B_0$ *with* $E_{C_0}(b_0) = 0$ *and* $c_1 \in C_1$ *with* $E_{C_0}(c_1) = 0$.

Moreover (i) to (vii) are equivalent with the analogous conditions obtained by interchanging B_0 *with* C_1.

<u>Proof.</u> Let p,q,r be three projections acting on some Hilbert space. The following are clearly equivalent:

(a) $pq = r$

(b) $pq = rq$ and $r \leq q$

(c) $pq = qp = r$.

As we may view the conditional expectations as projections on $L^2(B_1,tr)$, this shows the equivalence of (ii), (iii) and (iv). Obviously (ii) implies (i).

Assume (i) holds and let $b_0 \in B_0$. For all $c_0 \in C_0$ one has

$$tr(E_{C_0}(b_0)c_0) = tr(b_0c_0) = tr(E_{C_1}(b_0)c_0).$$

As $E_{C_1}(b_0) \in C_0$, this implies $E_{C_0}(b_0) = E_{C_1}(b_0)$, and (v) follows. As (v) implies (iii), conditions (i) to (v) are equivalent.

The equivalence of (vi) and (vii) follows from the formula

$$E_{C_0}\{(b_0-E_{C_0}(b_0))(c_1-E_{C_0}(c_1))\} = E_{C_0}(b_0c_1) - E_{C_0}(b_0)E_{C_0}(c_1)$$

for $b_0 \in B_0$ and $c_1 \in C_1$.

The next step is to show that (ii) and (vii) are equivalent. Observe first that one has

$$E_{C_1}(E_{B_0}(x) - E_{C_0}(x)) = E_{C_1}E_{B_0}(x) - E_{C_0}(x)$$

for any $x \in B_1$. Thus (ii) can be reformulated as

$$E_{B_0}(x) - E_{C_0}(x) \perp C_1 \quad \text{for all } x \in B_1.$$

Suppose (ii) holds. Then, in particular, $b_0 \perp C_1$ for $b_0 \in B_0$ with $E_{C_0}(b_0) = 0$. Consequently, for all $c_1 \in C_1$ and $c_0 \in C_0$, one has

$$tr(E_{C_0}(b_0c_1)c_0) = tr(b_0c_1c_0) = 0.$$

As tr is faithful on C_0, this implies $E_{C_0}(b_0c_1) = 0$ and (vii) holds.

Suppose (vii) holds. For all $x \in B_1$ and for all $c_1 \in C_1$, one has

$$\begin{aligned} &tr\{(E_{B_0}(x) - E_{C_0}(x))c_1\} \\ &= tr\{(E_{B_0}(x) - E_{C_0}(x))(c_1 - E_{C_0}(c_1))\} + tr\{E_{B_0}(x)E_{C_0}(c_1)\} - tr\{E_{C_0}(x)E_{C_0}(c_1)\} \\ &= 0 + tr\{E_{B_0}(xE_{C_0}(c_1))\} - tr\{E_{C_0}(xE_{C_0}(c_1))\}, \end{aligned}$$

which is zero, since the conditional expectations are trace preserving. Consequently (ii) holds.

Finally, as (iv) is symmetric with respect to B_0 and C_1, we may exchange B_0 and C_1 in any of the conditions (i) to (vii). #

It follows for example from (v) that in diagrams like

$$\begin{array}{ccc} C_2 \subset B_2 & & \\ \cup \quad \cup & & C_1 \subset B_1 \subset A_1 \\ C_1 \subset B_1 & \text{or} & \cup \quad \cup \quad \cup \\ \cup \quad \cup & & C_0 \subset B_0 \subset A_0 \\ C_0 \subset B_0 & & \end{array}$$

the "rectangles" are commuting squares as soon as the "small squares" are commuting squares.

A crucial point about commuting squares is their behavior with respect to fundamental construction defined in Section 3.6.

Proposition 4.2.2. *Consider a pair* $N \subset M$ *of finite von Neumann algebras, a finite faithful normal trace* tr *on* M, *and the algebra* $\langle M,e_N\rangle$ *obtained by the fundamental construction. Assume that* M *[respectively* N*] is generated as a von Neumann algebra by a nested sequence* $(B_j)_{j\geq 0}$ *[resp.* $(C_j)_{j\geq 0}$*] of von Neumann subalgebras in such a way that one has for each* $j \geq 0$ *a commuting square*

$$\begin{array}{ccc} C_{j+1} & \subset & B_{j+1} \\ \cup & & \cup \\ C_j & \subset & B_j \end{array}$$

and set $A_j = \{B_j,e_N\}''$. *Then*

(i) $e_N b e_N = E_{C_j}(b) e_N = e_N E_{C_j}(b)$ *for* $b \in B_j$, $j \geq 0$

(ii) *The algebras* $(A_j)_{j\geq 0}$ *generate* $\langle M,e_N\rangle$ *as a von Neumann algebra.*

Suppose moreover that tr *is a Markov trace of modulus* β *for the pair* $N \subset M$, *and denote the Markov extension of* tr *to* $\langle M,e_N\rangle$ *by* tr *again. (See Definition 3.7.1.) Then*

(iii) $\begin{array}{ccc} B_{j+1} & \subset & A_{j+1} \\ \cup & & \cup \\ B_j & \subset & A_j \end{array}$ *is a commuting square with respect to* $\mathrm{tr}|_{A_{j+1}}$.

Proof. (i) For each $j \geq 0$ and $k \geq 1$, the diagram $\begin{matrix} C_{j+k} & \subset & B_{j+k} \\ \cup & & \cup \\ C_j & \subset & B_j \end{matrix}$ is a commuting square, by induction on k. It follows that the limit diagram $\begin{matrix} N & \subset & M \\ \cup & & \cup \\ C_j & \subset & B_j \end{matrix}$ is also a commuting square, and thus for any $b \in B_j$ one has $e_N b e_N = E_N(b) e_N = E_{C_j}(b) e_N$. Since elements of N, and in particular $E_{C_j}(b)$, commute with e_N, this shows (i).

Claim (ii) is obvious.

(iii) One has $E_{B_{j+1}}(e_N) = \beta^{-1}$, because

$$\operatorname{tr}(E_{B_{j+1}}(e_N)x) = \operatorname{tr}(e_N x) = \beta^{-1}\operatorname{tr}(x)$$

for all $x \in B_{j+1}$. Consider now $y_0, y'_\alpha, y''_\alpha \in B_j$. Then

$$E_{B_{j+1}}(y_0 + \sum_\alpha y'_\alpha e_N y''_\alpha) = y_0 + \sum_\alpha y'_\alpha \beta^{-1} y''_\alpha \in B_j.$$

Thus $E_{B_{j+1}}(\tilde{A}_j) \subset B_j$ for a dense $*$-subalgebra $\tilde{A}_j$ of A_j.

Let $x \in A_j$. By the density theorem of Kaplansky, there exists a sequence $(x_k)_{k \geq 1}$, with $x_k \in \tilde{A}_j$ and $\|x_k\| \leq \|x\|$ for all $k \geq 1$, such that $x = \lim_{k\to\infty} x_k$ in the topology defined by the norm $\|\cdot\|_2$. It follows that $E_{B_{j+1}}(x) = \lim_{k\to\infty} E_{B_{j+1}}(x_k) \in B_j$. Thus $E_{B_{j+1}}(A_j) \subset B_j$ and this proves (iii). #

Corollary 4.2.3. *Consider a commuting square* $\begin{matrix} C_1 & \subset & B_1 \\ \cup & & \cup \\ C_0 & \subset & B_0 \end{matrix}$ *with respect to a trace* tr *which is a Markov trace for the pair* $B_0 \subset B_1$. *Let* $B_2 = \langle B_1, e_1 \rangle$ *be the von Neumann algebra obtained via the fundamental construction for* $B_0 \subset B_1$, *and let* $C_2 = \{C_1, e_1\}''$.

Then $\begin{matrix} C_2 & \subset & B_2 \\ \cup & & \cup \\ C_1 & \subset & B_1 \end{matrix}$ *is also a commuting square.*

Proof. This is the special case of 4.2.2 applied to $\begin{matrix} N=B_0 & \subset & M=B_1 & \subset & B_2 \\ \cup & & \cup & & \\ C_0 & \subset & C_1 & \subset & C_2 \end{matrix}$. #

Remark. Suppose moreover that B_0 and B_1 have finite dimensional centers. Then the fundamental construction iterates to give the tower $(B_j)_{j\geq 0}$ with $B_{j+1} = \langle B_j, e_j \rangle$ for all j. Define inductively $C_{j+1} = \{C_j, e_j\}''$ for $j \geq 1$. Then we obtain a ladder of commuting squares $\begin{matrix} C_{j+1} & \subset & B_{j+1} \\ \cup & & \cup \\ C_j & \subset & B_j \end{matrix}$. We are going to use this idea to construct examples of subfactors below, starting with a commuting square of finite dimensional algebras. The next two lemmas concern conditions which cause the inclusion matrices for the resulting ladder of finite dimensional algebras to be repeated with period 2.

Lemma 4.2.4. *Consider a commuting square* $\begin{matrix} C_1 & \subset & B_1 \\ \cup & & \cup \\ C_0 & \subset & B_0 \end{matrix}$ *of finite dimensional von Neumann algebras, with respect to some trace* tr *on* B_1. *Let* $B_2 = \langle B_1, e_1 \rangle$ *be the finite dimensional von Neumann algebra obtained via the fundamental construction for* $B_0 \subset B_1$ *and let* $C_2 = \mathrm{alg}\{C_1, e_1\}$.

Suppose that $C_2 = C_1 e_1 C_1$ *(or equivalently, by 2.6.9, that* $\sum x_i e_{C_0} y_i \overset{\sigma}{\longmapsto} \sum x_i e_1 y_i$ *is an isomorphism from the algebra* $\langle C_1, e_{C_0} \rangle$ *obtained by the fundamental construction for* $C_0 \subset C_1$ *onto* C_2*). Then*

(i) $\Lambda_{C_2}^{B_2} = \Lambda_{C_0}^{B_0}$. *More exactly, let* q, p *be minimal central projections in* C_0, B_0 *respectively. Let* $\tilde{q} = \sigma(J_{C_1} q J_{C_1})$ *and* $\tilde{p} = J_{B_1} p J_{B_1}$ *be the corresponding minimal central projections in* C_2, B_2 *respectively. Then* $[(B_0)_{pq} : (C_0)_{pq}] = [(B_2)_{\tilde{p}\tilde{q}} : (C_2)_{\tilde{p}\tilde{q}}]$.

Suppose in addition that tr *is a Markov trace with respect to* $B_0 \subset B_1$. *Let* $(B_j)_{j\geq 0}$ *be the tower obtained by iterating the fundamental construction for* $B_0 \subset B_1$, *with* $B_{j+1} = \langle B_j, e_j \rangle$, *and let* $C_{j+1} = \mathrm{alg}\{C_j, e_j\}$ *for all* $j \geq 1$. *Then*

(ii) *For all* $j \geq 1$, $C_j e_j C_j = C_{j+1}$ *and* $\Lambda_{C_{j+1}}^{B_{j+1}} = \Lambda_{C_{j-1}}^{B_{j-1}}$. *The inclusion matrices for* $C_{j-1} \subset C_j$ *are alternately* $\Lambda_{C_0}^{C_1}$ *and* $\Lambda_{C_1}^{C_2} = (\Lambda_{C_0}^{C_1})^t$.

Proof. (i) Let f be a minimal projection in $(C_0)q$ and let $pf = \sum_{i=1}^{n} g_i$ be a decomposition of pf into orthogonal minimal projections in $(B_0)p$ (so $n = [(B_0)_{pq} : (C_0)_{pq}]^{1/2}$). Then (by 2.6.4) $fe_1 = \sigma(fe_{C_0})$ is a minimal projection in $C_2\tilde{q}$ and

$$\begin{aligned} fe_1\tilde{p} &= fe_1 p && \text{(by 3.6.9)} \\ &= fpe_1 && \text{(because } p \in B_0) \\ &= pfe_1 && \text{(because } p \in Z(B_0)) \\ &= \sum_{i=1}^{n} g_i e_1. \end{aligned}$$

Thus $(fe_1)\tilde{p}$ is a sum of n orthogonal minimal projections in $(B_2)_{\tilde{p}\tilde{q}}$.

(ii) We are now supposing that tr is a Markov trace. The statement $C_{j+1} = C_j e_j C_j$ is valid for $j = 1$ by hypothesis. Suppose it is valid for some j. Then $C_{j+1}e_{j+1}C_{j+1}$ is an ideal in C_{j+2} containing $\beta e_j e_{j+1} e_j = e_j$, where β is the modulus of the Markov trace. Then $C_{j+1}e_{j+1}C_{j+1} \supset C_j e_j C_j \ni \mathbb{1}$, so $C_{j+1}e_{j+1}C_{j+1} = C_{j+2}$.

It follows that for all j, the tower $C_{j-1} \subset C_j \subset C_{j+1}$ is isomorphic to $C_{j-1} \subset C_j \subset \mathrm{End}_{C_{j-1}}(C_j)$, so the inclusion matrices $\Lambda_{C_{j-1}}^{C_j}$ are alternately $\Lambda_{C_0}^{C_1}$ and $(\Lambda_{C_0}^{C_1})^t$. Finally the statement regarding $\Lambda_{C_j}^{B_j}$ follows from (i) and induction. #

Lemma 4.2.5. *Consider a commuting square* $\begin{matrix} C_1 & \subset & B_1 \\ \cup & & \cup \\ C_0 & \subset & B_0 \end{matrix}$ *of finite dimensional von Neumann algebras, with respect to a trace* tr *on* B_1. *Suppose* $\Lambda_{C_0}^{C_1} = \Lambda_{C_0}^{B_0} = \Lambda^t$ *and* $\Lambda_{B_0}^{B_1} = \Lambda_{C_1}^{B_1} = \Lambda$ *for some* Λ. *Let* $B_2 = \langle B_1, e_1 \rangle$ *be the algebra obtained via the fundamental construction for* $B_0 \subset B_1$ *and let* $C_2 = \mathrm{alg}\{C_1, e_1\}$. *Then*

(i) $C_2 = C_1 e_1 C_1$, $\Lambda_{C_1}^{C_2} = \Lambda$, *and* $\Lambda_{C_2}^{B_2} = \Lambda^t$.

Suppose in addition that tr *is a Markov trace with respect to* $B_0 \subset B_1$. *Let* $(B_j)_{j\geq 0}$ *be the tower obtained by iterating the fundamental construction*, with $B_{j+1} = \langle B_j, e_j \rangle$, *and* set $C_{j+1} = \mathrm{alg}\{C_j, e_j\}$ *for all* $j \geq 1$. *Then*

(ii) *The chain* $C_{j-1} \subset C_j \subset C_{j+1}$ *is isomorphic to* $C_{j-1} \subset C_j \subset \mathrm{End}_{C_{j-1}}(C_j)$ *for all* j. *The inclusion matrices* $\Lambda_{C_j}^{C_{j+1}}$ *are alternately* Λ^t *and* Λ $(j \geq 0)$, *and the inclusion matrices* $\Lambda_{C_j}^{B_j}$ *are alternately* Λ^t *and* Λ $(j \geq 0)$.

Proof. (i) We have $C_2 = C_1 e_1 C_1 \oplus K$ and $\Lambda_{C_1}^{C_2} = \begin{bmatrix} \Lambda \\ \Omega_1 \end{bmatrix}$ for some matrix Ω_1, by 2.6.9. Also $\Lambda_{C_1 e_1 C_1}^{B_2} = \Lambda^t$ by the argument of 4.2.4(i), so $\Lambda_{C_2}^{B_2} = \begin{bmatrix} \Lambda^t & \Omega_2 \end{bmatrix}$ for some Ω_2. Therefore $\Lambda_{C_1}^{B_2} = \Lambda_{C_2}^{B_2} \Lambda_{C_1}^{C_2} = \Lambda^t \Lambda + \Omega_2 \Omega_1$. On the other hand $\Lambda_{C_1}^{B_2} = \Lambda_{C_1}^{B_2} \Lambda_{C_1}^{B_1}$ $= \Lambda^t \Lambda$. This is only possible if $K = (0)$, because otherwise $\Omega_2 \Omega_1 \neq 0$. The remainder of (i) and (ii) now follows from the previous lemma. #

The next result is that commuting squares are preserved under reduction by certain projections.

Proposition 4.2.6. *Consider a commuting square* $\begin{matrix} C_1 & \subset & B_1 \\ \cup & & \cup \\ C_0 & \subset & B_0 \end{matrix}$ *with respect to a trace* tr *on* B_1 *and a projection* $p \in B_0 \cap C_1'$, *not zero. Then*

$$\begin{matrix} pC_1 & \subset & pB_1p \\ \cup & & \cup \\ pC_0 & \subset & pB_0p \end{matrix}$$

is a commuting square with respect to $\mathrm{tr}|_{pB_1p}$.

Proof. Let $y \in pB_1p$. Then $E_{pB_0p}(y) = pE_{B_0}(y)p$ because one has $\mathrm{tr}(pE_{B_0}(y)pu) = \mathrm{tr}(pypu) = \mathrm{tr}(yu)$ for all $u \in pB_0p$. Consider $z \in pC_1$, say $z = pc$, with $c \in C_1$. Then

$$E_{pB_0p}(z) = pE_{B_0}(pcp)p = pE_{B_0}(c)p \in pC_0$$

and the claim follows. #

Remark. A similar result holds for reduction by projections in C_0.

Next we give some examples of commuting squares involving relative commutants, fixed–point algebras of groups, and crossed–products.

Proposition 4.2.7. *Let* $N \subset M$ *be a pair of von Neumann algebras, let* tr *be a finite faithful normal trace on* M, *and let* S *be a self–adjoint subset of* N. *Then*

$$\begin{array}{ccc} S' \cap M & \subset & M \\ \cup & & \cup \\ S' \cap N & \subset & N \end{array}$$

is a commuting square.

Proof. We may suppose that S is a von Neumann subalgebra of N. Choose $x \in M$. Denote by C the $\|\cdot\|_2$–closure of the convex hull of $\{uxu^* : u$ is unitary and $u \in S\}$ in $L^2(M,tr)$, and denote by y the projection of the origin onto C. Then $y \in M$ because the ball of radius $\|x\|$ in M is a $\|\cdot\|_2$–closed subset of $L^2(M,tr)$. Moreover, by the uniqueness of the projection onto a closed convex set, $uyu^* = y$ for any unitary $u \in S$. It follows that y is also in S'.

For any $z \in S' \cap M$ and for any unitary $u \in S$, one has $tr(uxu^*z) = tr(xu^*zu) = tr(xz)$, so that $E_{S'\cap M}(uxu^*) = E_{S'\cap M}(x)$. Consequently $E_{S'\cap M}(C) = E_{S'\cap M}(x)$, and $y = E_{S'\cap M}(y) = E_{S'\cap M}(x)$. In particular, if $x \in N$, then $C \subset N$ and $E_{S'\cap M}(x) = y \in S' \cap N$. #

Proposition 4.2.8. *Let* M *be a von Neumann algebra given with a finite faithful normal trace* tr. *Let* $G = H \rtimes K$ *be a semi–direct product group which acts on* M *and preserves* tr. *Assume that* K *is a compact group and that the restricted action of* K *on* M *is continuous. Denote by* M^G *the algebra of vectors in* M *fixed by* G, *and similarly for* M^H *and* M^K. *Then*

$$\begin{array}{ccc} M^K & \subset & M \\ \cup & & \cup \\ M^G & \subset & M^H \end{array}$$

is a commuting square.

Proof. For each $x \in M$, one has

$$E_{M^K}(x) = \int_K k(x)dk.$$

Suppose moreover that $x \in M^H$. Then $k(x) \in M^H$ for any $k \in K$, so that $E_{M^K}(x) \in M^H \cap M^K = M^G$. #

We leave it to the reader to formulate the details of a proposition involving the diagram

$$\begin{array}{ccc} M \rtimes H & \subset & M \rtimes G \\ \cup & & \cup \\ M & \subset & M \rtimes K \end{array}$$

where $\rtimes$ indicates now a crossed product.

We next describe three examples which are interesting in light of the connections between the theory of subfactors and that of the braid groups.

Example 4.2.9. Let $e_1, \cdots, e_n$ be a sequence of projections acting on some Hilbert space such that

$$\begin{array}{ll} \beta\, e_i e_j e_i = e_i & \text{if } |i-j| = 1 \\ e_i e_j = e_j e_i & \text{if } |i-j| \geq 2 \end{array}$$

for some real number $\beta \geq 1$ (see the last remark of Appendix IIc). Let tr be a normalized faithful trace on the algebra generated by the identity and the e_j's, and assume that the Markov relation

$$\beta\, \mathrm{tr}(we_j) = \mathrm{tr}(w), \quad w \in \mathrm{alg}\{\mathbb{1}, e_1, \cdots, e_{j-1}\}, \quad 1 \leq j \leq n$$

holds (see Section 3.4). Then the diagram

$$\begin{array}{ccc} C_1 = \mathrm{alg}\{\mathbb{1}, e_2, \cdots, e_{n-1}, e_n\} & \subset & B_1 = \mathrm{alg}\{\mathbb{1}, e_1, e_2, \cdots, e_{n-1}, e_n\} \\ \cup & & \cup \\ C_0 = \mathrm{alg}\{\mathbb{1}, e_2, \cdots, e_{n-1}\} & \subset & B_0 = \mathrm{alg}\{\mathbb{1}, e_1, e_2, \cdots, e_{n-1}\} \end{array}$$

is a commuting square.

Proof. Let us show that $E_{B_0}(x) \in C_0$ for any $x \in C_1$. This is obvious when $x \in C_0$. By Proposition 2.8.1, one may then assume without loss of generality that $x = ye_n z$ with $y,z \in C_0$. As $E_{B_0}(e_n) = \beta^{-1}$ (see the proof of 4.2.2.iii), one has $E_{B_0}(ye_n z) = y\beta^{-1}z \in C_0$.

#

Example 4.2.10. Let $N \subset M$ be a connected pair of finite von Neumann algebras with finite dimensional centers, of finite index (Definition 3.5.3). Let tr be the normalized Markov trace on $N \subset M$ (Corollary 3.7.4.i), and let $\beta = [M:N]$ be its modulus (Definition 3.7.5). Then tr has an extension to $\langle M,e_N\rangle$ which is again a Markov trace of modulus β on $M \subset \langle M,e_N\rangle$ (Corollary 3.7.4.ii), and that we denote by tr again.

Suppose moreover that $\beta \leq 4$, write $\beta = 2 + q + q^{-1}$, define

$$g = qe_N - (\mathbb{1}-e_N)$$

and observe that g is a unitary which commutes with N. Then

$$\begin{array}{ccc} gMg^{-1} & \subset & \langle M,e_N\rangle \\ \cup & & \cup \\ N & \subset & M \end{array}$$

is a commuting square.

Proof. Let $x \in gMg^{-1}$. If $y = g^{-1}xg \in M$, one has

$$x = \{(q+1)e_N - \mathbb{1}\}y\{(q^{-1}+\mathbb{1})e_N - \mathbb{1}\}.$$

Since $E_M(e_N) = \beta^{-1}$, we have

$$\begin{aligned} E_M(x) &= \beta\, E_M(e_N y e_N) + \{1 - (q+1)\beta^{-1} - (q^{-1}+1)\beta^{-1}\}y \\ &= \beta\, E_M(E_N(y)e_N) = E_N(y). \qquad \# \end{aligned}$$

Remarks.

(1) Up to scalars, g and g^{-1} are the only unitaries in $\mathrm{alg}\{\mathbb{1},e_N\}$ for which the above construction works. Observe that g is precisely the element involved in the braid group representation of [Jo2].

(2) This example is the basis for the examples of Section 4.4 below.

Example 4.2.11. Let $N \subset M$ be a pair of factors, of finite index β, and let tr denote the normalized trace on M. Assume that there exists a projection $e_0 \in M$ such that

$$e_0 \text{ and } N \text{ generate } M$$
$$tr(e_0 y) = \beta\, tr(y) \text{ for all } y \in N.$$

Let $(M_j)_{j\geq 0}$ be the tower and let $(e_j)_{j\geq 1}$ be as usual. (See Section 3.4; of course $M_1 = M$.) Let M_∞ denote the von Neumann algebra generated by $\bigcup_{j\geq 0} M_j$. Then

$$\begin{array}{ccc} \{\mathbb{1},e_0,e_1,\cdots\}'' & \subset & M_\infty \\ \cup & & \cup \\ \mathbb{C} & \subset & N \end{array}$$

is a commuting square.

Proof. We want to check that $tr(xy) = tr(x)tr(y)$ for all $x \in \{\mathbb{1},e_0,e_1,\cdots\}''$ and for all $y \in N$. Because of the density theorem of Kaplansky (see the proof of Proposition 4.2.2.iii), we may check this for all $x \in alg\{\mathbb{1},e_0,\cdots,e_n\}$ and for all $n \geq 0$. If $n = 0$, this follows from the hypothesis on e_0. To end the proof, we may assume that $n \geq 1$ and that the claim holds up to $n-1$.

For $a_\alpha, b_\alpha \in alg\{\mathbb{1},e_0,\cdots,e_{n-1}\}$ and $x = \sum_\alpha a_\alpha e_n b_\alpha$,

$$tr(xy) = tr(\sum_\alpha a_\alpha e_n b_\alpha y) = \beta^{-1} tr(\sum_\alpha a_\alpha b_\alpha y)$$

which is by induction

$$\beta^{-1} tr(\sum_\alpha a_\alpha b_\alpha) tr(y) = tr(x)tr(y).$$

This shows that the claim holds up to n. #

Remark. It would be interesting to have a systematic classification of commuting squares

$$\begin{array}{ccc} C_1 & \subset & (B_1,tr) \\ \cup & & \cup \\ C_0 & \subset & B_0 \end{array}$$

of finite dimensional von Neumann algebras.

4.3 Wenzl's index formula.

In this section we prove a formula due to H. Wenzl [Wen 2] for the index of a pair of factors generated by a ladder of commuting squares. The set up is as follows: We are given a chain $(B_j)_{j\geq 0}$ of finite dimensional von Neumann algebras and a faithful tracial state tr on $B_\infty = \bigcup_j B_j$. Since the GNS representation π of tr (on $\mathcal{H}_\pi$) is faithful, we regard B_∞ as a subalgebra of $B = \pi(\bigcup_j B_j)''$, a finite hyperfinite von Neumann algebra. We suppose we have a chain $(C_j)_{j\geq 0}$ of finite dimensional von Neumann algebras such that $\mathbb{1} \in C_j \subset B_j$ and:

Hypothesis (A). For each j, $\begin{matrix} C_{j+1} & \subset & B_{j+1} \\ \cup & & \cup \\ C_j & \subset & B_j \end{matrix}$ is a commuting square.

Then $C = (\bigcup_j C_j)''$ is a von Neumann subalgebra of B. In the *periodic* case which we consider below, tr is the unique tracial state on $\bigcup_j C_j$ and $\bigcup_j B_j$, so that C and B are factors.

If $E : B \to C$ and $E_j : B_j \to C_j$ denote the conditional expectations with respect to tr, then $E|_{B_j} = E_j$; that is $\begin{matrix} C & \subset & B \\ \cup & & \cup \\ C_j & \subset & B_j \end{matrix}$ is a commuting square for each j. Let $A = \langle B,e\rangle$ be the result of the fundamental construction for $C \subset B$ with respect to tr, and let $A_j = \{B_j,e\}''$ for each j. Then A_j is an E_j-extension of B_j, in the terminology of Section 2.6. Hence if $\langle B_j,f_j\rangle$ is the result of the fundamental construction for $C_j \subset B_j$, then the formula $\sigma_j(\sum_i a_i f_j b_i) = \sum_i a_i e b_i$ $(a_i,b_i \in B_j)$ defines an isomorphism from $\langle B_j,f_j\rangle$ onto the two sided ideal B_jeB_j generated by e in A_j, by 2.6.9.

Lemma 4.3.1.

(i) *The central support* z_j *of* e *in* A_j *is* $\sigma_j(\mathbb{1})$; *this is also the central support of the ideal* B_jeB_j.

(ii) $\lim_{j\to\infty} z_j = \mathbb{1}$ *in the strong operator topology.*

Proof. (i) This is straightforward, since the central support of f_j in $\langle B_j, f_j\rangle$ is $\mathbb{1}$, by 3.6.1(vi).

(ii) Since $(\bigcup_j A_j)'' = A$ and the central support of e in A is $\mathbb{1}$, one has $\bigvee_j [A_j e\mathcal{X}_\pi] = [Ae\mathcal{X}_\pi] = \mathcal{X}_\pi$. That is, z_j increases to $\mathbb{1}$. #

Next we introduce a very strong periodicity assumption on the inclusion data for the ladder of inclusions
$$\begin{array}{ccc} C_{j+1} & \subset & B_{j+1} \\ \cup & & \cup \\ C_j & \subset & B_j \end{array}.$$

Hypothesis (B). We assume there is a $j_0 \geq 0$ and a $p \geq 1$ and a suitable ordering of the factors in the B_j's and C_j's such that for all $j \geq j_0$:

(i) The inclusion matrix for $B_j \subset B_{j+1}$ is the same as that for $B_{j+p} \subset B_{j+p+1}$. Similarly for $C_j \subset C_{j+1}$ and $C_{j+p} \subset C_{j+p+1}$.

(ii) The inclusion matrices Φ_j for $B_j \subset B_{j+p}$ and Ψ_j for $C_j \subset C_{j+p}$ are primitive.

(iii) The inclusion matrix Λ_j for $C_j \subset B_j$ is the same as that for $C_{j+p} \subset B_{j+p}$.

We remark that under this hypothesis, tr is the unique tracial state on $\bigcup_j C_j$ and $\bigcup_j B_j$. In fact the trace vector $\overleftarrow{s}^{(j)}$ for B_j [resp. $\overleftarrow{t}^{(j)}$ for C_j] is a Perron–Frobenius (row) vector for Φ_j [resp. Ψ_j] for $j \geq j_0$, because $\overleftarrow{s}^{(j)} = \overleftarrow{s}^{(j+\ell p)}\Phi_j^\ell$ for all $\ell \geq 0$. Furthermore the dimension vectors $\vec{\mu}^{(j)}$ of B_j [resp. $\vec{\nu}^{(j)}$ of C_j] approach Perron–Frobenius eigenvectors of Φ_j [resp. Ψ_j]. More precisely, if φ_j is the spectral radius of Φ_j, then $\lim_{\ell\to\infty} \vec{\mu}^{(j+\ell p)}/\varphi_j^\ell = \lim_{\ell\to\infty}(\Phi_j^\ell \vec{\mu}^{(j)})/\varphi_j^\ell$ exists, and is a Perron–Frobenius eigenvector for Φ_j, and similarly for the vectors $\vec{\nu}^{(j)}$.

Lemma 4.3.2. *Assuming hypotheses* (A) *and* (B),

(i) B *and* C *are factors and* $[B:C] < \infty$.

(ii) $\overleftarrow{t}^{(j)}\Lambda_j^t\Lambda_j \leq [B:C]\,\overleftarrow{t}^{(j)}$ *for all* j, *the inequality holding component-by-component.*

(iii) *If* z_k *is the central support of* e *in* A_k *and* ψ_j *denotes the spectral radius of* Ψ_j, *then for* $j \geq j_0$,

$$\mathrm{tr}(\mathbb{1} - z_{j+\ell p}) = \langle \overleftarrow{t}^{(j)} - [B:C]^{-1}\overleftarrow{t}^{(j)}\Lambda_j^t\Lambda_j,\ \psi_j^{-\ell}\vec{\nu}^{(j+\ell p)}\rangle.$$

Proof. (i) That B and C are factors follows from the uniqueness of the trace on $\bigcup_j B_j$ and $\bigcup_j C_j$. We have to show that $A = \langle B,e \rangle$ is a finite factor. In any case A is semi-finite, so has a faithful normal semi-finite trace Tr; we have to show that $\mathrm{Tr}(\mathbb{1}) < \infty$. Now $eAe = Ce$ is isomorphic to C, which is a finite factor, so e is a finite projection and $\mathrm{Tr}(e) < \infty$. Fix some $j \geq j_0$ and let q_i be a minimal central projection in C_j, and $\tilde{q}_i = J_{B_j} q_i J_{B_j}$ and $\bar{q}_i = \sigma_j(\tilde{q}_i)$ the corresponding minimal central projections in $\langle B_j, f_j \rangle$ and $B_j e B_j = z_j A_j$. Then

$$\frac{\mathrm{Tr}(e\bar{q}_i)}{\mathrm{Tr}(\bar{q}_i)} = \frac{\mathrm{Tr}(eq_i)}{\mathrm{Tr}(\bar{q}_i)} \qquad \text{(using 3.6.9)}$$

$$= \nu_i^{(j)} / (\Lambda_j^t \Lambda_j \vec{\nu}^{(j)})_i,$$

because eq_i is the sum of $\nu_i^{(j)}$ minimal projections in $\bar{q}_i A_j$ (by 2.6.4) while $\bar{q}_i$ is the sum of $(\Lambda_j^t \Lambda_j \vec{\nu}^{(j)})_i$ minimal projections in $\bar{q}_i A_j$. Let $d_j = \min_i \{\nu_i^{(j)} / ((\Lambda_j^t \Lambda_j \vec{\nu}^{(j)})_i\}$. Then we have $d_j > 0$ and

$$\mathrm{Tr}(e) = \mathrm{Tr}(ez_j) = \sum_i \mathrm{Tr}(e\bar{q}_i) \geq d_j \sum_i \mathrm{Tr}(\bar{q}_i) = d_j \mathrm{Tr}(z_j).$$

Since $\vec{\nu}^{(j+\ell p)} / \psi_j^\ell$ converges to a Perron–Frobenius vector for Ψ_j, it follows that $d := \inf_{\ell \geq 0} d_{j+\ell p}$ is positive. We have

$$\mathrm{Tr}(e) \geq d_{j+\ell p} \mathrm{Tr}(z_{j+\ell p}) \geq d\, \mathrm{Tr}(z_{j+\ell p})$$

for all ℓ, and since $\lim_\ell z_{j+\ell p} = \mathbb{1}$, it follows that $\mathrm{Tr}(e) \geq d\, \mathrm{Tr}(\mathbb{1})$, and Tr is finite.

Since $\beta = [B:C] < \infty$, the normalized trace tr on A has the Markov property: $\mathrm{tr}(ex) = \beta^{-1} \mathrm{tr}(x)$ for $x \in B$. It follows from this and 2.6.4(c) that the weight vector of tr on $B_j e B_j = z_j A_j$ is $\beta^{-1} \vec{t}^{(j)}$.

(ii) It follows from 2.4.1(b) and 2.6.9 that the inclusion matrix of $B_j \subset A_j$ is of the form $\begin{bmatrix} \Lambda_j^t \\ \Omega_j \end{bmatrix}$ for some Ω_j, Λ_j^t being the inclusion matrix of $B_j \subset z_j A_j$. By the remark above, the weight vector of tr on A_j has the form $(\beta^{-1} \vec{t}^{(j)}, \vec{r}^{(j)})$, so that

$$\overleftarrow{t}^{(j)} = \overleftarrow{s}^{(j)}\Lambda_j = (\beta^{-1}\overleftarrow{t}^{(j)}, \overleftarrow{r}^{(j)})\begin{bmatrix}\Lambda_j^t\\ \Omega_j\end{bmatrix}\Lambda_j \geq \beta^{-1}\overleftarrow{t}^{(j)}\Lambda_j^t\Lambda_j.$$

(iii) First $\mathrm{tr}(z_j) = \langle \beta^{-1}\overleftarrow{t}^{(j)}, \Lambda_j^t\Lambda_j\vec{\nu}^{(j)}\rangle$, since the trace and dimension vectors on z_jA_j are respectively $\beta^{-1}\overleftarrow{t}^{(j)}$ and $\Lambda_j^t\Lambda_j\vec{\nu}^{(j)}$. Hence

(4.3.2.1) $$\mathrm{tr}(\mathbb{1}-z_j) = \langle \overleftarrow{t}_j - \beta^{-1}\overleftarrow{t}^{(j)}\Lambda_j^t\Lambda_j, \vec{\nu}^{(j)}\rangle.$$

Now apply this formula to $z_{j+\ell p}$ and use that $\overleftarrow{t}^{(j)} = \overleftarrow{t}^{(j+\ell p)}\Psi_j^\ell = \psi_j^\ell\, \overleftarrow{t}^{(j+\ell p)}$ and that $\Lambda_{j+\ell p} = \Lambda_j$ to get

(4.3.2.2) $$\begin{aligned}\mathrm{tr}(\mathbb{1}-z_{j+\ell p}) &= \langle \overleftarrow{t}^{(j+\ell p)} - \beta^{-1}\overleftarrow{t}^{(j+\ell p)}\Lambda_{j+\ell p}^t\Lambda_{j+\ell p}, \vec{\nu}^{(j+\ell p)}\rangle \\ &= \langle \overleftarrow{t}^{(j)} - \beta^{-1}\overleftarrow{t}^{(j)}\Lambda_j^t\Lambda_j, \psi_j^{-\ell}\, \vec{\nu}^{(j+\ell p)}\rangle. \qquad \#\end{aligned}$$

<u>Theorem 4.3.3.</u> *Assuming hypotheses* (A) *and* (B),

$$[B:C] = \|\overleftarrow{t}^{(j)}\|^2/\|\overleftarrow{s}^{(j)}\|^2 = \|\Lambda_j\|^2$$

for all $j \geq j_0$.

<u>Proof</u>. Fix $j \geq j_0$ and consider the formula (4.3.2.2) for $\mathrm{tr}(\mathbb{1}-z_{j+\ell p})$. Since $\lim_{\ell\to\infty} \mathrm{tr}(\mathbb{1}-z_{j+\ell p}) = 0$ and since $\psi_j^{-\ell}\, \vec{\nu}^{(j+\ell p)}$ converges to a strictly positive vector while $\overleftarrow{t}^{(j)} - \beta^{-1}\overleftarrow{t}^{(j)}\Lambda_j^t\Lambda_j$ is a non-negative vector, we must have $\overleftarrow{t}^{(j)} - \beta^{-1}\overleftarrow{t}^{(j)}\Lambda_j^t\Lambda_j = 0$. Therefore $z_j = \mathbb{1}$ by 4.3.2.1, and furthermore $\overleftarrow{t}^{(j)}$ is a Perron–Frobenius eigenvector for $\Lambda_j^t\Lambda_j$ with eigenvalue β, whence $\|\Lambda_j\|^2 = \beta$. Finally Λ_j^t is the inclusion matrix for $B_j \subset A_j$ and $\beta^{-1}\overleftarrow{t}^{(j)}$ is the trace vector of A_j; consequently $\overleftarrow{s}^{(j)} = \beta^{-1}\overleftarrow{t}^{(j)}\Lambda_j^t$ is a Perron–Frobenius eigenvector for $\Lambda_j\Lambda_j^t$ and

$$\|\overleftarrow{t}^{(j)}\|^2 = \langle \overleftarrow{s}^{(j)}\Lambda_j, \overleftarrow{s}^{(j)}\Lambda_j\rangle = \langle \overleftarrow{s}^{(j)}\Lambda_j\Lambda_j^t, \overleftarrow{s}^{(j)}\rangle = \beta\|\overleftarrow{s}^{(j)}\|^2. \qquad \#$$

<u>Remarks.</u>

(1) If it is known *a priori* that B and C are factors with $[B:C] < \infty$, Wenzl can obtain the index formula assuming only periodicity of the inclusion data for $(C_j)_{j\geq 0}$.

(2) We used periodicity of the inclusion data for $(B_j)_{j\geq 0}$ only to obtain that B is a factor.

4.4. Examples of irreducible pairs of factors of index less than 4, and a lemma of C. Skau.

We have shown in Chapter 2 that there are connected pairs $N \subset M$ of finite dimensional von Neumann algebras of index $4\cos^2(\pi/q)$ for any integer $q \geq 3$. One of the main results of [Jo1] shows that there are pairs of *factors* with the same indices. In the present section, we give a construction for such pairs which has been sketched in [Jo8]. These pairs are necessarily irreducible by 3.6.2(c). (Recall that $N \subset M$ is irreducible if the only elements of M which commute with N are the scalar multiples of the identity.)

We also present a theorem of C. Skau regarding irreducibility of certain subfactors.

Consider a connected pair $N \subset M$ of finite dimensional von Neumann algebras with inclusion matrix Λ, set $\beta = [M:N] = \|\Lambda\|^2$, and assume that $2 \leq \beta \leq 4$. Let tr denote both the normalized Markov trace of modulus β on this pair (see Theorem 2.7.3) and its Markov extension to $\langle M,e_N\rangle$. Set $g = qe_N - (\mathbb{1}-e_N)$, with $\beta = 2+q+q^{-1}$, and consider the commuting square

$$\begin{array}{ccc} C_1 = gMg^{-1} & \subset & \langle M,e_N\rangle = B_1 \\ \cup & & \cup \\ C_0 = N & \subset & M = B_0 \end{array}$$

of Example 4.2.10. Define inductively for each $j \geq 1$:

(i) The conditional expectation $B_j \to B_{j-1}$, denoted by e_j when viewed as an operator on $L^2(B_j,\mathrm{tr})$.

(ii) The algebra $B_{j+1} = \langle B_j,e_j\rangle$, together with the Markov extension of tr from B_j to B_{j+1}, again denoted by tr.

(iii) The subalgebra $C_{j+1} = \mathrm{alg}\{C_j,e_j\}$ generated in B_{j+1} by C_j and e_j.

These data satisfy Hypothesis (A) of section 4.3 because of Corollary 4.2.3. Before checking that these data also satisfy the periodicity Hypothesis (B), we need a preliminary proposition of independent interest.

Proposition 4.4.1. *There exists an endomorphism* Φ *of* B *with* $\Phi(B_j) = C_{j+1}$ *for* $j \geq 0$, *and consequently with* $\Phi(B) = C$. *Set* $e_0 = e_N$ *and* $g_0 = g$. *For* $j \geq 1$, *set* $g_j = (q+1)e_j - \mathbb{1}$, *so that* $g_jg_{j+1}g_j = g_{j+1}g_jg_{j+1}$ *for* $j \geq 0$. *Then*

$$\Phi(x) = \lim_{k\to\infty} g_0g_1\cdots g_{k-1}g_kxg_k^{-1}g_{k-1}^{-1}\cdots g_1^{-1}g_0^{-1}$$

for all $x \in B_\infty = \bigcup_{j\geq 0} B_j$.

Proof. Let $j \geq 0$ and let $x \in B_j$. The formula for $\Phi(x)$ makes sense, because $x \in B_j$ commutes with e_k, and thus also with g_k, for $k \geq j+1$. Observe for example that $g_{j-1} \in B_j$ so that one has, by using the braid relations:

$$\Phi(g_{j-1}) = g_0 \cdots g_{j-2}(g_{j-1}g_jg_{j-1}g_j^{-1}g_{j-1}^{-1})g_{j-2}^{-1} \cdots g_0^{-1}$$

$$= g_0 \cdots g_{j-2}g_jg_{j-2}^{-1} \cdots g_0^{-1} = g_j.$$

On $\bigcup_{j\geq 0} B_j$, the map Φ is clearly a $*$-endomorphism which preserves the trace and also the norm. Consequently this map extends to a (σ-weakly continuous) $*$-endomorphism of B, denoted by Φ again. As $\bigcup_{j\geq 0} C_j$ is strongly dense in C, the only thing left to be proved is that $\Phi(B_j) = C_{j+1}$ for $j \geq 0$.

For $j = 0$, this holds by definition of C_1. For $j \geq 1$ one has

$$B_j = \mathrm{alg}\{B_0,e_0,e_1,\cdots,e_{j-1}\} = \mathrm{alg}\{B_0,g_0,g_1,\cdots,g_{j-1}\}$$

and consequently, using the formula for $\Phi(g_k)$,

$$\Phi(B_j) = \mathrm{alg}\{C_1,g_1,\cdots,g_j\} = \mathrm{alg}\{C_1,e_1,\cdots,e_j\} = C_{j+1}$$

as wanted. #

Remark. For any complex number ω of absolute value 1 and for $j \geq 0$, set $g_j(\omega) = \omega e_j - (\mathbb{1} - e_j)$, and for $x \in \bigcup_{j\geq 0} B_j$, define

$$\Phi_\omega(x) = \lim_{k\to\infty} g_0(\omega)\cdots g_k(\omega)xg_k(\omega)^{-1}\cdots g_0(\omega)^{-1}.$$

The same argument as above shows that this defines an endomorphism Φ_ω of B, and the map

$$\begin{cases} \{\omega \in \mathbb{C} \mid |\omega| = 1\} \longrightarrow \mathrm{End}(B) \\ \qquad \omega \qquad\qquad \longmapsto \Phi_\omega \end{cases}$$

is pointwise strongly continuous. Moreover $\Phi_{-1} = \mathrm{id}$ and $\Phi_q = \Phi$. Thus C is connected to B by a continuous family of subfactors. It would be interesting to compute the index $[B : \Phi_\omega(B)]$ as a function of ω. We do it below for $\omega = q$ only.

Now we may check that the Hypothesis (B) of section 3 holds for the data of the present section, with $j_0 = 0$ and a period $p = 2$.

First, the inclusion matrices of $B_j \subset B_{j+1}$ are alternately Λ^t and Λ by Proposition 2.4.1.b, and those of $C_j \subset C_{j+1}$ are Λ and Λ^t by the Proposition above.

Second, as $N \subset M$ is a connected pair, the inclusion matrices for $B_j \subset B_{j+2}$ and $C_j \subset C_{j+2}$ are primitive for $j \geq 0$ by Lemma 1.3.2.

Third, as $\Phi|_{B_{j-1}} = \mathrm{Ad}(g_0 g_1 \cdots g_{j-1})$ for $j \geq 1$, the pair $C_j \subset B_j$ is the image of $B_{j-1} \subset B_j$ by an inner automorphism of B_j, so the inclusion matrices for $C_j \subset B_j$ are alternatively Λ and Λ^t (for $j \geq 0$).

Thus the Hypothesis (B) holds.

Theorem 4.4.2. *With the notation above, if* $\beta < 4$ *then the pair* $C \subset B$ *of factors of type* II_1 *is irreducible, and its index satisfies*

$$[B:C] = \beta \in \{4 \cos^2(\pi/q)\}_{q \geq 4}.$$

Proof. The index value follows from Wenzl's Theorem 4.3.3, and the irreducibility follows from 3.6.2(c). #

Remark. The construction of the pair of factors $C \subset B$ with $[B:C] = \beta$ still makes sense if $\beta = 4$, but the pair need not be irreducible. For example take $M = N \otimes \mathrm{Mat}_2$. Then

$$\begin{array}{ccc} g_0 M g_0^{-1} & \subset & \langle M, e_0 \rangle \\ \cup & & \cup \\ N & \subset & M \end{array}$$

is isomorphic to

$$\begin{array}{ccc} N \otimes \mathbb{1} \otimes \mathrm{Mat}_2 & \subset & N \otimes \mathrm{Mat}_2 \otimes \mathrm{Mat}_2 \\ \cup & & \cup \\ N & \subset & N \otimes \mathrm{Mat}_2 \otimes \mathbb{1} \end{array} .$$

Furthermore $C \subset B$ is isomorphic to $N \otimes \mathbb{1} \otimes (\overset{\infty}{\underset{1}{\otimes}} Mat_2) \subset N \otimes Mat_2 \otimes (\overset{\infty}{\underset{1}{\otimes}} Mat_2)$, so $C' \cap B \cong Mat_2$.

Let now $N \subset M$ be a pair of finite direct sums of finite factors, as in Chapter 3. Assume that N is of finite index in M and let tr be a Markov trace of modulus β on this pair. We consider as usual the tower

$$M_0 = N \subset M_1 = M \subset \cdots \subset M_k \subset M_{k+1} \subset \cdots$$

the projections $(e_k)_{k\geq 1}$ with $e_1 = e_N$ and the Markov extension of the trace on the finite von Neumann algebra M_∞ obtained by GNS–completion of $\bigcup_{j\geq 0} M_j$.

Theorem 4.4.3. (Skau's lemma). *If* $\beta \leq 4$ *then* $\{e_1,e_2,\cdots\}' \cap M_\infty = N$.

Proof. Set $\tilde{N} = \{e_1,e_2,\cdots\}' \cap M_\infty$. As $N \subset \tilde{N}$ is obvious, we have to show the opposite inclusion. For each $k \geq 1$, let F_k be the conditional expectation of M_∞ onto $\{e_k,e_{k+1},\cdots\}' \cap M_\infty$. Observe that $F_k F_\ell = F_{\min(k,\ell)}$. We have to show that $F_1(M_\infty) \subset N$.

It is enough to show that $F_2(M_\infty) \subset M$, because this and Proposition 3.6.1.iii show that $F_1 F_2(M_\infty) \subset N$.

Suppose we know that $F_{k+1}(e_k) \in \mathbb{C}$ for each $k \geq 1$. One has then for $\ell \geq 1$ and for $a,b \in M_\ell$

$$\begin{aligned} F_2(ae_\ell b) &= F_2 F_{\ell+1}(ae_\ell b) = F_2(aF_{\ell+1}(e_\ell)b) \\ &= F_{\ell+1}(e_\ell)F_2(ab) \in F_2(M_\ell). \end{aligned}$$

This implies $F_2(M_{\ell+1}) \subset F_2(M_\ell)$, and this implies in turn by induction that $F_2(M_\infty) \subset F_2(M_1) = M$, so that the proposition is proved.

We still have to prove that $F_{k+1}(e_k) \in \mathbb{C}$. The diagram

$$\begin{array}{ccc} \{e_{k+1},e_{k+2},\cdots\}' \cap M_\infty & \subset & M_\infty \\ \cup & & \cup \\ \{e_{k+1},e_{k+2},\cdots\}' \cap \{e_k,e_{k+1},\cdots\}'' & \subset & \{e_k,e_{k+1},\cdots\}'' \end{array}$$

is a commuting square by Proposition 4.2.7. As the pair $\{e_{k+1},e_{k+2},\cdots\}'' \subset \{e_k,e_{k+1},\cdots\}''$ is isomorphic to the pair $R_\beta \subset R$ of Lemma 3.4.5, we may write the commuting square above as

$$\begin{array}{ccc} R'_\beta \cap M_\infty & \subset & M_\infty \\ \cup & & \cup \\ R'_\beta \cap R & \subset & R \,. \end{array}$$

Let E denote the conditional expectation from R onto $R'_\beta \cap R$, and recall that F_{k+1} is the conditional expectation from M_∞ onto $R'_\beta \cap M_\infty$. As $e_k \in R$, one has $F_{k+1}(e_k) = E(e_k)$.

Now, if $\beta \leq 4$, the relative commutant $R'_\beta \cap R$ is trivial and E is just the trace. (by 3.6.2(c) when $\beta < 4$ and by [Jo1], §5.3 when $\beta = 4$.) Thus $F_{k+1}(e_k) \in \mathbb{C}$, and the proof is over. #

Remark. The condition $\beta \leq 4$ is necessary for Skau's lemma. For any pair $N \subset M$ of II_1-factors with $[M:N] > 4$, we claim that

$$\{e_1,e_2,\cdots\}' \cap M_\infty \underset{\neq}{\supset} N.$$

We will see in Section 4.7.f that there is a projection e_0 and a subfactor P of N such that $M = \langle N,e_0\rangle$ and M is obtained by the fundamental construction for $P \subset N$. Then by 4.7.5,

$$\{e_1,e_2,\cdots\}' \cap \{e_0,e_1,\cdots\}''$$

contains a non-scalar element x. Then $x \in \{e_1,e_2,\cdots\}' \cap M^\infty$ but $x \notin N$ by Example 4.2.11.

4.5. More examples of irreducible pairs of factors, and the index value $3+3^{1/2}$.

Consider a Coxeter graph Γ of finite type in one of the classes A,D,E, with a bicoloration involving m black vertices and n white vertices, and with a distinguished white vertex w_1. We shall associate to these data an irreducible subfactor C of the hyperfinite factor B of type II_1 and we shall compute the index $[B:C]$.

In particular, if $\Gamma = E_6$ with the vertex w_1 chosen as

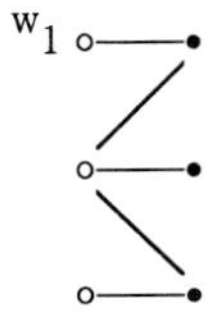

we shall find $[B:C] = 3 + 3^{1/2}$. At the time of writing, this is the smallest known value larger than 4 of the index of an irreducible subfactor.

Let M_0 denote the abelian von Neumann algebra $\mathbb{C}^n$. Let M_1 be a finite dimensional von Neumann algebra containing M_0 such that Γ is the Bratteli diagram of the inclusion $M_0 \subset M_1$. As Γ is connected, there is a unique normalized Markov trace tr on the pair $M_0 \subset M_1$. Form the tower $(M_j)_{j\geq 0}$ and let $(e_j)_{j\geq 1}$ be the usual sequence of projections. Let M be the factor of type II_1 obtained by completion of $\bigcup_{j\geq 0} M_j$ with respect to its unique positive normalized trace and let N be the subfactor of M generated by $\mathbb{1}$ and the e_j's.

Let h be the Coxeter number of Γ and set

$$\beta = [M_1:M_0] = \|\Gamma\|^2 = 4\cos^2(\pi/h).$$

Since $\beta < 4$, Skau's Lemma 4.4.3 applies and $M \cap N' = M_0$.

Let p be the minimal projection of M_0 corresponding to the vertex w_1 and set

$$C = pN \quad B = pMp.$$

(Observe that $pN = pNp$ because p commutes with e_j for each $j \geq 1$.) Then $C \subset B$ is a pair of factors with $C' \cap B = p(N' \cap M)p = pM_0p = \mathbb{C}p$; that is C has trivial commutant in B. Our goal is to compute $[B:C]$.

Number the vertices of Γ so that the distinguished white vertex is w_1. Departing somewhat from previous practice let Γ also denote the matrix of the bicolored graph Γ, which has m rows and n columns. Let $\overleftarrow{\xi}$ denote the unique Perron–Frobenius row vector for $\Gamma^t\Gamma$, normalized so that its first co-ordinate (corresponding to the distinguished white vertex) is 1. Thus $\xi > 0$, $\overleftarrow{\xi}\,\Gamma^t\Gamma = \beta\,\overleftarrow{\xi}$, and $\xi_1 = 1$.

Let Λ be the Coxeter graph of type A_{h-1}, with a bicoloration having at least one white end vertex; choose one white end vertex and label it as the first white vertex. Denote also by Λ the matrix of the graph Λ, and let $\overleftarrow{\eta}$ be the Perron–Frobenius row vector of $\Lambda^t\Lambda$, normalized so that its first co-ordinate (corresponding to the chosen white end vertex) is 1.

Theorem 4.5.1. $[B:C] = \|\vec{\eta}\|^2/\|\vec{\xi}\|^2$.

Proof. Define $N_0 = N_1 = \mathbb{C}$ and $N_{j+1} = \{\mathbb{1}, e_1, \cdots, e_j\}''$ for $j \geq 1$, so that $N = \overline{\bigcup_j N_j}$. Since

$$\begin{array}{ccc} N_1 & \subset & M_1 \\ \cup & & \cup \\ N_0 & \subset & M_0 \end{array}$$

is evidently a commuting square, so is

$$\begin{array}{ccc} N_{j+1} & \subset & M_{j+1} = \langle M_j, e_j \rangle \\ \cup & & \cup \\ N_j & \subset & M_j \end{array}$$

for all j, by 4.2.3, and induction. For each j, let $C_j = pN_j$ and $B_j = pM_jp$. Then since $p \in M_0 \subset N_j'$ for all j, Proposition 4.2.4 implies that

$$\begin{array}{ccc} C_{j+1} & \subset & B_{j+1} \\ \cup & & \cup \\ C_j & \subset & B_j \end{array}$$

is a commuting square for all j. Evidently $B = \overline{\bigcup_j B_j}$ and $C = \overline{\bigcup_j C_j}$. We will show that the inclusion data for these commuting squares are periodic with period 2 for large j.

First we need to describe the Bratteli diagram for the chain $(B_j)_{j \geq 0}$. The Bratteli diagram for $(M_j)_{j \geq 0}$ has n vertices each of dimension 1 in the 0th floor, and alternate stories given by Γ and Γ^t. To obtain the diagram for $(B_j)_{j \geq 0}$, take instead the dimension vector $\vec{\mu}^{(0)} = \begin{bmatrix} 1 \\ 0 \\ \vdots \\ 0 \end{bmatrix}$ on the 0th floor (that is introduce n-1 phantom vertices of dimension 0 on the 0th floor), and again form alternate stories by Γ and Γ^t. Compute the dimension vectors on each floor by $\vec{\mu}^{(2j)} = (\Gamma^t\Gamma)^j\vec{\mu}^{(0)}$ and $\vec{\mu}^{(2j+1)} = \Gamma(\Gamma^t\Gamma)^j\vec{\mu}^{(0)}$.

Finally erase all vertices of dimension 0 and all edges emanating from such a vertex. If ℓ_0 is the maximum distance from w_1 to any vertex of Γ, then for $j \geq \ell_0 - 1$

$$\Lambda_{B_j}^{B_{j+1}} = \begin{cases} \Gamma & \text{if } j \text{ is even} \\ \Gamma^t & \text{if } j \text{ is odd.} \end{cases}$$

For example if Γ is E_6 with the distinguished white vertex at the end then the diagrams are:

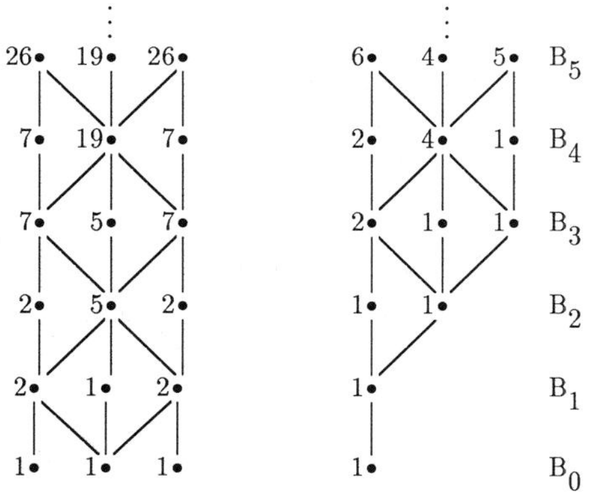

M B

If tr is the normalized trace on M, note that $\mathrm{Tr}(\cdot) = \mathrm{tr}(\cdot)/\mathrm{tr}(p)$ is the unique normalized trace on $\bigcup_j B_j$, and if $E_j : M_j \to M_{j-1}$ is the tr-preserving conditional expectation, then $E_j(B_j) = B_{j-1}$ (since $p \in M_0$) and $E_j|_{B_j}$ is Tr-preserving. Finally for $a \in B_j$, $e_j p a e_j p = E_j(a) e_j p$, so $\{B_j, e_j p\}''$ is an E_j-extension of B_j, in the terminology of Section 2.6. We have

$$B_{j+1} \supset \{B_j, e_j p\}'' \supset B_j,$$

and the inclusion matrix for $B_j \subset \{B_j, e_j p\}''$ is of the form $\begin{bmatrix} \Gamma_j^t \\ \Omega \end{bmatrix}$, where $\Gamma_j = \Lambda_{B_{j-1}}^{B_j}$ and Ω is some matrix, by 2.6.9. But if $j \geq \ell_0$, then $\Lambda_{B_j}^{B_{j+1}} = \Gamma_j^t$, and consequently

$$B_{j+1} = \{B_j, e_j p\}'' = B_j e_j p B_j \cong \langle B_j, e_{B_{j-1}} \rangle;$$

the chain $(B_j)_{j\geq \ell_0}$ is in fact (isomorphic to) a tower obtained by iterating the fundamental construction.

The Bratteli diagram for $(C_j)_{j\geq 0}$ is the same as that for $(N_j)_{j\geq 0}$, and is obtained from Λ, the Coxeter graph of type A_{h-1}, in exactly the same way as that of $(B_j)_{j\geq 0}$ is obtained from Γ; see section 2.9. In particular if $j \geq h-2$ then

$$\Lambda_{C_j}^{C_{j+1}} = \begin{cases} \Lambda & \text{if } j \text{ is even} \\ \Lambda^t & \text{if } j \text{ is odd}. \end{cases}$$

It now follows from Lemma 4.2.4 that for $j \geq j_0 = \max\{\ell_0, h-2\}$ the "horizontal" inclusion matrices $\Lambda_{C_j}^{B_j}$ are also periodic, $\Lambda_{C_{j-1}}^{B_{j-1}} = \Lambda_{C_{j+1}}^{B_{j+1}}$. Therefore by Wenzl's Theorem 4.3.3, $[B:C] = \|\overleftarrow{t}^{(j)}\|^2/\|\overleftarrow{s}^{(j)}\|^2$ for $j \geq j_0$, where $\overleftarrow{t}^{(j)}$ and $\overleftarrow{s}^{(j)}$ are respectively the weight vectors of Tr on C_j and B_j. Now for $2j \geq j_0$, $\overleftarrow{t}^{(2j)}$ [resp. $\overleftarrow{s}^{(2j)}$] is a Perron–Frobenius eigenvector for $\Lambda^t\Lambda$ [resp. $\Gamma^t\Gamma$] and so is proportional to $\overleftarrow{\eta}$ [resp. $\overleftarrow{\xi}$]. It remains only to establish the correct normalization of $\overleftarrow{s}^{(2j)}$ and $\overleftarrow{t}^{(2j)}$. We have

$$\begin{aligned} 1 &= \langle \overleftarrow{s}^{(2j)}, \vec{\mu}^{(2j)} \rangle = \langle \overleftarrow{s}^{(2j)}, (\Gamma^t\Gamma)^j \vec{\mu}^{(0)} \rangle = \langle \overleftarrow{s}^{(2j)}(\Gamma^t\Gamma)^j, \vec{\mu}^{(0)} \rangle \\ &= \beta^j \langle \overleftarrow{s}^{(2j)}, \vec{\mu}_0 \rangle = \beta^j s_1^{(2j)}, \end{aligned}$$

so the first component of $\overleftarrow{s}^{(2j)}$ is β^{-j} and $\overleftarrow{s}^{(2j)} = \beta^{-j}\overleftarrow{\xi}$. Similarly $\overleftarrow{t}^{(2j)} = \beta^{-j}\overleftarrow{\eta}$, and thus $[B:C] = \|\overleftarrow{t}^{(2j)}\|^2/\|\overleftarrow{s}^{(2j)}\|^2 = \|\overleftarrow{\eta}\|^2/\|\overleftarrow{\xi}\|^2$. #

Proposition 4.5.2. *The possible values of the index in Theorem 4.5.1 are as follows:*

For Γ *of type* A_ℓ $(\ell \geq 2)$:	$\sin^2[k\pi/(\ell+1)]/\sin^2[\pi/(\ell+1)]$,	$k = 1,2,\cdots,[(\ell+1)/2]$.
For Γ *of type* D_ℓ $(\ell \geq 4)$:	$2\sin^2[k\pi/(2\ell-2)]/\sin^2[\pi/(2\ell-2)]$,	$k = 1,2,\cdots,\ell-2$,
	$1/\sin^2[\pi/(2\ell-2)]$.	
For Γ *of type* E_6:	$3+\sqrt{3}$, $6+2\sqrt{3}$, $9+5\sqrt{3}$, $18+10\sqrt{3}$.	
For Γ *of type* E_7:	*seven values, the smallest being approximately* 7,759.	
For Γ *of type* E_8:	*eight values, the smallest being approximately* 19,48.	

Proof. The calculation, based on the data of 1.4.5, 1.4.8, and 1.4.9, is straightforward, and is left to the reader.

Remark: The only one of the values between 4 and 5 is $3+\sqrt{3}\cong 4{,}732$. The values between 5 and 10 are:

$[\sin^2 3\pi/\ell]/[\sin^2 \pi/\ell]$	for $\ell \geq 7$,
$8\cos^2 \pi/\ell$	for $\ell \geq 6$,
$3+\sqrt{5}\cong 5{,}236$	(D_6),
$6+2\sqrt{3}\cong 9{,}464$	(E_6),
$[\sin^2 4\pi/10]/[\sin^2 \pi/10]\cong 9{,}472$	(A_3),
ca. 7,759	(E_7).

4.6. The derived tower and the Coxeter invariant.

The results of this section, with the exception of 4.6.3(vi), were all known to V. Jones before the inception of A. Ocneanu's work on subfactors. Nonetheless, our exposition has been strongly influenced by conversations with Ocneanu, to whom we would like to record our gratitude.

The proof given in [Jo 1] for the restrictions on the possible values of the index of a pair $N \subset M$ of II_1-factors proceeded by constructing the tower associated to $N \subset M$ and then examining the algebra generated by the projections e_i in the construction, as in Chapters 2 and 3. It was of great importance that $\{e_1,e_2,\cdots,e_n\}''$ is finite dimensional for each $n \geq 1$. Here we will describe another chain $(Y_k)_{k\geq 0}$ of finite dimensional algebras associated to a pair $N \subset M$, such that Y_k contains $\{e_1,e_2,\cdots,e_{k-1}\}''$, but is in general strictly larger. It turns out that the chain $(Y_k)_{k\geq 0}$ is determined by a certain (possibly infinite) graph, called the principal graph of the pair $N \subset M$, which is a conjugacy invariant of the pair. In case $[M:N] < 4$, the principal graph is a Coxeter graph of type A, D, or E whose norm is the square root of index; this provides another proof of the restrictions on the index values, as well as a conjugacy invariant finer than index itself; these results were announced in [Jo8].

Consider a connected inclusion $N \subset M$ of finite von Neumann algebras with finite dimensional centers, with N of finite index in M. Let

$$N = M_0 \subset M = M_1 \subset M_2 \subset \cdots$$

be the tower obtained by iterating the fundamental construction, as in Chapter 3, with $M_{k+1} = \langle M_k, e_k\rangle$ for all k. Write $\beta = [M:N]$, and let tr be the unique trace on $\bigcup_k M_k$ with the β-Markov property with respect to each inclusion $M_{k-1} \subset M_k$.

Definition 4.6.1. The derived tower $\partial M/\partial N$ is the chain of algebras

$$\mathbb{C} = N' \cap N \subset N' \cap M \subset N' \cap M_2 \subset \cdots .$$

Lemma 4.6.2.

(i) $N' \cap M_k$ *is finite dimensional for all* k.

(ii) *If* N *and* M *are factors* $\dim(N' \cap M_k) \le [M:N]^k$.

(iii) *With respect to* tr,

$$\begin{array}{ccc} M_{k-1} & \subset & M_k \\ \cup & & \cup \\ N' \cap M_{k-1} & \subset & N' \cap M_k \end{array}$$

is a commuting square for all k.

Proof. (i) It follows by induction on k, using condition 3.5.4(iv) that N is of finite index in M_k for all k; hence by 3.6.2(a) $N' \cap M_k$ is finite dimensional.

(ii) If N and M are factors, then $[M_k:N] = [M:N]^k$, so the inequality follows from 3.6.2(b).

(iii) Follows from 4.2.7. #

We assume henceforth that N and M are factors and we denote $N' \cap M_k$ by Y_k, $E_{M_{k-1}}$ by E_k, and the inclusion matrix for $Y_k \subset Y_{k+1}$ by Λ_k. Since $e_k x e_k = E_k(x)e_k = E_{Y_{k-1}}(x)e_k$ for $x \in Y_k$ (by 4.6.2(iii)), $\{Y_k,e_k\}''$ is an $E_{Y_{k-1}}$–extension of Y_k, in the terminology of Section 2.6. Let $X_{k+1} = \langle Y_k, e_{Y_{k-1}} \rangle$ be the algebra obtained by the fundamental construction for $Y_{k-1} \subset Y_k$, with respect to tr, for $k \ge 1$. Give X_{k+1} the not necessarily normalized trace Tr defined by the weight vector $\beta^{-1}\vec{s}^{(k-1)}$, where $\vec{s}^{(k-1)}$ is the weight vector of tr on Y_{k-1}, and the minimal central projections of Y_{k-1} correspond to those of X_{k+1} as in 2.4.1. By 2.6.9, $\sigma_{k+1}(xe_{Y_{k-1}}x') = xe_kx'$ defines an injective algebra homomorphism from X_{k+1} into $\{Y_k,e_k\}'' \subset Y_{k+1}$ with image $Y_ke_kY_k$; it is evident that σ_{k+1} is a $*$–homomorphism. We will refer to $\sigma_{k+1}(X_{k+1})$ as "the old stuff" in Y_{k+1}, because it is determined by $Y_{k-1} \subset Y_k$. It turns out that the old stuff is an ideal, whose complementary ideal we call "the new stuff".

Theorem 4.6.3.

(i) *For* $k \geq 1$ *and* $a \in X_{k+1}$, $tr(\sigma_{k+1}(a)) = Tr(a)$.

(ii) *For* $k \geq 1$, $\sigma_{k+1}(X_{k+1}) = Y_k e_k Y_k$ *is an ideal in* Y_{k+1}.

(iii) *For* $k \geq 1$, $\sigma_{k+1}(\mathbf{1}_{X_{k+1}})$ *is the central support of* e_k *in* Y_{k+1}.

(iv) *For all* k, *the inclusion* $Y_k \subset Y_{k+1}$ *is connected.*

(v) *For all* k, $\|\Lambda_k\|^2 \leq [M:N]$.

(vi) ("*The new stuff comes only from the old new stuff.*") *For* $k \geq 2$, *if* $x \in Y_{k+1}$ *and* $x\sigma_{k+1}(X_{k+1}) = 0$, *then* $x\sigma_k(X_k) = 0$.

(vii) *For all* $k \geq 1$, *the following are equivalent:*

(a) $\sigma_{k+1}(X_{k+1}) = Y_{k+1}$.

(b) Tr *is normalized on* X_{k+1}.

(c) $\overleftarrow{s}^{(k-1)}$ *is an eigenvector of* $\Lambda_{k-1}^t \Lambda_{k-1}$ *with eigenvalue* β.

(d) $\Lambda_k = \Lambda_{k-1}^t$.

(e) $\|\Lambda_{k-1}\|^2 = [M:N]$.

(viii) *If the equivalent conditions of (vii) hold for* k, *then they also hold for* $k+1$.

Proof. (i) We have to show that $Tr(e_{Y_{k-1}} x) = tr(e_k x)$ for $x \in Y_k$. Since $Tr(e_{Y_{k-1}} x) = Tr(e_{Y_{k-1}} E_k(x))$ and $tr(e_k x) = tr(e_k E_k(x))$, it is enough to prove the equality for $x \in Y_{k-1}$, and since both $x \mapsto Tr(e_{Y_{k-1}} x)$ and $x \mapsto tr(e_k x)$ are traces on Y_{k-1}, it suffices to prove it for x a minimal projection in Y_{k-1}. But if x is a minimal projection in some direct summand of Y_{k-1}, then by 2.6.4(c), $e_{Y_{k-1}} x$ is a minimal projection in the corresponding direct summand of X_{k+1}, whose trace $Tr(e_{Y_{k-1}} x)$ is by definition $\beta^{-1} tr(x) = tr(e_k x)$.

(ii) We must show that if $a, b \in Y_k$ and $x \in Y_{k+1}$, then $xae_k b \in \sigma_{k+1}(X_{k+1})$. As in the proof of 3.6.3, $xae_k = \beta E_{k+1}(xae_k)e_k$, and by the N–N bilinearity of E_{k+1}, $E_{k+1}(xae_k) \in Y_k$. So $xae_k b \in Y_k e_k Y_k$.

(iii) This follows from (ii) and the fact that $e_{Y_{k-1}}$ has central support $\mathbf{1}$ in X_{k+1}.

(iv) An equivalent statement is that Λ_k is indecomposable for all k. This is evident for $k = 1$ since $Y_0 = \mathbb{C}$. By (ii) and 2.6.9, the matrix Λ_{k+1} has the form $\begin{bmatrix} \Lambda_k^t \\ \Gamma \end{bmatrix}$, where Γ has no row of zeroes; hence if Λ_k is indecomposable, so is Λ_{k+1}.

(v) If $\vec{\mu}^{(j)}$ denotes the dimension vector of Y_j, then for fixed i and for $\ell > 0$,

$$\vec{\mu}^{(i+\ell)} = \Lambda_{i+\ell-1}\Lambda_{i+\ell-2}\cdots\Lambda_i\vec{\mu}^{(i)}.$$

It follows from the form of Λ_j for $j \geq i$ (i.e., $\Lambda_{j+1} = \begin{bmatrix}\Lambda_j^t\\ \Gamma\end{bmatrix}$, that

$$\dim Y_{i+2p} = \|\vec{\mu}^{(i+2p)}\|^2 \geq \|(\Lambda_i^t\Lambda_i)^p\vec{\mu}^{(i)}\|^2.$$

Suppose that for some i, $[M:N] < \|\Lambda_i\|^2$, and choose $\epsilon > 0$ such that $\|\Lambda_i\|^2(1-\epsilon) > [M:N]$. Let $\vec{\nu}$ be a Perron–Frobenius vector for $\Lambda_i^t\Lambda_i$ (which exists due to (iv)), normalized so that $\vec{\mu}^{(i)} > \vec{\nu}$ (component-by-component inequality). Then for $p \geq 1$

$$\dim Y_{i+2p} \geq \|(\Lambda_i^t\Lambda_i)^p\vec{\nu}\|^2 = \|\Lambda_i\|^{4p}\|\vec{\nu}\|^2,$$

whence

$$\frac{(\dim Y_{i+2p})^{1/i+2p}}{\|\Lambda_i\|^2} \geq \left[\frac{\|\vec{\nu}\|^2}{\|\Lambda_i\|^{2i}}\right]^{1/i+2p}.$$

Since the right side converges to 1 as p increases, it follows that for some k,

$$(\dim Y_k)^{1/k} \geq (1-\epsilon)\|\Lambda_i\|^2 > [M:N]$$

which contradicts 4.6.2(ii).

(vi) Because of (ii) we can suppose x is a central projection in Y_{k+1}; then $xe_k = 0$ implies $xe_{k-1} = \beta\, e_{k-1}xe_ke_{k-1} = 0$, so that also $xae_{k-1}b = 0$ for all $a,b \in Y_k$.

(vii) Conditions (a) and (d) are equivalent since the inclusion matrix for $Y_k \subset X_{k+1}$ is Λ_{k-1}^t, and (a) is equivalent to (b) by (i). If (d) (and thus (b)) hold, then the vector of tr on Y_{k+1} is $\beta^{-1}\overleftarrow{s}^{(k-1)}$ and

$$\overleftarrow{s}^{(k-1)} = \overleftarrow{s}^{(k+1)}\Lambda_k\Lambda_{k-1} = \beta^{-1}\overleftarrow{s}^{(k-1)}\Lambda_{k-1}^t\Lambda_{k-1}.$$

Thus (d) implies (c). If (c) holds, then $\|\Lambda_{k-1}\|^2 \geq \beta$, and since the opposite inequality is true by (v), we have that (c) implies (e). If (e) holds, but not (d), then Λ_k is a non-negative matrix of the form $\begin{bmatrix}\Lambda_{k-1}^t\\ \Gamma\end{bmatrix}$, with $\Gamma \neq 0$, so $\|\Lambda_k\|^2 > \|\Lambda_{k-1}\|^2 = [M:N]$, contradicting (v).

(viii) This follows from (d) and (e) of (vii) together with $\|\Lambda\| = \|\Lambda^t\|$. #

Theorem 4.6.3 gives an interesting qualitative picture of the Bratteli diagram for the derived tower $\partial M/\partial N$. One is led to the following concepts which have been emphasized by Ocneanu:

Definition 4.6.4. A pair $N \subset M$ is said to have finite depth if there is a k for which the conditions of 4.6.3(viii) hold. In this case the smallest such k is called the depth of $N \subset M$.

Definition 4.6.5. The principal graph of $N \subset M$ is the bipartite multigraph constructed as follows: On the Bratteli diagram of the derived tower $\partial M/\partial N$ delete on each floor the vertices belonging to the old stuff, and the edges emanating from these vertices.

Since the new stuff is connected only to the old new stuff, the resulting graph Γ is connected. The principal graph Γ has a distinguished vertex $*$, the unique vertex on floor 0, and the distance of any vertex from $*$ is the number of its floor. The pair $N \subset M$ has finite depth if, and only if, Γ is finite, in which case the depth of $N \subset M$ is the maximum distance of any vertex from $*$. The Bratteli diagram of $\partial M/\partial N$ can be reconstructed from Γ (given together with the distinguished vertex $*$).

For purposes of illustration, let us give an example of what *might* be the Bratteli diagram of $\partial M/\partial N$ and the principal graph for a pair $N \subset M$ of finite depth. (We are *not* claiming that this example actually occurs; this is a much more delicate question!)

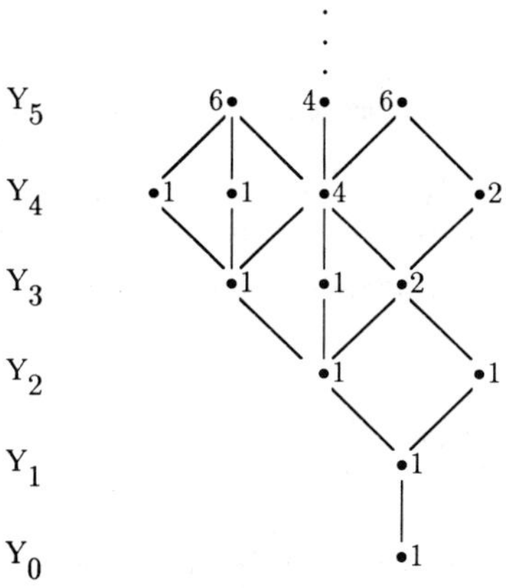

B

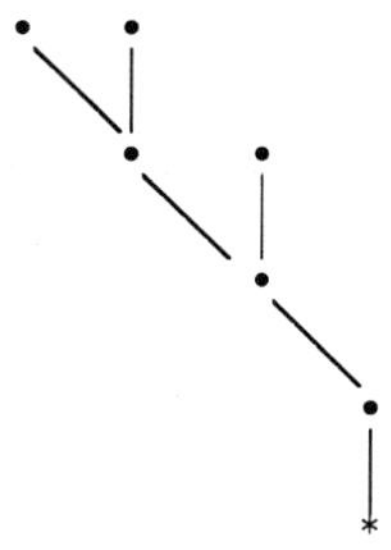

Γ

The pair $N \subset M$ would have depth 4 according to our conventions. The Bratteli diagram of $\partial M/\partial N$ for a finite depth pair $N \subset M$ will always exhibit a growth in complexity up to a certain level, after which the remaining structure is obtained by reflecting. Note that if $N \subset M$ has depth k, then the Bratteli diagram for $Y_{k-1} \subset Y_k$ is isomorphic to the principal graph.

We can now record the following consequence of Theorem 4.6.3 and the work of Chapter 1.

Corollary 4.6.6. *Suppose* $N \subset M$ *is a pair of* II_1*-factors with* $[M:N] < 4$. *Then*

(a) $[M:N] = 4\cos^2\pi/h$ *for some* $h \geq 3$.

(b) *The depth of* $N \subset M$ *is no greater than* $h-2$.

(c) *The principal graph of* $N \subset M$ *is a Coxeter graph of type* A, D, *or* E, *whose norm is* $[M:N]^{1/2}$.

Proof. If $N \subset M$ were not of finite depth, then one would have a sequence Λ_i of non-negative integer matrices with $\|\Lambda_i\|^2 < \|\Lambda_{i+1}\|^2 \leq [M:N] < 4$, which is impossible by the classification of Chapter 1. If k is the depth of $N \subset M$, then for $j \geq k-1$ the graph of Λ_j is isomorphic to the principal graph Γ, and $\|\Lambda_j\|^2 = \|\Gamma\|^2 = [M:N] < 4$. Therefore, by Chapter 1, Γ is a Coxeter graph of type A, D, or E and $[M:N] = 4\cos^2\pi/h$, where h is the Coxeter number. Furthermore $k \leq \operatorname{diam}(\Gamma) \leq h-2$. #

This completes the proof of the restriction on index values outlined in [Jo8].

Corollary 4.6.7. *Suppose* $N \subset M$ *is a pair of* II_1*-factors with* $[M:N] = 4$.

(a) *If* $N \subset M$ *is of finite depth, then the principal graph* Γ *is a completed Coxeter graph of type* A, D, *or* E, *i.e., one of the graphs in Table 1.4.6.*

(b) *If* $N \subset M$ *is of infinite depth, then* Γ *is one of the following:*

A_∞ 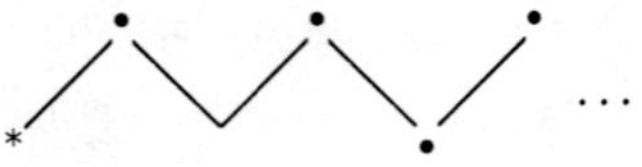

$A_{n,\infty}$ 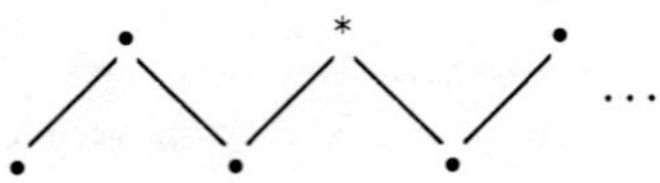 *(end vertex at distance* n *from* $*$*)*

$A_{\infty,\infty} = \mathbb{T}$

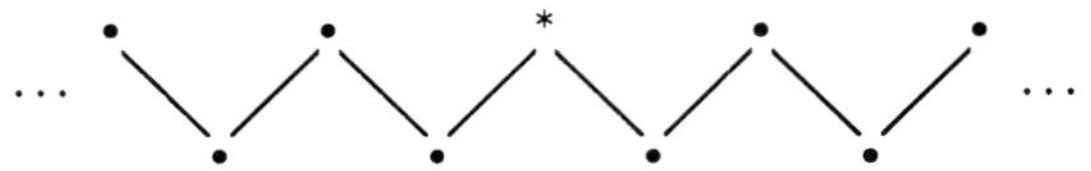

(doubly infinite linear graph)

D_{∞}

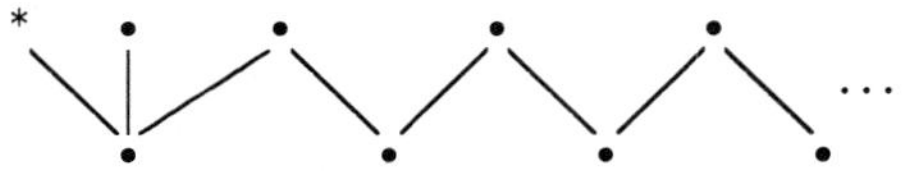

$D_{n,\infty}$

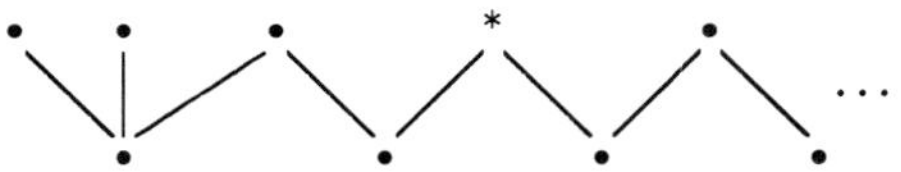

(end vertex at distance n *from* $*$*)*

Proof. (a) Follows from 4.6.3 and 1.4.3.

(b) Let Γ_k denote the subgraph of Γ containing vertices of distance no greater than k from $*$; then Γ_k is also isomorphic to the Bratteli diagram for $Y_{k-1} \subset Y_k$, so $\|\Gamma_k\|^2 < 4$ for all k. Thus each Γ_k is a Coxeter graph of type A, D, or E; furthermore Γ_{k+1} is obtained from Γ_k by addition of one or more vertices at distance $k+1$ from $*$. It is easy to verify that the only possibilities are those listed. #

We will see in Section 4.7 that subfactors of finite depth and of infinite depth do occur. In particular all possibilities allowed by 4.6.7 do actually occur except possibly $A_{n,\infty}$ and $D_{n,\infty}$; we will also see why $A_{\infty,\infty}$ is also labelled $\mathbb{T}$.

Finally, we cannot resist saying a few words about the truly exciting results of Ocneanu, who has added to the structure described here something we have completely neglected in our treatment, namely the involutions J_i coming from each basic construction in the tower M_i. He shows that they combine to define an endomorphism of $\partial M/\partial N$ and, together with the e_i's and the principal graph, seem to completely determine $N \subset M$ in many cases. In particular, he can show that there are only finitely many subfactors of the hyperfinite II_1-factor (of index < 4) for each Coxeter graph, up to conjugation by automorphisms, and he determines which Coxeter graphs are allowed.

4.7. Examples of derived towers.

4.7.a. Finite group actions. We shall analyze the derived tower for a pair $N \subset M$, when N is the fixed point algebra M^G for an outer action of a finite group G on a II_1 factor M, and also when M is the crossed product $N \rtimes G$ of N by an outer action of a finite group G on N.

(i) $N = M^G$. In this case we know from [Jo1], that $\langle M,e_N \rangle = M_2$ can be identified with the crossed product $M \rtimes G$, so that $Y_2 = M_2 \cap N'$ contains the group (von Neumann) algebra $\mathbb{C}G$. The inclusion matrix for $\mathbb{C} \subset \mathbb{C}G$ is $[n_0 = 1, n_1, \cdots, n_k]$, where the n_i are the degrees of the irreducible representations of G. Thus $\|\Lambda_1\|^2 \geq \sum n_i^2 = |G|$. On the other hand, by [Jo1], $[M:N] = |G|$. Hence by 4.6.3, $Y_2 = \mathbb{C}G$ and $N \subset M$ is of depth 2. Note that the derived tower is independent of M or the action of G; for example, in case $G = S_3$ the Bratteli diagram for $\partial M/\partial N$ is

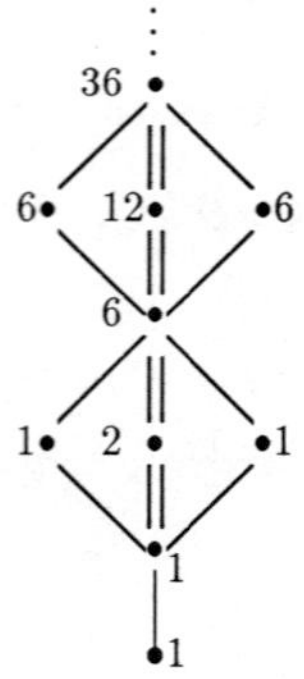

Remark. According to Ocneanu, depth = 2 and $N' \cap M = \mathbb{C}$ characterizes fixed point algebras of outer actions of finite dimensional Kac algebras.

(ii) $M = N \rtimes G$. In this case $\langle M,e_N \rangle$ is known to be the crossed product of M by the "dual action" of $\hat{G}$ (see [NT]). To be more concrete, denote by u_g the canonical unitaries of the crossed product. On $L^2(M)$, each of the projections $e_g = u_g e_N u_g^*$ onto the closure of $Nu_g = u_g N$ is evidently in $\langle M,e_N \rangle \cap N' = Y_2$. By the same reasoning as in case (i), the (commutative) algebra which they generate is equal to Y_2, and one always has the following Bratteli diagram for $\partial M/\partial N$.

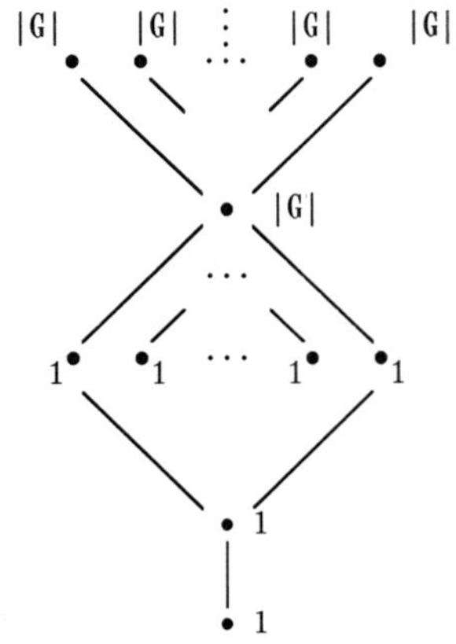

Remarks. (1) Ocneanu's endomorphism allows one to reconstruct the multiplication table for G, once the elements have been put into bijection with the factors on the third line!

(2) By choosing $G = \mathbb{Z}/2\mathbb{Z}$ one obtains the Coxeter graph A_3 as the principal graph, and $G = \mathbb{Z}/3\mathbb{Z}$ gives D_4 (for either the fixed point algebra or crossed product case). One can also check that the pair $N \rtimes \mathfrak{S}_2 \subset N \rtimes \mathfrak{S}_3$ or $M^{\mathfrak{S}_3} \subset M^{\mathfrak{S}_2}$ has principal graph A_5. ($\mathfrak{S}_n$ denotes the symmetric group.)

4.7.b. The A_n Coxeter graphs. Let $\{e_i\}_{i\geq 0}$ be a sequence of self-adjoint projections satisfying the relations

$$e_i e_{i\pm 1} e_i = \beta^{-1} e_i$$
$$e_i e_j = e_j e_i \qquad \text{for } |i-j| \geq 2,$$

for some $\beta > 0$. Realize the hyperfinite II_1 factor R as $R = \{\mathbb{1}, e_0, e_1, \ldots\}''$ and write R_β for $\{\mathbb{1}, e_1, e_2, \ldots\}''$. We have computed that $[R:R_\beta] = \beta$ (Proposition 3.4.4).

When $\beta = 4\cos^2 \pi/n$ for some $n \geq 3$, it follows easily from the proof of Skau's lemma (4.4.3) that the principal graph of $R_\beta \subset R$ is the Coxeter graph A_{n-1}. In fact, write $e_{-1}, e_{-2}, \ldots$ for the projections occurring in the tower construction for $R_\beta \subset R$, so that $M_i = \{\mathbb{1}, e_{-i+1}, \ldots, e_1, e_2, \ldots\}''$ $(i \geq 0)$. The proof of Skau's lemma shows that $N' \cap M_i = \{\mathbb{1}, e_{-i+1}, \ldots, e_{-1}\}''$, and we saw in Section 2.9 that the chain $(N' \cap M_i)_{i\geq 1}$ has the appropriate Bratteli diagram. (The *statement* of Skau's lemma does not apply since $\{e_{-i+1}, \ldots, e_0, e_1\}''$ is *not* isomorphic to the result of the fundamental construction for

$$\{e_{-i+1},\ldots,e_{-1}\}'' \subset \{e_{-i+1},\ldots,e_0\}''$$

unless $i \geq n-3$.)

Since Skau's lemma also is valid for $\beta = 4$, the same argument shows that the principal graph for the pair $R_\beta \subset R$ when $\beta = 4$ is

$A_\infty =$ 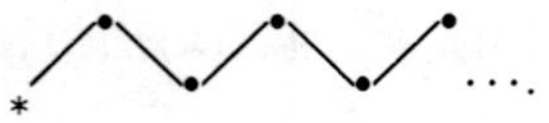

4.7.c. A general method. The following result is useful as it allows the computation of the derived tower in many examples coming from group actions. A more powerful result has been discovered and exploited independently by A. Wassermann [Wa].

Lemma 4.7.1. *Let* $N \subset M \subset P$ *be von Neumann algebras,* φ *a faithful normal state of* P, e *a projection in* $N' \cap P$, *and* G *a group of automorphisms of* P *preserving* N, M, φ, *and* e. *Suppose:*

(i) $eMe = Ne$,

(ii) $\varphi(xe) = \varphi(e)\varphi(x) \quad \forall\, x \in M$,

(iii) $tr = \varphi|_{P^G}$ *is a trace,*

(iv) $N^G \subset M^G \subset P^G$ *are* II_1 *factors with* $[M^G:N^G] = [P^G:M^G] = \varphi(e)^{-1}$.

Then there is an isomorphism Φ *of* $\langle M^G, e_{N^G}\rangle$ *onto* P^G *such that* $\Phi(e_{N^G}) = e$ *and* $\Phi(x) = x$ *for all* $x \in M^G$.

Proof. Let us first show that $exe = E_{N^G}(x)e$ for $x \in M^G$. By hypothesis, $exe = ae$ for some $a \in N$, and for all $g \in G$ we have $ae = exe = g(exe) = g(ae) = g(a)e$. But then $\varphi(e)\varphi((a-g(a))^*(a-g(a))) = \varphi((a-g(a))^*(a-g(a))e) = 0$, so $a \in N^G$ since φ is faithful. Also if $y \in N^G$, then $\varphi(e)tr(yx) = tr(yxe) = tr(eyxe) = tr(yexe) = tr(yae) = \varphi(e)tr(ya)$, so $a = E_{N^G}(x)$.

Now by 2.6.9 and 3.6.4 there is a $*$-algebra isomorphism Φ of $\langle M^G, e_{N^G}\rangle$ into P^G such that $\Phi(xe_{N^G}y) = xey$ for $x,y \in M^G$. The map Φ is trace preserving (since $tr(e_{N^G}) = [M^G:N^G]^{-1} = tr(e)$), so normal and unital. We need only show that Φ is surjective. But the image of Φ is a II_1 factor containing M^G (since Φ is unital) as a subfactor of index $tr(e)^{-1} = [P^G:M^G]$, so the image is P^G. #

This lemma will be used repeatedly in 4.7.d and 4.7.f to calculate derived towers by calculating it in a simple situation and then passing to the fixed point algebra of some group action. See also [PP3]. Wassermann calls the lemma the invariance principle, since in cases where $[M:N]$ makes sense it should also be $\varphi(e)^{-1}$.

4.7.d. Some examples of derived towers for index 4 subfactors. We realize the hyperfinite II_1 factor R as the completion, with respect to the unique tracial state tr, of the infinite tensor product of $Mat_2(\mathbb{C})$, $R = (\overset{\infty}{\otimes} Mat_2(\mathbb{C}))^{-}$. Any closed subgroup G of $SU(2)$ acts on $\overset{\infty}{\otimes} Mat_2(\mathbb{C})$ by the infinite tensor product of its action by conjugation on $Mat_2(\mathbb{C})$. The action preserves the trace, so extends to an action on R. The group G acts in the same way on $R \otimes Mat_2(\mathbb{C})$, so one has a commuting square

$$(*) \qquad \begin{array}{ccc} R & \subset & R \otimes Mat_2(\mathbb{C}) \\ \cup & & \cup \\ R^G & \subset & (R \otimes Mat_2(\mathbb{C}))^G . \end{array}$$

Now $R^G \supset R^{SU(2)}$ and $R^{SU(2)}$ contains in its unitary group a copy of the infinite symmetric group $\mathfrak{S}_\infty$, which acts ergodically on R. So $(R^G)' \cap R = \mathbb{C}\mathbf{1}$, and, in particular, R^G is a factor. The projection

$$e_1 = \tfrac{1}{2}\mathbf{1}_R \otimes \{e_{11} \otimes {}_{22} - e_{12} \otimes e_{21} - e_{21} \otimes e_{12} + e_{22} \otimes e_{11}\}$$

satisfies $e_1 x e_1 = E_R(x) e_1$ and $(R \otimes Mat_2(\mathbb{C})) e_1 (R \otimes Mat_2(\mathbb{C})) = R \otimes Mat_2(\mathbb{C}) \otimes Mat_2(\mathbb{C})$. Hence, using 2.6.9 and 3.6.4, the result of the fundamental construction for $R \subset R \otimes Mat_2(\mathbb{C})$ can be identified with $R \otimes Mat_2(\mathbb{C}) \otimes Mat_2(\mathbb{C})$, the projection e_R being identified with e_1. Hence the tower for $R \subset R \otimes Mat_2(\mathbb{C})$ is identified with

$$R \subset R \otimes Mat_2(\mathbb{C}) \subset \cdots \subset R \otimes (\overset{k}{\otimes} Mat_2(\mathbb{C})) \subset \cdots,$$

The projections e_i being identified with $\mathbf{1}_R \otimes \alpha^{i-1}(e_1)$, with α the shift endomorphism of $\overset{\infty}{\otimes} Mat_2(\mathbb{C})$. Note that the projections e_i are $SU(2)$–invariant.

In case G is finite, it follows from the multiplicativity of the index that $[(R \otimes Mat_2(\mathbb{C}))^G : R^G] = 4$, and then 4.7.1 gives an isomorphism of

$\langle (R \otimes \mathrm{Mat}_2(\mathbb{C}))^G, e_{R^G}\rangle$ onto $(R \otimes \mathrm{Mat}_2(\mathbb{C})) \otimes \mathrm{Mat}_2(\mathbb{C}))^G$ taking e_{R^G} to e_1. If G is infinite, we have to do a little more work to reach the same conclusion.

Identify R (G–equivariantly) with $R_0 \otimes \mathrm{Mat}_2(\mathbb{C})$, where $R_0 \cong R$. Set $e_0 = \frac{1}{2} \mathbf{1}_{R_0} \otimes \{e_{11} \otimes e_{22} - e_{12} \otimes e_{21} - e_{21} \otimes e_{12} + e_{22} \otimes e_{11}\}$. Thus $(R \otimes \mathrm{Mat}_2(\mathbb{C}))e_0 = Re_0$ (by 3.6.3 or by direct computation). If $x \in (R \otimes \mathrm{Mat}_2(\mathbb{C}))^G$, then there is an $x_0 \in R$ such that $xe_0 = x_0e_0$, and

$$xe_0 = E_{(R\otimes \mathrm{Mat}_2(\mathbb{C}))^G}(xe_0) = R_{R^G}(x_0)e_0,$$

so $(R \otimes \mathrm{Mat}_2(\mathbb{C}))^G e_0 = R^G e_0$. Therefore,

$$R^G e_0 R^G = (R \otimes \mathrm{Mat}_2(\mathbb{C}))^G e_0 (R \otimes \mathrm{Mat}_2(\mathbb{C}))^G = (R \otimes \mathrm{Mat}_2(\mathbb{C}))^G,$$

the last equality because $(R \otimes \mathrm{Mat}_2(\mathbb{C}))^G$ is a factor. If e denotes the orthogonal projection of $L^2((R \otimes \mathrm{Mat}_2(\mathbb{C}))^G)$ onto $L^2(R^G)$, then $e\, e_0\, e = \frac{1}{4} e$. We claim that also $e_0\, e\, e_0 = \frac{1}{4} e_0$. Because $R^G\, e_0\, R^G = (R \otimes \mathrm{Mat}_2(\mathbb{C}))^G$, it suffices to check this equality on vectors $xe_0y\Omega$, where $x,y \in R^G$ and Ω is the trace vector in $L^2((R \otimes \mathrm{Mat}_2(\mathbb{C}))^G)$. But

$$\begin{aligned} e_0\, e\, e_0(x\, e_0\, y\, \Omega) &= e_0\, e\, e_0\, E_{R_0^G}(x)y\Omega \\ &= e_0\, E_{R^G}(e_0 E_{R_0^G}(x)y)\Omega \\ &= \tfrac{1}{4}\, e_0\, E_{R_0^G}(x)y\Omega \\ &= \tfrac{1}{4}\, e_0(x\, e_0\, y\Omega). \end{aligned}$$

It follows that $e_0 \sim e$ in $\langle (R \otimes \mathrm{Mat}_2(\mathbb{C}))^G, e\rangle$ and since e is a finite projection by 3.6.1(v), it follows that e_0 is finite in $\langle (R \otimes \mathrm{Mat}_2(\mathbb{C}))^G, e\rangle$. But $\mathbf{1}$ is the sum of finitely many projections, each equivalent to a subprojection of e_0, so $\langle (R \otimes \mathrm{Mat}_2(\mathbb{C}))^G, e\rangle$ is a finite factor. Therefore $[(R \otimes \mathrm{Mat}_2(\mathbb{C}))^G : R^G] = \mathrm{tr}(e)^{-1} = \mathrm{tr}(e_0)^{-1} = 4$.

We can now conclude from 4.7.1 and induction that the tower for $R^G \subset (R \otimes \mathrm{Mat}_2(\mathbb{C}))^G$ is

$$R^G \subset (R \otimes \mathrm{Mat}_2(\mathbb{C}))^G \subset \cdots \subset (R \otimes (\overset{k}{\otimes} \mathrm{Mat}_2(\mathbb{C})))^G \subset \cdots,$$

and since $(R^G)' \cap R = \mathbb{1C}$, we have $(R^G)' \cap R \otimes (\overset{k}{\otimes} \mathrm{Mat}_2(\mathbb{C})) = \mathbb{1}_R \otimes (\overset{k}{\otimes} \mathrm{Mat}_2(\mathbb{C}))$ and

$$(R^G)' \cap (R \otimes (\overset{k}{\otimes} \mathrm{Mat}_2(\mathbb{C}))^G = \mathbb{1}_R \otimes (\overset{k}{\otimes} \mathrm{Mat}_2(\mathbb{C}))^G.$$

Thus we have identified the derived tower for $R^G \subset (R \otimes \mathrm{Mat}_2(\mathbb{C}))^G$ with the sequence of finite dimensional von Neumann algebras $Y_0 = \mathbb{C}$, $Y_1 = (\mathrm{Mat}_2(\mathbb{C}))^G$, $Y_2 = [\mathrm{Mat}_2(\mathbb{C}) \otimes \mathrm{Mat}_2(\mathbb{C})]^G$, etc.

It is fairly evident that these algebras are just the commutants of the tensor powers of the fundamental representation of G on $\mathbb{C}^2$ determined by its inclusion in SU(2). We can now use the McKay correspondence between finite subgroups of SU(2) and affine Coxeter graphs (see [Slo] or [Jo 4]) to calculate the Bratteli diagrams or principal graphs when G is finite. The correspondence is as follows. Let Γ be the matrix with rows and columns indexed by irreducible representations of G, whose (i,j)–entry is the multiplicity of j in the tensor product of i with the fundamental representation of G on $\mathbb{C}^2$. Then Γ is the adjacency matrix of an affine Coxeter graph of type $A_n^{(1)}$, $D_n^{(1)}$, $E_6^{(1)}$, $E_7^{(1)}$, or $E_8^{(1)}$. (In fact, $A_n^{(1)}$ corresponds to a cyclic group, $D_n^{(1)}$ a dihedral group, $E_6^{(1)}$ the tetrahedron group, $E_7^{(1)}$ the cube group, and $E_8^{(1)}$ the icosohedron group.) The method of constructing the Bratteli diagram is clear from the representation–theoretic interpretation of Γ: use Γ $(= \Gamma^t)$ as the inclusion matrix and start with the dimension vector $[1,0,0,\cdots]^t$ on the 0th floor (as for example in the calculation of Bratteli diagrams in Section 4.5). The resulting principal graphs are $A_n^{(1)}$ $(n \geq 2)$, $D_n^{(1)}$ $(n \geq 4)$, and $E_6^{(1)}$, $E_7^{(1)}$, $E_8^{(1)}$ (see Table 1.4.6).

For G the maximal torus $\mathbb{T}$, the principal graph for $R^G \subset (R \otimes \mathrm{Mat}_2(\mathbb{C}))^G$ is the graph $A_{\infty,\infty}$ of Corollary 4.6.7. For G the infinite dihedral group D_∞ the graph is D_∞, and for $G = SU(2)$, the graph is A_∞.

Of course, the method we have described here is quite general and also applies in dimension greater than 2.

<u>4.7.e The tunnel construction.</u> We describe the tunnel construction of V. Jones ([Jo 1]) for a pair $N \subset M$ of finite factors with finite index. This is a sort of mirror image inside N of the tower construction of Chapter 3. The essential observation is that there is a representation of M on $L^2(N,tr)$ extending the standard representation of N, although not a canonical one. Start with the representation of M on $H = L^2(M,tr)$ and choose any projection $p \in M'$ with $tr_{M'}(p) = [M:N]^{-1}$. Then $\dim_N(pH) = 1$ by 3.2.5(e), so that the N–module pH is isomorphic to $L^2(N,tr)$ by 3.2.4(a); if

$u : L^2(N,tr) \longrightarrow pH$ is a unitary N–module map then $x \longmapsto u^* x|_{pH} u$ is the desired representation.

Represent the pair $N \subset M$ on $L^2(N,tr)$ and let J denote the conjugate linear isometry of $L^2(N,tr)$ extending the map $x \longmapsto x^*$ on N. Write M_{-1} for $JM'J$ and let e_0 be the projection of $L^2(N,tr)$ onto $L^2(M_{-1},tr)$. Then $[N:M_{-1}] = [N':M'] = [N:M]$, and $JMJ = M'_{-1} = \langle N',e_0\rangle$, so $M = JM'_{-1}J = \langle N,e_0\rangle$. That is, the pair $N \subset M$ is the result of the fundamental construction for the pair $M_{-1} \subset N$.

Iterating this construction, one obtains a decreasing "tunnel" of II_1 factors

$$\cdots \subset M_{-k} \subset M_{-k+1} \cdots \subset M_{-1} \subset N = M_0 \subset M = M_1$$

along with projections $\{e_0, e_{-1}, e_{-2}, \cdots\}$ such that $\langle M_{-k}, e_k\rangle = M_{-k+1}$. The projections $\{e_i : i \le 0\}$ satisfy the usual relations with $\beta = [M:N]$.

The tunnel construction has been exploited systematically by Ocneanu in his classification of subfactors.

4.7.f The derived tower for $R \supset R_\beta$ when $\beta > 4$. In this section we will compute the derived tower for $R \supset R_\beta$ when $\beta > 4$; compare Section 4.7.b for the cases $\beta < 4$ and $\beta = 4$. The computation uses a beautiful representation of the e_i's due to Pimsner and Popa [PP 1] as well as a theorem of Popa on the tunnel construction for certain pairs of factors. We begin with a preliminary lemma from [Jo 1].

Lemma 4.7.2. *Suppose* M *is a* II_1 *factor containing a projection* f *such that there is an isomorphism* $\theta : fMf \longrightarrow (\mathbb{1}-f)M(\mathbb{1}-f)$. *Let* $N = \{x+\theta(x) : x \in fMf\}$. *Then* N *is a subfactor with* $[M:N] = tr(f)^{-1} + tr(\mathbb{1}-f)^{-1} = tr(f)^{-1} tr(\mathbb{1}-f)^{-1}$.

Proof. The algebra N is isomorphic to fMf, so is a subfactor. Since $f \in N' \cap M$, $[M:N] = \dim_N(H) = \dim_N(fH) + \dim_N((\mathbb{1}-f)H)$, where $H = L^2(M)$. Since $Nf = fMf$,

$$\begin{aligned} \dim_N(fH) &= \dim_{fMf}(fH) \\ &= tr(f)^{-1} \dim_M(H) \quad \text{by 3.2.5(h).} \\ &= tr(f)^{-1}, \end{aligned}$$

and similarly for $\mathbb{1}-f$. #

The following theorem is due to Popa, who has kindly showed us the proof and allowed us to present it here.

Theorem 4.7.3 (Popa). *Suppose* M *is a* II_1 *factor containing a projection* f, *with* $tr(f) < \frac{1}{2}$, *such that there is an isomorphism* $\theta : fMf \longrightarrow (\mathbb{1}-f)M(\mathbb{1}-f)$. *Let* $N = \{x+\theta(x) : x \in fMf\}$. (*By 4.7.2,* $[M:N] = tr(f)^{-1} + tr(\mathbb{1}-f)^{-1} := \beta$; *note that* $\beta > 4$.) *Let* $\{e_0,e_{-1},e_{-2},\cdots\}$ *be projections in the tunnel construction for the pair* $N \subset M$, *set* $R = \{e_0,e_{-1},\cdots\}''$ *and* $R_\beta = \{e_{-1},e_{-2},\cdots\}''$. *Then*

(1) $f \in R_\beta' \cap R$,

(2) $fRf = R_\beta f$ *and* $(\mathbb{1}-f)R(\mathbb{1}-f) = R_\beta(\mathbb{1}-f)$.

Proof. Write $t = tr(f)$, so $\beta = t^{-1} + (1-t)^{-1}$. Since $E_N(e_0) = \beta\mathbb{1}$,

$$\begin{array}{ccc} R & \subset & M \\ \cup & & \cup \\ R_\beta & \subset & N \end{array}$$

is a commuting square, and $E_N E_R = E_R E_N = E_{R_\beta}$.

We first claim that $R_\beta' \cap R \underset{\neq}{\supset} \mathbb{C}\mathbb{1}$. Let p be any projection with $p \leq f$ and $tr(p) = \beta^{-1} = t(1-t)$, and set $q = \mathbb{1}-f-\theta(p)$, so $tr(q) = \beta^{-1}$ as well. Let v be a partial isometry in M such that $v^*v = q$ and $vv^* = p$ and define

$$e = (1-t)p + tq + \sqrt{t(1+t)}\,(v+v^*).$$

Then e is a projection. One can check that for any $x \in M$,

$$\begin{aligned} E_N(x) &= E_N(fxf) + E_N((\mathbb{1}-f)x(\mathbb{1}-f)) \\ &= t(fxf + \theta(fxf)) + (1-t)(\theta^{-1}((\mathbb{1}-f)x(\mathbb{1}-f)) + (\mathbb{1}-f)x(\mathbb{1}-f)). \end{aligned}$$

In particular, $E_N(e) = \beta^{-1}\mathbb{1}$. By [PP 1], there is an automorphism of M leaving N fixed pointwise and carrying e_0 to e, so we can assume $e_0 = e$.

Since $f \in M \cap N'$, $E_R(f) \in R \cap (N \cap R)' = R \cap R_\beta'$. Suppose that $E_R(f)$ is a scalar, that is, $E_R(f) = t\mathbb{1}$. Then

$$\begin{aligned} (1-t)\beta^{-1} &= tr(fe_0f) = tr(e_0f) \\ &= tr(E_R(e_0f)) = tr(e_0E_R(f)) \\ &= t\, tr(e_0) = t\beta^{-1}, \end{aligned}$$

which is impossible since $t < \frac{1}{2}$. Thus $E_R(f) \in R'_\beta \cap R \backslash \mathbb{C}\mathbb{1}$.

Now let f_0 be any projection in $R'_\beta \cap R$ with $0 < tr(f_0) \le \frac{1}{2}$. Ultimately we will show that $f = f_0$. Since

$$\frac{1}{t} + \frac{1}{1-t} = \beta = [R:R_\beta] \ge \frac{1}{tr(f_0)} + \frac{1}{tr(\mathbb{1}-f_0)} \quad \text{(by the proof of 3.6.2)},$$

it follows by calculus that $tr(f_0) \ge t$. Assume $tr(f_0) > t$. If $g := f_0 \wedge (\mathbf{1}-f) = 0$, then $f_0 = f_0 - f_0 \wedge (\mathbf{1}-f) \sim f_0 \vee (\mathbf{1}-f)-(\mathbf{1}-f) \le f$, so $tr(f_0) \le t$, contradicting our assumption. Since $E_N(f_0) \in N \cap R \cap R'_\beta = R_\beta \cap R'_\beta = \mathbb{C}\mathbb{1}$, it follows that $E_N(f_0) = tr(f_0)$. Thus

$$\begin{aligned} \tfrac{1}{2}\mathbf{1} \ge tr(f_0)\mathbb{1} = E_N(f_0) &\ge E_N(g) && \text{(since } f_0 \ge g) \\ &= (1-t)(\theta^{-1}(g)+g) && \text{(since } (\mathbf{1}-f) \ge g). \end{aligned}$$

Since $g \ne 0$, this implies $\frac{1}{2} \ge 1-t$, contradicting the definition of t. It follows that $tr(f_0) = t$.

Define $f_1 = f - f \wedge f_0$ and $f_2 = s((\mathbb{1}-f)f_0(\mathbb{1}-f))$, where $s(\cdot)$ denote support. Remark that

$$s((f_0-f\wedge f_0)(\mathbb{1}-f)(f_0-f\wedge f_0)) = f_0-f\wedge f_0$$

because $(f_0-f\wedge f_0) \wedge f = 0$. Hence

$$\begin{aligned} tr(f_2) &= tr[s((\mathbb{1}-f)(f_0-f\wedge f_0)(\mathbb{1}-f))] \\ &= tr[s((f_0-f\wedge f_0)(\mathbb{1}-f)(f_0-f\wedge f_0))] \\ &= (f_0-f\wedge f_0) = t - tr(f\wedge f_0) \\ &= tr(f_1). \end{aligned}$$

Write $h = \theta(f\wedge f_0)$. We have

$$\begin{aligned} t\mathbf{1} &= E_N(f_0) \\ &= E_N(ff_0f + (\mathbb{1}-f)f_0(\mathbb{1}-f)) \\ &\ge E_N(f\wedge f_0 + (\mathbb{1}-f)f_0(\mathbb{1}-f) && \text{since } E_N((1-f)f_0f) = 0 \end{aligned}$$

$$= t(f\wedge f_0+h) + (1-t)(\theta^{-1}(\mathbb{1}-f)f_0(\mathbb{1}-f) + (\mathbb{1}-f)f_0(\mathbb{1}-f))$$
$$\geq th + (1-t)(\mathbb{1}-f)f_0(\mathbb{1}-f).$$

Hence

$$th \geq th + (1-t)h(\mathbb{1}-f)f_0(\mathbb{1}-f)h,$$

so $h(\mathbb{1}-f)f_0(\mathbb{1}-f)h = 0$. It follows that

$$0 = hs((\mathbb{1}-f)f_0(\mathbb{1}-f)) = hf_2,$$

so $\theta(f\wedge f_0) \leq \mathbb{1}-f-f_2$. On the other hand, multiplying the equation

$$t\mathbb{1} = E_N(f_0) = t[ff_0f+(ff_0f)] + (1-t)[\theta^{-1}((\mathbb{1}-f)f_0(\mathbb{1}-f)) + (\mathbb{1}-f)f_0(\mathbb{1}-f)]$$

by $\mathbb{1}-f-f_2$ gives

$$t(\mathbb{1}-f-f_2) = t\theta(ff_0f)(\mathbb{1}-f-f_2),$$

which implies that $\theta(f_0\wedge f) \geq (\mathbb{1}-f-f_2)$. Hence $\theta(f_0\wedge f) = \mathbb{1}-f-f_2$.

Finally we compute

$$\begin{aligned} tr(f_0\wedge f) &= \tfrac{t}{1-t}\, tr\ \theta(f_0\wedge f) = \tfrac{t}{1-t}\, tr(\mathbb{1}-f-f_2) \\ &= \tfrac{t}{1-t}(1-t-tr(f_1)) \\ &= \tfrac{t}{1-t}[1-2t + tr(f_0\wedge f)]. \end{aligned}$$

Solving gives $(1-2t)\ tr(f_0\wedge f) = (1-2t)t$, or, since $1-2t > 0$, $tr(f_0\wedge f) = t = tr(f) = tr(f_0)$. It follows that $f = f\wedge f_0 = f_0$ and $f \in R \cap R'_\beta$, as was to be shown.

If $x \in R$, then $t^{-1}E_N(xfx) = fxf + \theta(fxf)$ lies in $R \cap N = R_\beta$, and $fxf = t^{-1}E_N(cfc)f \in R_\beta f$. Thus $fRf = R_\beta f$. Similarly $(\mathbb{1}-f)R(\mathbb{1}-f) = R_\beta(\mathbb{1}-f)$. #

<u>Corollary 4.7.4.</u> [Jo1] Let $\beta > 4$, and let $R \supset R_\beta$ be the pair of factors in Proposition 3.4.4. Then $R \cap R'_\beta \supsetneq \mathbb{C}\mathbb{1}$.

<u>Proof.</u> Let M be the hyperfinite II_1 factor, and let $f \in M$ be a projection of trace $t = \frac{1}{2} - \frac{1}{2}\sqrt{1-4\beta^{-1}}$, so that $t^{-1}(1-t)^{-1} = \beta$. Then there is an isomorphism $\theta : fMf \to (\mathbb{1}-f)M(\mathbb{1}-f)$. Define the factor N and the projections $\{e_0, e_{-1}, e_{-2}, \cdots\}$ as in Theorem

4.7.3. Then $f \in \{e_0, e_{-1}, e_{-2}, \cdots\}'' \cap \{e_{-1}, e_{-2}, \cdots\}'$. But the the pair $R \supset R_\beta$ is isomorphic the the pair $\{e_0, e_{-1}, e_{-2}, \cdots\}'' \supset \{e_{-1}, e_{-2}, \cdots\}''$. #

We now describe a representation of the e_i's due to Pimsner and Popa [PP 1]. Let A be the infinite tensor product of $\mathrm{Mat}_2(\mathbb{C})$, that is, the inductive limit of the algebras $\bigotimes_1^n \mathrm{Mat}_2(\mathbb{C})$, each imbedded into the next via $x \longmapsto x \otimes \mathbb{1}$. It is well known that A has a unique C^*-norm and that $\overline{A}$, the C^*-completion of A, is a simple C^*-algebra.

Fix $\beta > 4$ and let $\lambda \in]0,1[$ satisfy $2+\lambda+\lambda^{-1} = \beta$. Define a state η on $\mathrm{Mat}_2(\mathbb{C})$ by $\eta(x) = \frac{1}{1+\lambda} \mathrm{tr}\left[\begin{bmatrix} 1 & 0 \\ 0 & \lambda \end{bmatrix} x\right]$ and let φ be the corresponding Powers state [P] on $\overline{A}$ defined on A by $\varphi(x_1 \otimes x_2 \otimes \cdots \otimes x_n) = \prod_{i=1}^n \eta(x_i)$. Let N be the weak closure of $\overline{A}$ in the GNS-representation corresponding to φ, the Powers factor of type III_λ.

There is an obvious shift endomorphism on A,

$$\alpha : x_1 \otimes x_2 \otimes \cdots \otimes x_n \mapsto \mathbb{1} \otimes x_1 \otimes \cdots \otimes x_n$$

which preserves φ and so extends to N. Define $e_1 \in \mathrm{Mat}_2(\mathbb{C}) \otimes \mathrm{Mat}_2(\mathbb{C})$ by

$$e_1 = \frac{1}{1+\lambda}(e_{11} \otimes e_{22} + \sqrt{\lambda}\,(e_{12} \otimes e_{21} + e_{21} \otimes e_{12}) + \lambda\, e_{22} \otimes e_{11}),$$

or

$$e_1 = \frac{1}{1+\lambda}\begin{bmatrix} 0 & 0 & 0 & 0 \\ 0 & 1 & \sqrt{\lambda} & 0 \\ 0 & \sqrt{\lambda} & \lambda & 0 \\ 0 & 0 & 0 & 0 \end{bmatrix}$$

under an appropriate identification of $\mathrm{Mat}_2(\mathbb{C}) \otimes \mathrm{Mat}_2(\mathbb{C})$ with $\mathrm{Mat}_4(\mathbb{C})$. Let $e_i = \alpha^{i-1}(e_1)$ for $i \geq 2$. It is a bit tedious but straightforward to check that $\{e_i : i \geq 1\}$ satisfy

$$\beta\, e_i e_{i\pm1} e_i = e_i,$$
$$e_i e_j = e_j e_i, \qquad |i-j| \geq 2,$$

and that $\varphi|_{\mathrm{alg}\{\mathbb{1}, e_1, e_2, \ldots\}}$ is the Markov trace of modulus β. The circle group acts on A by the infinite tensor product of its action $z \longmapsto \mathrm{Ad}\begin{bmatrix} z & 0 \\ 0 & z^{-1} \end{bmatrix}$ on $\mathrm{Mat}_2(\mathbb{C})$; this action preserves φ and extends to an action on N. One checks that the projections e_i are $\mathbb{T}$-invariant. Equivalently (since the $\mathbb{T}$-action is actually the modular automorphism group for φ) the e_i's are in the normalizer of φ, $\varphi(e_i y) = \varphi(y e_i)$ for all $y \in N$. We shall apply Lemma 4.7.1 with $M = N \otimes \mathrm{Mat}_2(\mathbb{C})$, $P = N \otimes \mathrm{Mat}_2(\mathbb{C}) \otimes \mathrm{Mat}_2(\mathbb{C})$, $G = \mathbb{T}$, and the state $\varphi \otimes \eta \otimes \eta$ on P, in order to compute the derived tower for $R_\beta \subset R$. The key result is the following, from [PP1].

Theorem 4.7.5. *With the notation above,* $\{e_1, e_2, \cdots\}''$ *is the fixed-point algebra* $N^{\mathbb{T}}$ *for the* $\mathbb{T}$ *action on* N.

Proof. A proof of this result using the Connes–Stormer relative entropy is in [PP1]. We shall give a simplified proof due to Popa which is based on his Theorem 4.7.3.

First we observe that $\{\mathbb{1}, e_1, e_2, \cdots\}$ together with the projection $f = e_{11} \in \mathrm{Mat}_2(\mathbb{C})$ generate $N^{\mathbb{T}}$. In fact $f e_1 f = \frac{1}{1+\lambda} e_{11} \otimes e_{22}$ and $(\mathbb{1}-f) e_1 (\mathbb{1}-f) = \frac{\lambda}{1+\lambda} e_{22} \otimes e_{11}$, so that $\mathrm{alg}\{\mathbb{1}, f, e_1\}$ contains the self-adjoint unitary $\sigma = \sum_{i,j=1}^{2} e_{i,j} \otimes e_{j,i}$ as well as the diagonal algebra $\{d_1 \otimes d_2 : d_i$ is a diagonal 2-by-2 matrix$\}$, and contains in particular $\alpha(f) = \mathbb{1} \otimes e_{11}$. By induction, $\mathrm{alg}\{\mathbb{1}, f, e_1, e_2, \cdots\}$ contains the infinite tensor product of the 2-by-2 diagonal algebra, as well as $\{\alpha^k(\sigma) : k \geq 0\}$. It is a well known fact of invariant theory that, on the Fock space $\bigoplus_{k=0}^{\infty} (\otimes^k \mathbb{C}^2)$ over $\mathbb{C}^2$, the commutant of the direct sum of the tensor powers of the representation $z \mapsto \begin{bmatrix} z & 0 \\ 0 & z^{-1} \end{bmatrix}$ of $\mathbb{T}$ on $\mathbb{C}^2$ is the algebra generated by the permutations $\{\alpha^k(\sigma) : k \geq 0\}$ together with the infinite diagonal algebra. This implies that $\overline{A}^{\mathbb{T}} = C^*\{\mathbb{1}, f, e_1, e_2, \cdots\}$ and $N^{\mathbb{T}} = \{\mathbb{1}, f, e_1, e_2, \cdots\}''$. Therefore it suffices to prove that $f \in \{e_1, e_2, \cdots\}''$.

The equality $N^{\mathbb{T}} = \{\mathbb{1}, f, e_1, e_2, \cdots\}''$ implies that $f N^{\mathbb{T}} f = \alpha(N^{\mathbb{T}}) f$ and $(\mathbb{1}-f) N^{\mathbb{T}} (\mathbb{1}-f) = \alpha(N^{\mathbb{T}})(\mathbb{1}-f)$, so that $\theta : xf \mapsto x(\mathbb{1}-f)$ $(x \in \alpha(N^{\mathbb{T}}))$ is an isomorphism of $f N^{\mathbb{T}} f$ onto $(\mathbb{1}-f) N^{\mathbb{T}} (\mathbb{1}-f)$ and $\alpha(N^{\mathbb{T}}) = \{x + \theta(x) : x \in f N^{\mathbb{T}} f\}$. Thus the pair $\alpha(N^{\mathbb{T}}) \subset N^{\mathbb{T}}$ is of the type discussed in Lemma 4.7.2, and therefore $[N^{\mathbb{T}} : \alpha(N^{\mathbb{T}})] = \mathrm{tr}(f)^{-1} + \mathrm{tr}(\mathbb{1}-f)^{-1} = 2 + \lambda + \lambda^{-1} = \beta$.

We can now verify that the conditions of Lemma 4.7.1 are satisfied by the algebras $\alpha^2(N) \subset \alpha(N) \subset N$, the state φ, the projection e_1, and the group $\mathbb{T}$, and therefore we can identify the inclusion $\alpha(N^{\mathbb{T}}) \subset N^{\mathbb{T}}$ as coming from the fundamental construction applied to $\alpha^2(N^{\mathbb{T}}) \subset \alpha(N^{\mathbb{T}})$, the projection e_1 being identified with $e_{\alpha^2(N^{\mathbb{T}})}$. Applying the endomorphism α repeatedly, we have that $\alpha^k(N^{\mathbb{T}}) = \langle \alpha^{k+1}(N^{\mathbb{T}}), e_{k+1} \rangle$ is isomorphic to the result of the fundamental construction for $\alpha^{k+2}(N^{\mathbb{T}}) \subset \alpha^{k+1}(N^{\mathbb{T}})$, and the projections $\{e_1,e_2,\cdots\}$ are those of a tunnel construction for $\alpha(N^{\mathbf{T}}) \subset N^{\mathbf{T}}$. Therefore the conclusion $f \in \{e_1,e_2,\cdots\}''$ follows from Popa's theorem 4.7.3. #

There is no further difficulty in computing the entire derived tower for $R_\beta \subset R$. Using 4.7.1 and 4.7.5, the tower for the pair $R_\beta \subset R$ can be identified with

$$N^{\mathbf{T}} \subset (N \otimes \mathrm{Mat}_2(\mathbb{C}))^{\mathbf{T}} \subset (N \otimes \mathrm{Mat}_2(\mathbb{C}) \otimes \mathrm{Mat}_2(\mathbb{C}))^{\mathbf{T}} \subset \cdots .$$

It is well known that $(N^{\mathbf{T}})' \cap N = \mathbb{C}\mathbf{1}$. In fact, the infinite symmetric group $\mathfrak{S}_\infty$ acts ergodically on N (by permutation of the tensorands), and this action is implemented by unitaries in $N^{\mathbf{T}}$; the image of $\mathfrak{S}_\infty$ in $N^{\mathbf{T}}$ is the group generated by $\{\alpha^k(\sigma) : k \geq 0\}$. Therefore $(N^{\mathbf{T}})' \cap (N \otimes (\overset{k}{\otimes} \mathrm{Mat}_2(\mathbb{C})) = \mathbf{1} \otimes (\overset{k}{\otimes} \mathrm{Mat}_2(\mathbb{C}))$ and $(N^{\mathbf{T}})' \cap (N \otimes (\overset{k}{\otimes} \mathrm{Mat}_2(\mathbb{C}))^{\mathbf{T}} = \mathbf{1} \otimes (\overset{k}{\otimes} \mathrm{Mat}_2(\mathbb{C}))^{\mathbf{T}}$. That is, the derived tower is precisely the sequence of fixed-point algebras for the tensor product action of $\mathbf{T}$ on $\overset{k}{\otimes} \mathrm{Mat}_2(\mathbb{C})$ $(k \geq 0)$. The Bratteli diagram is Pascal's triangle [Bra]:

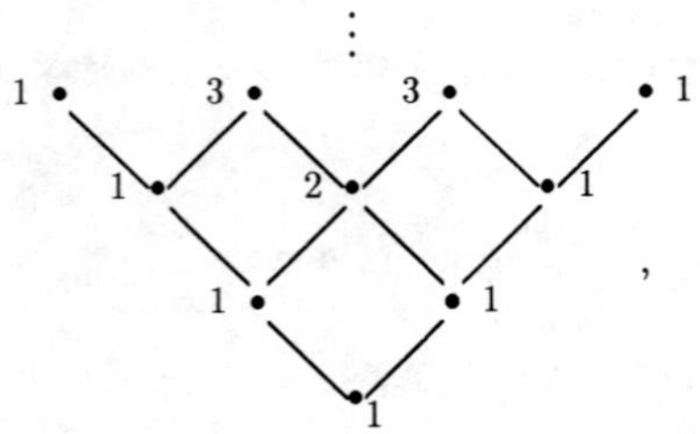

and the principal graph is

$A_{\infty,\infty} =$ 

Note that it differs from the graph for $\beta = 4$ (Section 4.7.b).

APPENDIX I

Classification of Coxeter graphs with spectral radius just beyond the Kronecker range

I.1. The results.

The first purpose of this appendix is to complement section 1.4 and to classify finite connected graphs satisfying $2 < \|\Gamma\| \le \lambda_\infty$ where we set

$$\lambda_\infty = \sqrt{\sqrt{5} + 2} \approx 2{,}058171.$$

The results are those of Cvetković, Doob and Gutman [CDG]. Then we also classify Coxeter graphs with norms satisfying the same inequality.

The following graphs enter the classification. First, the $T_{p,q,r}$'s already introduced in Section 1.4. Next, for an integer $m \ge 3$, we set

$T_{p,q,r}(m)$

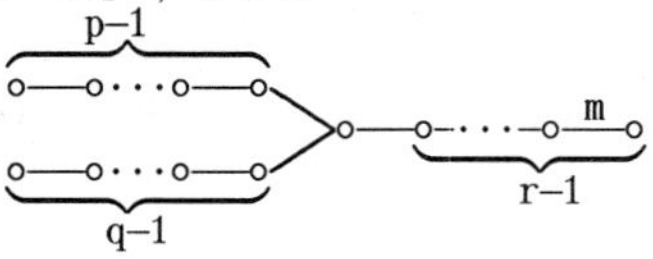

As usually for Coxeter graphs, edges to which the associated integer is 3 are left unmarked, so that

$$T_{p,q,r}(3) = T_{p,q,r}.$$

Given integers p, q, r, m, m′ with $2 \le p \le q \ge 3$, $r \ge 5$ and $m,m' \ge 3$, we also define the H-shaped Coxeter tree

$H_{p,q,r}(m,m')$

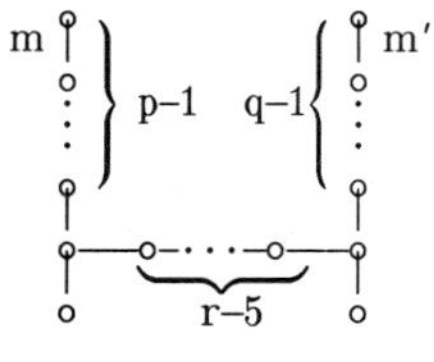

with $p+q+r-3$ vertices.

The constant λ_∞ makes its first appearance in a result of Hoffman (Proposition 3.7 of [Hof]):

Proposition I.1.1. *Let* Γ *be a finite connected graph which is not a cycle and which contains a cycle. Then* $\|\Gamma\| > \lambda_\infty$. *Moreover* λ_∞ *is the largest constant for which this holds.*

One important result about the set $\mathbf{E}$ of norms of graphs (see section 1.5) is also essentially due to Hoffman [Hof], and can be stated in terms of a sequence $(\lambda_q)_{q\geq 2}$ of numbers defined as follows. Let ν_q be the largest real root of the polynomial

$$L_q(\nu) = \nu^q - (\nu^{q-2} + \nu^{q-3} + \cdots + \nu + 1).$$

We set

$$\lambda_q = \nu_q^{1/2} + \nu_q^{-1/2}.$$

If $\nu_\infty = \frac{1}{2}(5^{1/2}+1)$ is the golden mean, namely, the positive root of $\nu^2-\nu-1$, it follows by induction on q from the identity $L_{q+1}(\nu) - L_q(\nu) = \nu^{q-1}(\nu^2-\nu-1)$ that

$$1 = \nu_2 < \nu_3 < \cdots < \nu_q < \nu_{q+1} < \cdots < \nu_\infty \approx 1{,}618034.$$

As $\lambda_\infty = \nu_\infty^{1/2} + \nu_\infty^{-1/2}$ one has also

$$2 = \lambda_2 < \lambda_3 < \cdots < \lambda_q < \lambda_{q+1} < \cdots < \lambda_\infty \approx 2{,}058171.$$

Theorem I.1.2. *The accumulation points of* $\mathbf{E} \cap [0,\lambda_\infty]$ *are precisely the* λ_q's *for* $q = 2,3,\cdots,\infty$ *and* λ_3 *is maximal among real numbers* c *such that* $\mathbf{E} \cap [0,c]$ *is well ordered. Moreover,*

(i) $\|A_k\|$ *and* $\|D_k\|$ *both increase strictly with* k *and converge to* $\lambda_2 = 2$. *One has* $\mathbf{E} \cap\,]2,d[\, = \phi$ *where* $d = \|T_{2,3,7}\|$ *is the square root of the largest root of the polynomial* $\rho^5 - 9\rho^4 + 27\rho^3 - 31\rho^2 + 12\rho - 1$.

(ii) *For* $q \geq 3$ *and* p *with* $2 \leq p \leq q$:

$\|T_{2,q,r}\|$ *increases strictly with* r *and converges to* λ_q,

$\|T_{3,3,r}\|$ *increases strictly with* r *and converges to* λ_∞,

$\|H_{p,q,r}\|$ *decreases strictly with* r *and converges to* λ_q.

To state the corresponding classification of Coxeter graphs, we set

F_4^{++} 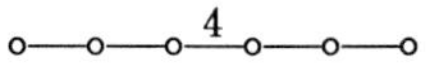

F_k 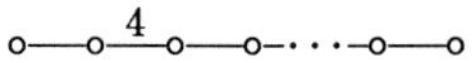 k vertices

H_k 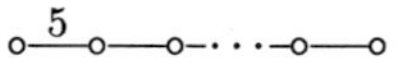 k vertices

Theorem I.1.3. *Let* Γ *be a finite connected Coxeter graph with* $2 < \|\Gamma\| \leq \lambda_\infty$. *Then* Γ *is one of the following.*

F_4^{++} *with norm approximately* 2,053;

F_k *with* $k \geq 6$; *the norms* $\|F_k\|$ *increase and tend to* λ_∞;

H_k *with* $k \geq 5$; *the norms* $\|H_k\|$ *increase and tend to* λ_∞;

$T_{2,q,r}$ *with* $3 \leq q \leq r$, *and* $r \geq 7$ *if* $q = 3$, *and* $r \geq 5$ *if* $q = 4$; *the norms increase and tend to* λ_q;

$T_{2,q,r}(4)$ *with* $3 \leq q \leq r$ *and* r *large enough; the norms decrease and tend to* λ_q;

$T_{3,3,r}$ *with* $r \geq 4$; *the norms increase and tend to* λ_∞;

$T_{3,4,4}$ *with norm approximately* 2,053;

$H_{p,q,r}(m,m')$ *with* $2 \leq p \leq q \geq 3$ *and* r *large enough, with* $m,m' \in \{3,4\}$ *and* $m = 3$ *if* $p = 2$; *the norms decrease and tend to* λ_q.

The classification of matrices $X \in M_{fin}(\mathbb{E})$ with $\|X\| \leq \lambda_\infty$ (up to pseudo-equivalence) follows from Theorems 1.1.3 and I.1.3.

Remarks.

(1) The number $\|F_4^{++}\|^2 = \|T_{3,4,4}\|^2$ is the largest root of $\rho^3 - 6\rho^2 + 8\rho - 2$.

(2) The polynomials $\frac{L_2(\nu)}{\nu+1}$, $L_3(\nu)$, $\frac{L_4(\nu)}{\nu+1}$ are irreducible in $\mathbb{Z}[\nu]$. We have not checked whether $\frac{L_{2p}(\nu)}{\nu+1}$ and $L_{2p+1}(\nu)$ are irreducible for larger p's.

(3) The list of theorem I.1.3 is neither including nor included in other lists of Coxeter graphs "just larger" than those of theorem 1.1.3, such as the list of hyperbolic graphs in [Che] and [Kos], or the list of trees with Lorentzian associated quadratic form in [Mxl].

(4) The consideration of infinite graphs brings no surprise [Tor].

The proofs which follow are very elementary, though rather tedious. Many of the partial results have been checked numerically with a computer. It is a pleasure to thank F. Ronga, G. Wanner, and E. Hairer for crucial help with the computations.

I.2. Computations of characteristic polynomials for ordinary graphs.

Most computations below are by induction, based on the following.

<u>Lemma I.2.1</u>. *Let* Γ *be a graph and let* v *be a vertex of* Γ. *Let* Γ_- *be the graph obtained from* Γ *by deleting* v *and all edges ending at* v; *let* Γ_+ *be the graph obtained from* Γ *by adding a vertex* v_+ *and an edge between* v *and* v_+. *Denote by* P, P_- *and* P_+ *the characteristic polynomials of* Γ, Γ_- *and* Γ_+. *Then*

$$P_+(\lambda) = \lambda P(\lambda) - P_-(\lambda).$$

<u>Proof</u>: By standard expansion of determinants; more similar lemmas in [Sch1]. #

One example is provided by

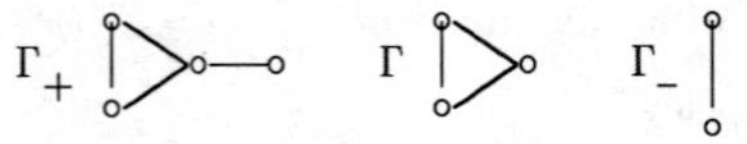

Let $P \in \mathbb{Z}[\lambda]$ be the characteristic polynomial of some graph Γ. For our computations, it helps to consider larger rings

$$\mathbb{Z}[\lambda] \subset \mathbb{Z}[\lambda, \sqrt{\lambda^2-4}] \subset \mathbb{Z}[\mu, \mu^{-1}]$$

with

$$\lambda = \mu + \mu^{-1} \qquad 2\mu = \lambda + (\lambda^2-4)^{1/2}$$

$$(\lambda^2-4)^{1/2} = \mu - \mu^{-1} \qquad 2\mu^{-1} = \lambda - (\lambda^2-4)^{1/2}$$

$$\mu^2 - \lambda\mu + 1 = 0.$$

For example, we write $P(\lambda) = \mu^3 + \mu^{-3} - 2$ to mean $P(\lambda) = \lambda^3 - 3\lambda - 2$. The function

$$\begin{cases} [1, \infty[\longrightarrow [2, \infty[\\ \mu \longrightarrow \mu + \mu^{-1} \end{cases}$$

is strictly increasing and bijective. When $P \in \mathbb{Z}[\lambda]$ is given by a function f of μ, this has the following consequence: for $\mu_0 \in [1, \infty[$, one has $f(\mu_0) = 0$ if and only if $P(\lambda_0) = 0$, with $\lambda_0 = \mu_0 + \mu_0^{-1}$; and $\|\Gamma\| = \lambda_0$ if and only if μ_0 is the largest root of f. This will be used constantly below.

For any integer $k \geq 1$, we denote by P_k^A the characteristic polynomial of A_k.

Proposition I.2.2. *One has* $P_k^A(\lambda) = (\mu-\mu^{-1})^{-1}(\mu^{k+1}-\mu^{-k-1}) =$

$$\prod_{1\leq j\leq k}(\lambda\text{-}2\cos(\pi j/(k+1))) \quad \textit{and} \quad \|A_k\| = 2\cos(\pi/(k+1)).$$

Proof. From the previous lemma one has the difference equation

$$P_{k+1}^A(\lambda) - \lambda P_k^A(\lambda) + P_{k-1}^A(\lambda) = 0$$

which holds also for $k = 1$ if $P_0^A(\lambda) = 1$. The indicial equation being $\mu^2 - \lambda\mu + 1 = 0$, the general solution of the difference equation is $P_k(\lambda) = C\mu^k + D\mu^{-k}$, where C and D are constants in $\mathbb{Z}[\mu,\mu^{-1}]$ independent of k. Adjustment of C and D to fit P_0^A and P_1^A provide the formula with μ's. Roots of $P_k^A(\lambda) = \mu^{-k}(\mu^2-1)^{-1}(\mu^{2k+2}-1)$ are given by $\mu^2 = \exp(2i\pi j/(k+1))$, namely by

$$\lambda = \mu + \mu^{-1} = 2\cos(\pi j/(k+1))$$

for $j = 1,\cdots,k$. The largest of these is $2\cos(\pi/(k+1))$. #

Observe that the roots of P_k^A are $2\cos(\pi m j/h)$ where $h = k+1$ is the Coxeter number of A_k and where $m_1,\cdots,m_k$ are its Coxeter exponents $1,2,\cdots,k$.

For any integer $k \geq 2$, we denote by $P_k^{A,1}$ the characteristic polynomial of $A_k^{(1)}$ (the cycle with $k+1$ vertices) by $A_k^{(1,1)}$ the graph with $k+2$ vertices

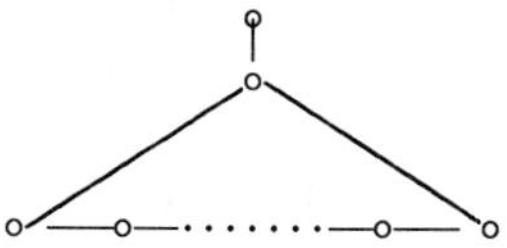

and by $P_k^{A,1,1}$ its characteristic polynomial.

Lemma I.2.3. *One has* $P_k^{A,1}(\lambda) = \mu^{k+1} + \mu^{-k-1} - 2$ *and* $\|A_k^{(1)}\| = 2$. *One has*

$$P_k^{A,1,1}(\lambda) = \frac{\mu^{k-1}-\mu^{-2}}{\mu-\mu^{-1}}\{\mu^4-\mu^2-1-\mu^{-k-1}(\mu^4+\mu^2-1)\}$$

so that $\|A_k^{(1,1)}\|$ *decreases strictly with* k *and converges to* $\lambda_\infty = (5^{1/2}+2)^{1/2}$.

Proof. Expansion of the determinant defining $P_k^{A,1}$ along the first column gives $P_k^{A,1}(\lambda)-\lambda P_k^A(\lambda)$ as a sum of two k-by-k determinants; expansions of these along their first lines give

$$P_k^{A,1}(\lambda) = \lambda P_k^A(\lambda) - 2P_{k-1}^A(\lambda)-2$$

and the formula for $P_k^{A,1}$. The largest root of $P_k^{A,1}$ is given by $\mu = 1$, namely by $\lambda = 2$ (see also lemma 1.4.1).

The formula for $P_k^{A,1,1}$ follows by lemma I.2.1. Set $\mu_\infty = (\frac{1}{2}(5^{1/2}+1))^{1/2} = \nu_\infty^{1/2}$ (with ν_∞ as in I.1). If $\mu = \mu_\infty$ one has $\mu^4-\mu^2-1 = 0$ and

$$\mu^4-\mu^2-1-\mu^{-k-1}(\mu^4+\mu^2-1) = -2\mu_\infty^{-k+1} < 0.$$

Consequently the left-hand side has a root larger than μ_∞, and $\|A_k^{(1,1)}\| > \lambda_\infty = \mu_\infty + \mu_\infty^{-1}$ for all $k \geq 2$. Let $\mu > \mu_\infty$; then $\mu^4-\mu^2-1 > 0$ and $P_k^{A,1,1}(\mu+\mu^{-1}) > 0$ for k large enough. Consequently $\lim_{k\to\infty}\|A_k^{(1,1)}\| = \lambda_\infty$. It is clear that $\|A_k^{(1,1)}\|$ is strictly decreasing when k increases; this is a particular case of the next lemma. #

Proposition I.1.1 follows from this lemma and from 1.4.2.

The following lemma is reproduced from [HS]. Let $\ell = (u,w)$ be an edge of a connected graph Γ; we denote by $\hat\Gamma_{(u,w)}$ or simply by $\hat\Gamma$ the graph obtained from Γ as follows: delete ℓ, add a vertex v, add two edges $\ell_- = (u,v)$ and $\ell_+ = (v,w)$. For example,

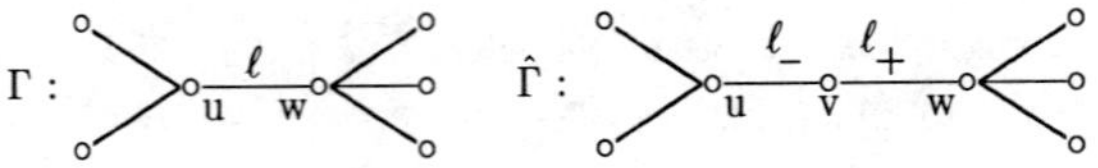

Say that ℓ is internal if there exist an integer $n \geq 1$ and a sequence of vertices $x_0,\cdots,x_j = u, x_{j+1} = w,\cdots,x_n$ such that all x_i are distinct (except possibly $x_0 = x_n$), where the degrees $d(x_i)$ satisfy

$$d(x_0) \geq 3, \quad d(x_1) = \cdots = d(x_{n-1}) = 2, \quad d(x_n) \geq 3$$

and where x_{i-1} is adjacent to x_i for $i = 1,\cdots,n$.

Lemma I.2.4. *Let* Γ *be a connected graph with an internal edge* $\ell = (u,w)$; *assume that* Γ *is not one of the* $D_k^{(1)}$*'s. Then* $\|\hat{\Gamma}_{(u,w)}\| < \|\Gamma\|$.

Proof. Let $x_0,\cdots,x_n$ be as above. We shall assume that $x_0 \neq x_n$ (the case $x_0 = x_n$, substantially easier, is left as an exercise, and is also done with details in [HS]). Let $x_{-m},\cdots,x_{-1}$ be vertices adjacent to x_0 other than x_1; by hypothesis $m \geq 2$. Let ξ be a Perron–Frobenius vector for the adjacency matrix Y of Γ; one has $Y\xi = \|\Gamma\|\xi$; also $\|\Gamma\| > 2$ because there exists an integer k such that $D_k^{(1)}$ is a proper subgraph of Γ. We write $\hat{\Gamma}$ instead of $\hat{\Gamma}_{(u,w)}$; we denote by $\hat{V}$ the set of vertices of $\hat{\Gamma}$ and by $\hat{Y}$ its adjacency matrix. Suppose one can find a positive vector $\hat{\xi} \in \mathbb{R}^{\hat{V}}$ with $\hat{Y}\hat{\xi} \lneqq \|\Gamma\|\hat{\xi}$; it will then follow from Perron–Frobenius theory that $\|\hat{\Gamma}\| < \|\Gamma\|$; see lemma 2 and remark 4 in § XIII.2 of [Gan]. We distinguish three cases.

Assume first that there exists $t \in \{1,\cdots,n-1\}$ with $\xi_t \leq \xi_i$ for $i = 0,\cdots,n$. There is no loss of generality in assuming $x_t = u$ and $x_{t+1} = w$. One has

$$(Y\xi)_t = \xi_{t-1} + \xi_{t+1} = \|\Gamma\|\xi_t > 2\xi_t. \qquad (*)$$

Define $\hat{\xi}$ to be the vector with coordinate corresponding to v given by $\hat{\xi}_v = \xi_t$ and with other coordinates as those of ξ. One has

$$(\hat{Y}\hat{\xi})_u = \xi_{t-1} + \hat{\xi}_v \leq \xi_{t-1} + \xi_{t+1} = \|\Gamma\|\hat{\xi}_u$$

$$(\hat{Y}\hat{\xi})_v = \xi_t + \xi_{t+1} \leq \xi_{t-1} + \xi_{t+1} = \|\Gamma\|\hat{\xi}_v$$

with at least one inequality being strict by (*). Hence $\hat{Y}\hat{\xi} \leq \|\Gamma\|\hat{\xi}$ and $\hat{Y}\hat{\xi} \neq \|\Gamma\|\hat{\xi}$.

For the two next cases, $\xi_0 \leq \xi_i$ for $i = 0,\cdots,n$, and we assume $x_0 = u$, $x_1 = w$ without loss of generality.

Assume now that $\sum_{i=1}^{m} \xi_{-i} \geq \xi_0$. One has

$$(Y\xi)_0 = \sum_{i=1}^{m} \xi_{-i} + \xi_1 = \|\Gamma\|\xi_0 > 2\xi_0. \qquad (**)$$

Define $\hat{\xi}$ by $\hat{\xi}_v = \xi_0$, the other coordinates being as those of ξ. One has

$$(\hat{Y}\hat{\xi})_u = \sum_{i=1}^{m} \xi_{-i} + \xi_0 \leq \sum_{i=1}^{m} \xi_{-i} + \xi_1 = \|\Gamma\|\xi_0$$

$$(\hat{Y}\hat{\xi})_v = \xi_0 + \xi_1 \leq \sum_{i=1}^{m} \xi_{-i} + \xi_1 = \|\Gamma\|\xi_0$$

with at least one inequality being strict by (**). The conclusion follows.

Assume finally that $\sum_{i=1}^{m} \xi_{-i} < \xi_0$. One has $\xi_0 \leq \|\Gamma\|\xi_{-i}$ for $i = 1,\cdots,m$. Since $m \geq 2$, this implies $2\xi_0 \leq m\xi_0 \leq \|\Gamma\| \sum_{i=1}^{m} \xi_{-i}$ and consequently

$$\sum_{i=1}^{m} \xi_{-i} + \xi_0 < \|\Gamma\| \sum_{i=1}^{m} \xi_{-i}. \qquad (***)$$

Define $\hat{\xi}$ by $\hat{\xi}_u = \sum_{i=1}^{m} \xi_{-i}$ and $\hat{\xi}_v = \xi_0$, the other coordinates being as those of ξ. One has

$$(\hat{Y}\hat{\xi})_u = \sum_{i=1}^{m} \xi_{-i} + \xi_0 < \|\Gamma\| \sum_{i=1}^{m} \xi_{-i}$$

by (***). The proposition follows. #

One may also apply lemma I.2.4 to Coxeter graphs such that, notations being as above, edges (x_{i-1},x_i) for $i = 1,\cdots,n$ are marked with $m = 3$.

For q,r with $2 \leq q \leq r$, we denote by $P^T_{2,q,r}$ the characteristic polynomial of the graph $T_{2,q,r}$. For $k \geq 4$ and $k \geq 6$, one has respectively $P^D_k = P^T_{2,2,k-2}$ and $P^E_k = P^T_{2,3,k-3}$.

Proposition I.2.5. *With* $\nu = \mu^2$ *one has*

$$P^T_{2,q,r}(\lambda) = (\mu-\mu^{-1})^{-1}\{\mu^{r-q+1}[\nu^q-(\nu^{q-2}+\cdots+1)] - \mu^{-r+q-1}[\nu^{-q}-(\nu^{-q+2}+\cdots+1)]\}.$$

In particular,

$$P^D_k(\lambda) = (\mu+\mu^{-1})(\mu^{k-1}+\mu^{-k+1}) = \prod_{1\leq j\leq k}(\lambda-2\cos(\pi m_j/h))$$

with $h = 2k-2$ and $m_1,\cdots,m_k = 1,3,\cdots,2k-3,k-1$;

$$P^E_6(\lambda) = \prod_{1\leq j\leq 6}(\lambda-2\cos(\pi m_j/12)) \quad \text{with } m_j = 1,4,5,7,8,11;$$

$$P^E_7(\lambda) = \prod_{1\leq j\leq 7}(\lambda-2\cos(\pi m_j/18)) \quad \text{with } m_j = 1,5,7,9,11,13,17;$$

$$P^E_8(\lambda) = \prod_{1\leq j\leq 8}(\lambda-2\cos(\pi m_j/30)) \quad \text{with } m_j = 1,7,11,13,17,19,23,29;$$

so that

$$\|D_k\| = 2\cos(\pi/(2k-2)) \qquad \|E_6\| = 2\cos(\pi/12)$$

$$\|E_7\| = 2\cos(\pi/18) \qquad \|E_8\| = 2\cos(\pi/30).$$

Proof. The formula for $P^T_{2,q,r}$ follows from I.2.1 and I.2.2. Roots of $(\mu+\mu^{-1})(\mu^{k-1}+\mu^{-k+1})$ are given by $\mu = \pm i$ and $\mu = \exp\left\{\frac{i\pi}{2k-2}(1+2j)\right\}$ for $j = 0,1,\cdots,2k-3$. As one value of μ and its inverse correspond to the same value of λ, roots of P^D_k are given by $\lambda = 0$ and $\lambda = 2\cos\left\{\frac{\pi}{2k-2}(1+2j)\right\}$ for $j = 0,1,\cdots,k-2$; the product formula for P^D_k follows.

From the formula for P^T or from a direct computation,

$$P^E_6(\lambda) = \lambda^6 - 5\lambda^4 + 5\lambda^2 - 1$$

$$P^E_7(\lambda) = \lambda(\lambda^6 - 6\lambda^4 + 9\lambda^2 - 3)$$

$$P^E_8(\lambda) = \lambda^8 - 7\lambda^6 + 14\lambda^4 - 8\lambda^2 + 1.$$

The product formulas for the P^E's may be checked as follows. Firstly, the roots of P^E_6 are $\lambda = \pm 1$ and $\lambda = \pm\frac{3^{1/2}\pm 1}{2^{1/2}}$. Secondly, for P^E_7, set $\lambda = 2\cos\theta$, so that

$\lambda^2 = 2(1+\cos 2\theta)$ and

$$\lambda^6 - 6\lambda^4 + 9\lambda^2 - 3 = 8\cos^3(2\theta) - 6\cos(2\theta) - 1 = 2\cos(6\theta) - 1;$$

the roots of P_7^E are given by $\lambda = 0$ and $\cos(6\theta) = \frac{1}{2}$, namely by $\lambda = 0$ and $\lambda = 2\cos(\frac{\pi}{18} + j\frac{\pi}{3})$ for $j = 0,1,2,3,4,5$. Finally, set $Q(\lambda) = \lambda^4 - 7\lambda^3 + 14\lambda^2 - 8\lambda + 1$; a straightforward computation shows that

$$Q(T+2+T^{-1}) = T^{-4}\Phi_{30}(T)$$

where $\Phi_{30}(T) = \prod_{1 \leq j \leq 8}(T-T_j)$ is the cyclotomic polynomial with roots the primitive 30th roots of unity $T_j = \exp(im_j 2\pi/30)$, where $m_j = 1,7,\cdots,29$. As $T_j + 2 + T_j^{-1} = 4\cos^2(m_j\pi/30)$, the roots of Q are $4\cos^2(m_j\pi/30)$ for $m_j = 1,7,11,13$, and the formula for P_8^E follows. #

Corollary I.2.6. *For* $q \geq 3$, *the sequence* $(\|T_{2,q,r}\|)_{r \geq q}$ *is strictly increasing and converges to* $\lambda_q = \nu_q^{1/2} + \nu_q^{-1/2}$, where ν_q *is the largest root of the polynomial*

$$L_q(\nu) = \nu^q - (\nu^{q-2} + \nu^{q-3} + \cdots + \nu + 1).$$

Similarly, the sequence $(\|T_{3,3,r}\|)_{r \geq 3}$ *is strictly increasing and converges to* λ_∞.

Proof. The sequences are strictly increasing by lemma 1.4.2. One has $-\mu^{-r+q-1} L_q(\nu^{-1}) > 0$ for $\mu > 1$ (and $\nu = \mu^2 > 1$), so that $\|T_{2,q,r}\| > \lambda_q$ for all $r \geq q$ by proposition I.2.5. The argument for $\lim_{r \to \infty}\|T_{2,q,r}\| = \lambda_q$ is as in the proof of I.2.3.

For the graphs $T_{3,3,r}$, one computes first

$$\begin{aligned} P^T_{3,3,r}(\lambda) &= \lambda P^T_{2,3,r}(\lambda) - P^A_{r+2}(\lambda) \\ &= (\mu^2+1+\mu^{-2})(\mu-\mu^{-1})^{-1}\mu^{r-1}\{\mu^4-\mu^2-1+\mu^{-2r-2}(\mu^4+\mu^2-1)\} \end{aligned}$$

and proceeds then in the same way. #

For integers p,q,r with $2 \leq p \leq q \geq 3$ and $r \geq 5$, we denote by $P^H_{p,q,r}$ the characteristic polynomial of the graph $H_{p,q,r} = H_{p,q,r}(3,3)$ defined in I.1.

Proposition I.2.7. *With* L_q *as in corollary* I.2.6, *one has*

$$P^H_{p,q,r}(\lambda) = (\mu-\mu^{-1})^{-1}\{\mu^{r-p-q-2}L_p(\nu)L_q(\nu)-\mu^{-r+p+q+2}L_p(\nu^{-1})L_q(\nu^{-1})\}$$

where $\nu=\mu^2$. *The sequence* $(\|H_{p,q,r}\|)_{r\geq 5}$ *is strictly decreasing and converges to*

$$\lambda_q = \nu_q^{1/2}+\nu_q^{-1/2}.$$

Proof. By lemma I.2.1,

$$\begin{aligned} P^H_{p,q,r}(\lambda) &= \lambda P^T_{2,p,q+r-4} - P^A_{q-1}P^T_{2,p,r-4} \\ &= \lambda^2 P^A_{p+q+r-5} - \lambda(P^A_{p-1}P^A_{q+r-5} + P^A_{q-1}P^A_{p+r-5}) + P^A_{p-1}P^A_{q-1}P^A_{r-5}. \end{aligned}$$

By proposition I.2.2 and routine manipulation

$$P^H_{p,q,r}(\lambda) = (\mu-\mu^{-1})^{-3}\{F(\mu)-F(\mu^{-1})\}$$

with

$$F(\mu) = \mu^{r-p-q-4}\{\mu^{2p+2q}(\mu^2-1-\mu^{-2})^2 + (\mu^{2p}+\mu^{2q})(\mu^2-1-\mu^{-2}) + 1\}.$$

As μ_∞ is the largest root of $\mu^2-1-\mu^{-2}$, one has for $\mu \geq \mu_\infty$

$$P^H_{p,q,r}(\lambda) \geq (\mu-\mu^{-1})^{-3}\{\mu^{r-p-q-4} - F(\mu^{-1})\}$$

which is positive for r large enough; otherwise said $\|H_{p,q,r}\| < \lambda_\infty$ for r large enough. One has also $F(\mu) = \mu^{r-p-q-4}f(\nu)$ with $\nu=\mu^2$ and

$$\begin{aligned} f(\nu) &= \nu^{p+q+2} - 2\nu^{p+q+1} - \nu^{p+q} + 2\nu^{p+q-1} + \nu^{p+q-2} + \nu^{p+1} - \nu^{p-1} \\ &\qquad + \nu^{q+1} - \nu^{q-1} - \nu^p - \nu^q + 1 \\ &= (\nu-1)^2 L_p(\nu)L_q(\nu). \end{aligned}$$

The formula given for $P^H_{p,q,r}(\lambda)$ follows, and the other claims are checked as in earlier proofs of the present section. #

I.3. Proofs of theorems I.1.2 and I.1.3.

We know from proposition I.1.1 that the spectral radius of a connected graph which contains strictly a cycle is strictly larger than λ_∞. There are other conditions which imply the same inequality.

Lemma I.3.1. *Let* Γ *be one of the graphs*

$$H_{2,k+1,k+3} \qquad k \geq 3$$

$$H_{2,k+1,k+2} \qquad k \geq 3$$

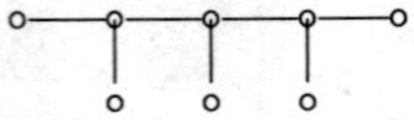

Then $\|\Gamma\| > \lambda_\infty$.

Proof. By proposition I.2.7 one has

$$P^H_{2,k+1,k+3}(\lambda) = (\mu-\mu^{-1})^{-1}\{(\mu^2-\mu^{-2})L_{k+1}(\nu)-(\mu^{-2}-\mu^2)L_{k+1}(\nu^{-1})\}$$

$$= \frac{\mu+\mu^{-1}}{\mu-\mu^{-1}}\{\mu^{2k-1}(\mu^4-\mu^2-1) +\mu^{-2k-3}(\mu^4+\mu^2-1) +\mu^{-1} +\mu\}.$$

It follows that

$$P^H_{2,k+1,k+3}(\lambda_\infty) = \frac{\mu_\infty+\mu_\infty^{-1}}{\mu_\infty-\mu_\infty^{-1}}\{2\mu_\infty^{-2k+1} -\mu_\infty^3\}.$$

The first factor is positive and the second is

$$2\mu_\infty^{-2k+1} -\mu_\infty^3 \leq -\mu_\infty^{-7}(\mu_\infty^4-2) < 0$$

for $k \geq 3$. Consequently $P^H_{2,k-1,k+3}(\lambda_\infty) < 0$ and $\|H_{2,k+1,k+3}\| > \lambda_\infty$ for $k \geq 3$.

One has similarly

$$P^H_{2,k+1,k+2}(\lambda) = \frac{\mu+\mu^{-1}}{\mu-\mu^{-1}}\{\nu^k(\nu^2-\nu-1) + \nu^{-k-2}(\nu^2+\nu-1) + \nu^{-1}-\nu\}$$

and $\|H_{2,k+1,k+2}\| > \lambda_\infty$ for $k \geq 3$. This follows also from the first computation and from I.2.4.

By a direct computation, the norm of the last graph in the lemma is $(\frac{1}{2}(5+17^{1/2})^{1/2} \approx 2{,}136$, and is thus larger than λ_∞. #

Proposition I.3.2. *Let* Γ *be a connected graph. Each of the following conditions implies* $\|\Gamma\| > \lambda_\infty$.

(i) Γ *contains a vertex of degree* $d \geq 5$;

(ii) Γ *contains a vertex of degree* 4 *and* $\Gamma \neq D_4^{(1)}$;

(iii) Γ *contains three vertices of degree* 3;

(iv) Γ *contains the graph*

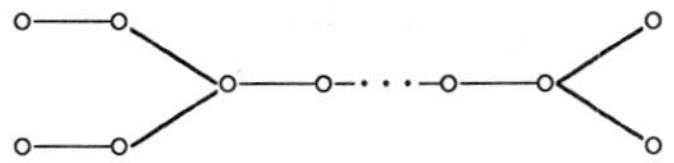

with $k \geq 8$ *vertices.*

Proof. (i) The adjacency matrix of  is $Y = \begin{bmatrix} 0 & X \\ X^t & 0 \end{bmatrix}$ with $X = (1\ 1\ 1\ 1\ 1) \in M_{1,5}(\mathbb{N})$. As $XX^t = 5$ one has $\|Y\| = 5^{1/2} > \lambda_\infty$.

(ii) The adjacency matrix of 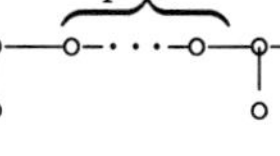 is $Y = \begin{bmatrix} 0 & X \\ X^t & 0 \end{bmatrix}$ with $X = \begin{bmatrix} 1 & 1 & 1 & 1 \\ 0 & 0 & 0 & 1 \end{bmatrix}$.

as $XX^t = \begin{bmatrix} 4 & 1 \\ 1 & 1 \end{bmatrix}$ has norm $\frac{1}{2}(5+13^{1/2}) > \lambda_\infty^2$ one has $\|Y\| > \lambda_\infty$.

(iii) If Γ contains three vertices of degree 3, then Γ contains

$\Gamma_{p,q}$:

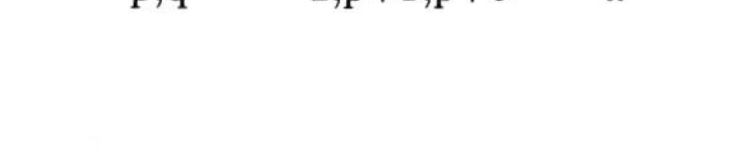

for some integers with $2 \leq p \leq q$. If $p \geq 3$, one has

$$\|\Gamma\| \geq \|\Gamma_{p,q}\| > \|H_{2,p+1,p+3}\| > \lambda_\infty$$

and if $p = 2$, $q \geq 3$ one has

$$\|\Gamma\| \geq \|\Gamma_{2,q}\| > \|H_{2,q+1,5}\| \geq \|H_{2,4,5}\| > \lambda_\infty$$

by the previous lemma. If $p = q = 2$, one has also

$$\|\Gamma\| \geq \|\Gamma_{2,2}\| > \lambda_\infty$$

by the same lemma.

(iv) The characteristic polynomial P_k of the graph in (iv) is given by

$$P_k(\lambda) = \lambda P^T_{3,3,k-5}(\lambda) - \lambda P^T_{3,3,k-7}(\lambda)$$
$$=(\mu^2-\mu^{-2})(\mu^2-1)(\mu^2+1+\mu^{-2})\{\mu^{k-8}(\mu^4-\mu^2-1)-\mu^{-k+2}(\mu^4+\mu^2-1)\}.$$

It follows that the norms of these graphs decrease and converge to λ_∞ when k increases. #

Corollary I.3.3. *Let Γ be a connected graph with*

$$2 < \|\Gamma\| \leq \lambda_\infty.$$

Then Γ is either some $T_{p,q,r}$ or some $H_{p,q,r}$.

Proof. Clear from the previous proposition. #

The adjacency matrix of

$T_{2,3,7}$ 

is

$$Y = \begin{pmatrix} 0 & X \\ X^t & 0 \end{pmatrix} \qquad X = \begin{pmatrix} 1 & 0 & 0 & 0 & 0 \\ 1 & 1 & 1 & 0 & 0 \\ 0 & 0 & 1 & 1 & 0 \\ 0 & 0 & 0 & 1 & 1 \\ 0 & 0 & 0 & 0 & 1 \end{pmatrix}.$$

The characteristic polynomial of XX^t is

$$\rho^5 - 9\rho^4 + 27\rho^3 - 31\rho^2 + 12\rho - 1.$$

Its largest root ρ_0 is near 4,0264; consequently $\|T_{2,3,7}\| = d$ with $d = \rho_0^{1/2} \approx 2,0066$.

Proposition I.3.4. *Let Γ be a connected graph with $2 < \|\Gamma\| \leq d$. Then $\Gamma = T_{2,3,7}$.*

Proof. By I.3.3, the graph Γ is some of $T_{p,q,r}$ or of $H_{p,q,r}$, because $d < \lambda_\infty$. By I.2.7, it cannot be some of $H_{p,q,r}$, because $d < \lambda_3$; indeed $\lambda_3 = \nu_3^{1/2} + \nu_3^{-1/2}$ where ν_3 is the largest root of $\nu^3 - \nu - 1$, so that $\nu_3 \approx 1{,}325$ and $\lambda_3 \approx 2{,}0198$. Hence $\Gamma = T_{p,q,r}$ with $2 \leq p \leq q \leq r$. Moreover the triples

$$(2,2,r) \text{ with } r \geq 2 \qquad (2,4,4)$$

$$(2,3,r) \text{ with } r \leq 6 \qquad (3,3,3)$$

are ruled out, because $\|\Gamma\| > 2$. This and lemma 1.4.2 imply that Γ must be one of

$$T_{2,3,7} \quad T_{2,4,5} \quad T_{3,3,4}.$$

By I.2.5 one has

$$P^T_{2,4,5}(\lambda) = (\mu+\mu^{-1})(\nu^4-\nu-1-\nu^{-1}+\nu^{-4}) \qquad \nu = \mu^2$$

$$P^T_{3,3,4}(\lambda) = (\mu^2+1+\mu^{-2})(\mu-\mu^{-1})^2(\nu^3-\nu-1-\nu^{-1}+\nu^{-3})$$

and by direct computation

$$2{,}0153 \approx \|T_{2,4,5}\| > d \text{ and } 2{,}0285 \approx \|T_{3,3,4}\| > d$$

It follows that $\Gamma = T_{2,3,7}$. #

Remarks.

(1) The proof of theorem I.1.2 is now complete.

(2) Proposition I.3.4 calls to mind various extremal properties of the triple (2,3,7). One is that $T_{2,3,7}$ is a hyperbolic Coxeter graph of highest possible rank, namely 10; see [Kos]. Another one is about Hurwitz' group, of presentation

$$\{u,v;\ u^2 = v^3 = (uv)^7 = 1\}$$

which is "the largest" group of automorphisms of a Riemann surface with genus $g \geq 2$; see the discussion in [Magl], page 103. This group has a quotient which is the simple group $PSL_2(\mathbb{Z}/7\mathbb{Z})$ of order 168, namely the finite simple group which is neither cyclic nor alternating and of smallest possible order. (See [Bur], in particular, section 166 and note N.)

Given $f = \lambda^n + a_1\lambda^{n-1} + \cdots + a_n \in \mathbb{Z}[\lambda]$, let $\Omega(f)$ be the product of the roots of f outside the unit disc. Lehmer [Le 1] has asked about f with $\Omega(f)$ minimal. The best polynomial found by Lehmer (and known today) is $L(\lambda) = \lambda^{10} + \lambda^9 - \lambda^7 - \lambda^6 - \lambda^5 - \lambda^4 - \lambda^3 + \lambda + 1$. Let $P(\lambda) = \lambda^5 - 9\lambda^4 + 27\lambda^3 - 31\lambda^2 + 12\lambda - 1$, so that $P^T_{2,3,7}(\lambda) = P(\lambda^2)$. Then $L(\lambda) = \lambda^5 P(\lambda + 1/\lambda)$. We know this from Misiurewicz [Mis].

Lemma I.3.5. *One has* $\|T_{3,4,5}\| > \lambda_\infty > \|T_{3,4,4}\|$.

Proof. The adjacency matrix of $T_{3,4,5}$ is $\begin{bmatrix} 0 & X \\ X^t & 0 \end{bmatrix}$ with

$$X = \begin{bmatrix} 1 & 1 & 0 & 0 & 0 \\ 0 & 1 & 1 & 1 & 0 \\ 0 & 0 & 1 & 0 & 0 \\ 0 & 0 & 0 & 1 & 1 \\ 0 & 0 & 0 & 0 & 1 \end{bmatrix}.$$

The characteristic polynomial of XX^t is

$$P(\rho) = \rho^5 - 9\rho^4 + 27\rho^3 - 32\rho^2 + 13\rho - 1.$$

If $\rho_\infty = \lambda_\infty^2 = 5^{1/2} + 2$ one has $P(\rho_\infty) = -4 + 5^{1/2} < 0$; hence $\|T_{3,4,5}\| = \|XX^t\|^{1/2} > \lambda_\infty$. The adjacency matrix of $T_{3,4,4}$ is $\begin{bmatrix} 0 & Z \\ Z^t & 0 \end{bmatrix}$ for

$$Z = \begin{bmatrix} 1 & 1 & 0 & 0 & 0 \\ 0 & 1 & 1 & 1 & 0 \\ 0 & 0 & 1 & 0 & 0 \\ 0 & 0 & 0 & 1 & 1 \end{bmatrix}.$$

The characteristic polynomial of ZZ^t is

$$Q(\rho) = (\rho - 2)f(\rho), \quad f(\rho) = \rho^3 - 6\rho^2 + 8\rho - 2.$$

One has

$$f(0) < 0, \quad f(1) > 0, \quad f(2) < 0, \quad f(5^{1/2} + 2) = -2 + 5^{1/2} > 0.$$

Hence all roots of Q are in $]0, \rho_\infty[$ and $\|T_{3,4,4}\| = \|ZZ^t\|^{1/2} < \lambda_\infty$. #

Proposition I.3.6. *Let* Γ *be one of the* $T_{p,q,r}$*'s. Then* $2 < \|\Gamma\| \leq \lambda_\infty$ *if and only if* (p,q,r) *is one of the following:*

$(2,3,r)$ *with* $r \geq 7$ $\qquad$ $(3,3,r)$ *with* $r \geq 4$

$(2,4,r)$ *with* $r \geq 5$ $\qquad$ $(3,4,4)$.

$(2,q,r)$ *with* $r \geq q \geq 5$.

Proof. It follows from lemma 1.4.1, corollary I.2.6 and lemma I.3.5. #

The proof of theorem I.1.3 for ordinary graphs is now complete. One may add minor refinements. For example, $\|H_{2,q,r}\| < \lambda_\infty$ implies $r \geq q+2$ by I.3.1 (and a direct computation if $q = 3$); but this is not sharp, because one has for example $\|H_{2,4,r}\| \leq \lambda_\infty$ if and only if $r \geq 7$.

We observe that $\lambda_3 \approx 2{,}0198$ is in the closure of $\mathbb{E}$ but not in $\mathbb{E}$. One of the estimates in the proof of I.3.4 shows that $\lambda_3 \neq \|T_{2,4,5}\|$. Similar estimates show that λ_3 is strictly smaller than

$$\|T_{2,4,6}\| = \text{the largest root of } \lambda^8 - 9\lambda^6 + 27\lambda^4 - 31\lambda^2 + 11 \approx 2{,}0237$$

as well as

$$\|T_{2,5,5}\| = \text{the largest root of } \lambda^6 - 6\lambda^4 + 8\lambda^2 - 1 \approx 2{,}0285$$

Thus if follows from Theorem I.1.3 that $\lambda_3 \notin \mathbb{E}$, and in particular that $\mathbb{E}$ is not closed.

Similarly, Theorem I.1.3 shows that, if $\lambda_\infty \in \mathbb{E}$, then $\lambda_\infty = \|\Gamma\|$, with $\Gamma = H_{p,q,r}$ for some triple of integers satisfying $2 \leq p \leq q$, $3 \leq q$, and $q \leq r$. A.E. Brouwer (private communication) has checked that this is not the case, so that $\lambda_\infty \notin \mathbb{E}$. We have not checked whether $\lambda_q \in \mathbb{E}$ for $q \geq 4$.

Lemma I.3.7. *For* $s \in \mathbb{R}_+$ *with* $s > 1$ *and for* $\ell \in \mathbb{N}$ *with* $\ell \geq 1$, *consider the symmetric matrices*

$$Y_\ell = \begin{bmatrix} s & 1 & 0 & \cdots & 0 & 0 \\ 1 & 0 & 1 & \cdots & 0 & 0 \\ \cdots & \cdots & \cdots & \cdots & \cdots & \cdots \\ 0 & 0 & 0 & \cdots & 0 & 1 \\ 0 & 0 & 0 & \cdots & 1 & 0 \end{bmatrix}, \quad Z_\ell = \begin{bmatrix} 0 & s & 0 & \cdots & 0 & 0 \\ s & 0 & 1 & \cdots & 0 & 0 \\ \cdots & \cdots & \cdots & \cdots & \cdots & \cdots \\ 0 & 0 & 0 & \cdots & 0 & 1 \\ 0 & 0 & 0 & \cdots & 1 & 0 \end{bmatrix}$$

in $M_\ell(\{0,1,s\})$. *When* ℓ *increases,*

$\|Y_\ell\|$ *increases and converges to* $s+s^{-1}$

$\|Z_\ell\|$ *increases and converges to* $s^2(s^2-1)^{-1/2}$.

Proof. Observe that Y_ℓ and Z_ℓ differ by one and two entries only from the example in section 1.2.

Let P_ℓ be the characteristic polynomial of Y_ℓ and let Q_ℓ be that of Z_ℓ. As in proposition I.2.2, one has $P_{\ell+1} = \lambda P_\ell - P_{\ell-1}$, so that

$$P_\ell(\lambda) = (\mu-\mu^{-1})^{-1}\{(\mu P_2 - P_1)\mu^{\ell-2} - (\mu^{-1}P_2 - P_1)\mu^{-\ell+2}\}.$$

When ℓ increases, the largest root $\|Y_\ell\|$ of P_ℓ increases and converges to a limit y which is the largest root of $\mu P_2 - P_1$. As $\mu P_2 - P_1 = \mu^2(\mu - s)$ one has $y = s+s^{-1}$.

Similarly the largest root $\|Z_\ell\|$ of Q_ℓ increases and converges to a limit z which is the largest root of $\mu Q_2 - Q_1 = \mu(\mu^2+1-s^2)$, so that $z = (s^2-1)^{1/2} + (s^2-1)^{-1/2}$.

The proof is that of lemma 3.5 in [Hof]. #

Proposition I.3.8. *Let* Γ *be a Coxeter graph.*

(i) *Suppose that the underlying graph of* Γ *is a segment. Then* $2 < \|\Gamma\| \leq \lambda_\infty$ *if and only if* Γ *is one of the following.*

H_ℓ o—5—o——o—· · ·—o——o *with* ℓ *vertices,* $\ell \geq 5$

F_ℓ o——o—4—o——o—· · ·—o——o *with* ℓ *vertices,* $\ell \geq 6$

F_4^{++} o——o——o—4—o——o——o *with 6 vertices.*

(ii) *Suppose that* $2 < \|\Gamma\| \leq \lambda_\infty$. *Then either* $\Gamma = H_\ell$ *as in* (i), *or* Γ *is a Coxeter graph where marked edges are marked with* $m = 4$.

Proof. The graph o—7—o——o has characteristic polynomial

$$\begin{vmatrix} \lambda & -2\cos(\pi/7) & 0 \\ -2\cos(\pi/7) & \lambda & -1 \\ 0 & -1 & \lambda \end{vmatrix} = \lambda\{\lambda^2-1-4\cos^2(\pi/7)\}.$$

Its norm is near 2,0608, and thus strictly larger than λ_∞. As the norm of $\circ\overset{m}{\text{——}}\circ$ is $2\cos(\pi/m) < 2$ for all $m \geq 3$, a Coxeter graph with $2 < \|\Gamma\| \leq \lambda_\infty$ cannot have any edge marked with some $m \geq 7$.

The graph $G_2^{(1)}$: $\circ\overset{6}{\text{——}}\circ\text{——}\circ$ is of norm 2. The graph $\circ\overset{6}{\text{——}}\circ\text{——}\circ\text{——}\circ$ has characteristic polynomial $\lambda^4-5\lambda^2+3$ and its norm is $2{,}074 \approx (\frac{1}{2}(5+13^{1/2}))^{1/2} > \lambda_\infty$. The graph $\circ\overset{6}{\text{——}}\circ\overset{4}{\text{——}}\circ$ and the graph $\circ\overset{6}{\text{——}}\circ\!<^{\circ}_{\circ}$ both have norm $5^{1/2} > \lambda_\infty$. Finally the graph $\circ\text{——}\circ\overset{6}{\text{——}}\circ\text{——}\circ$ has norm $(\frac{1}{2}(5+19^{1/2}))^{1/2} > \lambda_\infty$. Hence a Coxeter graph with $2 < \|\Gamma\| \leq \lambda_\infty$ cannot have any edge marked with $m = 6$.

The graph $\circ\text{——}\circ\overset{5}{\text{——}}\circ\text{——}\circ$ has characteristic polynomial $\lambda^4 - \frac{1}{2}(7+5^{1/2})\lambda^2 + 1$; its norm is near 2,095, and thus strictly larger than λ_∞. Consequently, if a Coxeter graph Γ with $\|\Gamma\| \leq \lambda_\infty$ has an edge marked with $m = 5$, this edge must be a free edge (namely one at an end of Γ).

The previous lemma shows that, for $m \geq 4$, the norms of

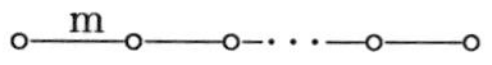

ℓ vertices

increase with ℓ and converge to $4\cos^2(\pi/m)(4\cos^2(\pi/m)-1)^{-1/2}$.
For $m = 5$, these are the H_ℓ's and the limit is λ_∞; observe that $\|H_4\| = 2$. Now the norms of

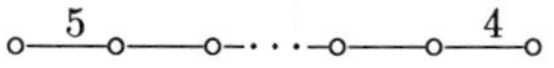

ℓ vertices

decrease and converge to λ_∞ by lemma I.2.4; the same holds for

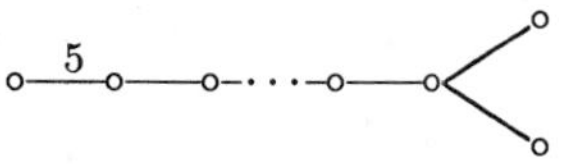

and this ends the proof of (ii).

The same argument as in I.3.7 shows that the norms of F_ℓ increase and converge to the largest root x of μP_2-P_1 with

$$P_1(\lambda) = \begin{vmatrix} \lambda & -1 & 0 \\ -1 & \lambda & -2^{1/2} \\ 0 & -2^{1/2} & \lambda \end{vmatrix}, \quad P_2(\lambda) = \begin{vmatrix} \lambda & -1 & 0 & 0 \\ -1 & \lambda & -2^{1/2} & 0 \\ 0 & -2^{1/2} & \lambda & -1 \\ 0 & 0 & -1 & \lambda \end{vmatrix}.$$

As $\mu P_2 - P_1 = \mu(\mu^4 - \mu^2 - 1)$ one has $x = \lambda_\infty$. Observe that $F_5 = F_4^{(1)}$ has norm 2. As above, the norms of

K_ℓ  ℓ vertices

decrease and converge to λ_∞.

Consider now a Coxeter graph Γ with underlying graph a segment and which has exactly one edge marked with $m = 4$ (all other edges, being unmarked, correspond to $m = 3$). If this is a free edge, $\Gamma = B_\ell$ for some $\ell \geq 2$ and $\|\Gamma\| < 2$. If this is next to a free edge, $\Gamma = F_\ell$ as considered above. In the other cases, either $\Gamma = F_4^{++}$ or Γ contains

L_7 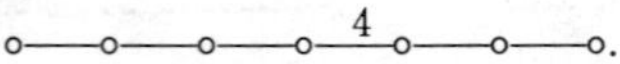

Direct computations show that $\|F_4^{++}\|^2$ is the largest root of $\rho^3 - 6\rho^2 + 8\rho - 2$ and that $\|L_7\|^2$ is the largest root of $(\rho-2)(\rho^2-5\rho+3)$. Hence $\|F_4^{++}\| < \lambda_\infty$ and $\|L_7\| = (\frac{1}{2}(5+13^{1/2}))^{1/2} > \lambda_\infty$.

Consider finally a Coxeter graph Γ with underlying graph a segment and which has at least two edges marked with $m = 4$. If there are the two free edges, $\Gamma = C_\ell$ for some $\ell \geq 3$ and $\|\Gamma\| = 2$. In the other cases one has $\|\Gamma\| \geq \|K_\ell\| > \lambda_\infty$ for some $\ell \geq 4$. #

<u>Proposition I.3.9</u>. *Let* Γ *be a Coxeter graph, the underlying graph of which is some* $T_{p,q,r}$ *with* $2 \leq p \leq q \leq r$. *Then* $2 < \|\Gamma\| \leq \lambda_\infty$ *if and only if* Γ *is one of the Coxeter graphs on the list of* I.3.6, *or* $T_{2,q,r}(4)$ *for some* $q \geq 3$ *and for* r *large enough (see* I.1 *for a picture of* $T_{2,q,r}(4)$).

<u>Proof</u>. Suppose first that Γ has underlying graph D_k for some $k \geq 4$. The norm of

$B_\ell^{(1)}$ 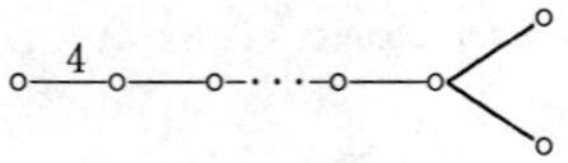 $\ell+1$ vertices, $\ell \geq 3$

is 2. The norms of

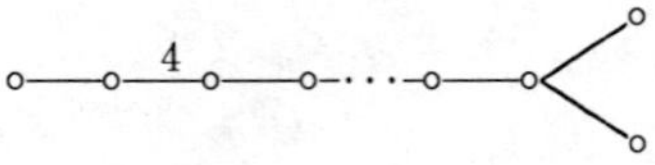 k vertices

decrease and converge to λ_∞ by lemma I.2.4. The graph

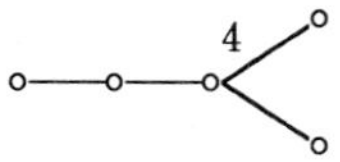

has characteristic polynomial $\lambda(\lambda^4-5\lambda^2+3)$ and its norm, equal to that of  , is larger than λ_∞. Hence Γ cannot satisfy $2 < \|\Gamma\| \leq \lambda_\infty$.

Suppose now that the Coxeter graph Γ has underlying graph $T_{2,q,r}$ with $3 \leq q \leq r$. From lemma I.2.1 and proposition I.2.5, the characteristic polynomial of

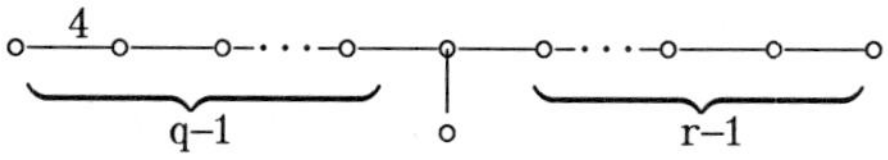

is given by

$$\lambda P^T_{2,q-1,r} - 2P^T_{2,q-2,r} = (\mu-\mu^{-1})^{-1}\{\mu^{r-q+1}f(\mu)-\mu^{-r+q-1}f(\mu^{-1})\}$$

$$f(\mu) = \mu^{2q}-\mu^{2q-2}-\mu^{2q-4}-1.$$

For $(q,r) \neq (3,3)$, it is easy to check that the value for $\lambda = \lambda_\infty$ of this polynomial is strictly negative; consequently the norm of the graph is strictly larger than λ_∞. Lemma I.2.4 shows that $\|T_{2,q,r}(4)\|$ decreases and converges to λ_q when r increases.

Also the norms of

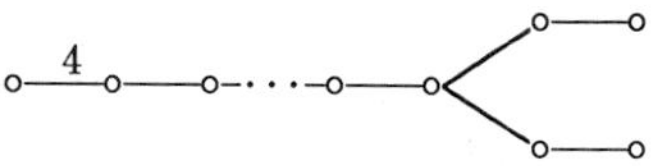

k vectors, $k \geq 7$

decrease and converge to λ_∞. The proposition follows. #

A final computation of the same kind for Coxeter graphs $H_{p,q,r}(m,m')$, also based on I.2.4, ends the proof of theorem I.1.3.

APPENDIX II.a

Complex semisimple algebras and finite dimensional C^*–algebras

Let M be a *complex* algebra. An <u>involution</u> on M is a $\mathbb{R}$–linear map $\alpha : M \to M$ satisfying

$$\alpha(xy) = \alpha(y)\alpha(x) \qquad \alpha(ix) = -i\alpha(x) \qquad \alpha(\alpha(x)) = x$$

for all $x,y \in M$; such an involution is <u>positive</u> if moreover

$$\alpha(x)x = 0 \quad \text{implies} \quad x = 0.$$

In particular, let $M = \mathrm{End}_{\mathbb{C}}(V)$ for some finite dimensional complex vector space V. Let $\langle\ |\ \rangle$ be a hermitian product on V. For $x \in M$, denote by x^* the adjoint endomorphism, defined by $\langle x^*\xi|\eta\rangle = \langle\xi|x\eta\rangle$ for all $\xi,\eta \in V$. Then $x \mapsto x^*$ is the standard example of a positive involution on M. Let α be any involution on M. It follows from Skolem–Noether's theorem that $\alpha(x) = gx^*g^{-1}$ for some $g \in GL(V)$, and $\alpha(\alpha(x)) = x$ implies that gg^{-1*} is central, hence a scalar. One may always choose g hermitian: $g^* = g$ and all eigenvalues of g are real. Then α is positive if and only if all eigenvalues of g have the same sign. If $\mathrm{tr} : M \to \mathbb{C}$ is the usual trace, we leave it to the reader to check that α is positive if and only if $\mathrm{tr}(\alpha(x)x) \geq 0$ for all $x \in M$; hence "positive" has the same meaning above as in [Wei].

<u>Proposition II.1</u>. *Let* M *be a finite dimensional complex algebra.*

(a) *If* M *has a positive involution, then* M *is semisimple.*

(b) *If* M *is semisimple, then* M *has a positive involution* $x \mapsto x^*$. *Moreover, for any other positive involution* α *on* M, *there exists an invertible element* $g \in M$ *with* $g^* = g$ *and with positive spectrum, such that* $\alpha(x) = gx^*g^{-1}$ *for all* $x \in M$.

(c) *Let* M *and* $x \mapsto x^*$ *be as in* (b) *and let* p *be a central idempotent of* M. *Then* $p^* = p$.

(d) *Let* $N \subset M$ *be a pair of semisimple algebras and let* β *be a positive involution on* N. *Then there exists a positive involution* α *on* M *which extends* β.

<u>Proof</u>. (a) Let $x \mapsto x^*$ be a positive involution on M. Assume that the Jacobson radical J of M contains some $x_1 \neq 0$; then $x_2 = x_1^*x_1$ is not zero, so that $x_3 = x_2^*x_2 \neq 0$, and so on; but this contradicts the fact that J is nilpotent. Hence $J = 0$ and M is semisimple.

(b) If M is semisimple, M is a direct sum of factors, and thus has a positive involution by the example before the proposition.

Let V be as in this same example. A straightforward application of Skolem–Noether's theorem is that any involution α on $\mathrm{End}(V) \times \mathrm{End}(V)$ which exchanges the factors is of the form $\alpha(x,y) = (gy^*g^{-1}, g^*x^*g^{*-1})$ for some $g \in GL(V)$; such an involution is not positive because $\alpha(1,0)(1,0) = (0,0)$. Now let M be an arbitrary semisimple algebra. What precedes implies that factors in M are invariant by any positive involution; claims (b) and (c) follow.

(d) Assume first that M is a factor, so that one may set $M = \mathrm{End}(V)$ as above. Let $U(N)$ be the unitary group $\{h \in N \mid \beta(h)h = 1\}$, which is compact, and let dh denote its Haar measure of total mass 1. Define a new scalar product on V by

$$\langle\langle \xi | \eta \rangle\rangle = \int_{U(N)} \langle h\xi | h\eta \rangle \, dh, \qquad \xi,\eta \in V.$$

Then $\langle\langle \beta(y)\xi | \eta \rangle\rangle = \langle\langle \xi | y\eta \rangle\rangle$ for all $\xi,\eta \in V$ and $y \in N$; for $y \in U(N)$, this follows from invariance of Haar measure, but any $Y \in N$ is a linear combination of unitary elements. Define the involution α on M by $\langle\langle \alpha(x)\xi | \eta \rangle\rangle = \langle\langle \xi | x\eta \rangle\rangle$ for all $\xi,\eta \in V$ and $x \in M$.

In the general case, one has $M = \bigoplus_{j=1}^{n} \mathrm{End}(V_j)$. One defines $\alpha(x_1,\cdots,x_m) = (\alpha_1(x_1),\cdots,\alpha_m(x_m))$ where each α_i is defined from $\beta|_{(N \cap \mathrm{End}(V_i))}$ as in the factorial case. #

The proposition holds also for real algebras; see [Wei], Proposition 1 and corollary.

Example II.2. Let G be a finite group. Its group algebra $\mathbb{C}[G]$ is made of complex functions on G, the product being convolution

$$(xy)(g) = \sum_{h \in G} x(h)y(h^{-1}g), \qquad g \in G.$$

There is a natural involution defined on $\mathbb{C}[G]$ by

$$x^*(g) = \overline{x(g^{-1})}, \qquad g \in G.$$

This is positive, because $x^*x = 0$ implies

$$(x^*x)(1) = \sum_{h \in G} |x(h)|^2 = 0$$

and thus $x = 0$. Thus proposition II.1.a provides one proof of (one formulation of) Maschke's theorem.

Example II.3. Consider the two-dimensional truncated polynomial algebra $M = \mathbb{C}[x]/(x^2)$, with basis $\{1,x\}$ and with product defined by $x^2 = 0$. Then M has two involutions which map 1 to 1, and x to x and $-x$ respectively. Of course M is not semisimple and these involutions are not positive.

Recall that a C^*-algebra is a complex algebra M (possibly infinite dimensional, possibly without unit) furnished with an involution $x \mapsto x^*$ and a norm $x \mapsto \|x\|$ such that

(i) M is a Banach space for $\| \ \|$;

(ii) $\|xy\| \leq \|x\|\|y\|$ for all $x,y \in M$.

(iii) $\|x^*x\| = \|x\|^2$ for all $x \in M$.

In particular, (iii) implies that the involution is positive. If M is finite dimensional, (i) is of course automatic, but moreover the norm is uniquely determined by the involution:

Proposition II.4. (a) *Let* M *be a semisimple complex algebra (of finite dimension) and let* $x \mapsto x^*$ *be a positive involution on* M. *For* $x \in M$, *define*

$$\|x\| = \mathrm{Sup}\{t \in \mathbb{R}_+ : t^2 - x^*x \textit{ is not invertible on } M\}.$$

Then $x \mapsto \|x\|$ *is a norm which, together with* $x \mapsto x^*$, *make* M *a* C^**-algebra.*

(b) *Let* M,M' *be two* C^**-algebras and let* $\varphi : M \to M'$ *be an injective linear map such that*

$$\varphi(xy) = \varphi(x)\varphi(y), \qquad \varphi(x^*) = \varphi(x)^*, \quad \textit{for all } x,y \in M.$$

Then φ *is an isometry:*

$$\|\varphi(x)\| = \|x\| \qquad \textit{for all } x \in M.$$

Proof. (a) If $M = \mathrm{End}(V)$ with V a hermitian vector space, one may also define the norm by

$$\|x\| = \mathrm{Sup}\{\|x\xi\| : \xi \in V \text{ and } \|\xi\| \leq 1\}.$$

Then $\|x^*x\| = \|x\|^2$ as in lemma 1.2.4, and the two definitions of $\|x\|$ coincide by the spectral theorem for hermitian matrices. The general case follows from that of factors.

(b) This claim is proved by elementary functional calculus, and comes early in any book about C^*-algebras, for example on page 13 of [Arv]. This claim would not hold without the completeness requirement (i) in the definition of a C^*-algebra. #

Let M be an algebra (possibly infinite dimensional) with a positive involution $x \mapsto x^*$. Recall from section 2.5 that a trace on M is a linear map $\mathrm{tr} : M \to \mathbb{C}$ such that $\mathrm{tr}(xy) = \mathrm{tr}(yx)$ for all $x,y \in M$; say it is positive if $\mathrm{tr}(x^*x) \geq 0$ for any $x \in M$. In case

$\dim_{\mathbb{C}}(M) < \infty$, this definition is the same as that of Section 2.5 by II.1.c. It follows from the Cauchy–Schwarz inequality that a positive trace is faithful if and only if $tr(x^*x) > 0$ for all $x \neq 0$ in M.

In the context of C^**–algebras and* W^**–algebras, it is usual that "trace" means "positive trace." We shall take up this habit in Chapter 3.*

Then, in the same way as for measures on locally compact spaces, positivity makes it possible (and quite useful) to consider infinite traces; we shall return to this in Section 3.2.

Let N be a sub–C^*–algebra of a C^*–algebra M. A conditional expectation $E : M \to N$ is <u>self–adjoint</u> if $E(x^*) = E(x)^*$ for all $x \in M$; it is <u>positive</u> if, for every $x \in M$, there exists $y \in N$ with $E(x^*x) = y^*y$. It is an important point that the construction of Proposition 2.6.2 provides positive conditional expectations from positive traces.

<u>Proposition II.5</u>. *Let* $N \subset M$ *and* tr *be as in Proposition* 2.6.2 (*with* $\mathbb{K} = \mathbb{C}$); *assume moreover that* M *is given a positive involution* $x \mapsto x^*$ *and that* the *faithful trace* tr *is positive. Then the faithful conditional expectation* $E : M \to N$ *of Proposition* 2.6.2 *is self–adjoint and positive.*

<u>Proof</u>. Observe that $tr(x^*) = \overline{tr(x)}$ for any $x \in M$; indeed, this is clear when x is positive, and one has in general $x = x_1 - x_2 + ix_3 - ix_4$ with $x_1, \cdots, x_4$ positive.

Since tr is positive, both tr and $tr|_N$ are faithful. We check then that E is self–adjoint. For any $z \in N$ one has

$$\begin{aligned} tr(E(x^*)z) &= tr(E(x^*z)) = tr(x^*z) = \overline{tr(z^*x)} \\ &= \overline{tr(z^*E(x))} = tr(E(x)^*z) \end{aligned}$$

and the claim follows because $tr|_N$ is faithful.

Consider now $x \in M$. As $E(x^*x)$ is self–adjoint, there exist positive elements $z_1, z_2 \in N$ with $E(x^*x) = z_1^*z_1 - z_2^*z_2$ and $z_1z_2^* = 0$. If $z_2 \neq 0$ there would exist a self–adjoint idempotent $e \in N$ with $z_1^*z_1e = 0$ and $ez_2^*z_2e \neq 0$, so that

$$tr((xe)^*(xe)) = tr(E(x^*x)e) = -tr((z_2e)^*(z_2e)) < 0$$

contradicting the positivity of tr. Hence $E(x^*x) = z_1^*z_1$. #

It can be shown in general that a positive conditional expectation is faithful if and only if $E(x^*x) > 0$ for any non–zero x. See 9.2 in [Str]. This is evident for conditional expectations defined by a faithful positive trace, for $E(x^*x) = 0$ implies that $tr(x^*x) = tr(E(x^*x) = 0$, and therefore $x = 0$ by faithfulness of the trace.

In the context of C^**-algebras and* W^**-algebras, it is usual that "conditional expectation" means "positive conditional expectation". We will adopt this habit in Chapter 3.*

Conditional expectations are very important in the study of operator algebras; see, e.g. [Str], §9.

In the remainder of this appendix, we comment on C^* versions of the various constructions of Chapter 2. All of the results of Sections 2.2 and 2.3 have C^* versions. For example, in 2.2.3 and 2.2.5, let F be the factor $End_{\mathbb{C}}(V)$, where V is a finite dimensional complex Hilbert space, M a C^*-subalgebra, and in 2.2.5 take q to be a self-adjoint projection. Then $C_F(M)$ and qMq are also C^*-subalgebras. Versions of 2.3.9 and 2.3.10 are valid for C^*-algebras, with the isomorphisms respecting the involutions. We have already remarked and the end of Section 2.3 that one can associate to a Bratteli diagram a chain of C^*-algebras, via the path model.

Now let $N \subset M$ be a pair of finite dimensional C^*-algebras and set $L = End_N^r(M)$. Then L has a unique C^* structure such that the inclusion of M in L is a $*$-isomorphism. In fact, from the Bratteli diagram for $N \subset M \subset L$, we can impose such a C^* structure on L, using the C^* version of 2.3.9, or the path model. A more natural way to go about this, however, is to take any positive faithful trace tr on M, and to give M the hermitian inner product $\langle x|y\rangle = tr(xy^*)$. Then the left regular representation of M on (M,tr) is a faithful $*$-representation, and the right regular representation of N is a $*$-anti-representation. Hence $L := \rho(N)'$ is naturally a C^*-algebra, by the C^* version of 2.2.3. (Here the prime ($'$) denotes centralizer in $End_{\mathbb{C}}(M,tr)$.)

Furthermore, the (faithful, positive) conditional expectation $E : M \to N$ determined by the trace tr is also the self-adjoint projection of M onto N, with respect the trace inner product. (See the proof of 2.6.2.) Therefore the subalgebra $\langle M,E\rangle$ of $End_{\mathbb{C}}(M,tr)$ generated by $\lambda(M)$ and E is a $*$-subalgebra, so equal to its own bicentralizer by the C^* version of 2.2.3. But the centralizer of $\langle M,E\rangle$ is easily seen to be $\rho(N)$, so that $\langle M,E\rangle = \langle M,E\rangle'' = \rho(N)' = L$. Finally, if z is a central projection in L orthogonal to the ideal $\lambda(M)E\lambda(M)$, then for all $y,x \in M$, one has $0 = (E\lambda(y)z)(x) = E(yz(x))$; hence $z(x) = 0$ by faithfulness of E, and so $z = 0$. Thus $L = \lambda(M)E\lambda(M)$. This is the C^* version of Corollary 2.6.4.

Suppose now that $N \subset M$ is a connected pair and choose for tr the unique positive Markov trace of modulus $\beta = [M:N]$ (Theorem 2.7.3). Then the unique extension of tr to L with the Markov property, $\beta tr(xE) = tr(x)$ for $x \in M$, is also positive, since the weights of the trace on L are β^{-1} times the corresponding weights on N. Iterating the fundamental construction thus yields a chain of finite dimensional C^* algebras carrying a faithful positive trace tr and a sequence of self-adjoint projections $(E_i)_{i\geq 1}$ satisfying the conditions of Proposition 2.7.5.

We have already noted in Sections 2.8 and 2.9 that, for $\mathbb{K} = \mathbb{C}$ and for $\beta \geq 4$ or $\beta = 4\cos^2(\pi/q)$ for some integer $q \geq 3$, the algebras $B_{\beta,k}$ constructed there have a C^* structure making the generators $\{e_j\}$ self-adjoint projections. Furthermore, 2.8.5 or 2.9.6(e) implies that $B_{\beta,k}$ is, up to $*$-isomorphism, the unique C^*-algebra generated by the identity and self-adjoint projections $\{e_1, \cdots e_{k-1}\}$ satisfying the relations of 2.8.4 (ii), with a faithful trace tr satisfying $\beta tr(we_j) = tr(w)$ for $w \in alg\,\{\mathbb{1}, e_1, \cdots e_{j-1}\}$.

APPENDIX II.b

The algebras $\mathcal{A}_{\beta,k}$ in statistical mechanics

The first recorded occurrence of the algebras of Section 2.8 arose in a work of H.N.V. Temperley and E.H. Lieb (1971) about microscopic models for magnets. The purpose of this appendix is to introduce the reader to this circle of ideas. For a systematic account see [Bax].

Generalities on the partition function.

It is an idea going back to Daniel Bernoulli (1738) to compute the pressure of a gas in a container from the change in momentum of the molecules impinging on the walls. During the last century, physicists made this computation on several occasions and compared the outcome with the equation of state of a perfect gas $pV = nRT$. One result of this comparison is that the mean kinetic energy of a simple (i.e., monatomic) molecule in a gas at temperature T is

$$E_{\text{mean}} = \tfrac{3}{2}kT \quad \text{(II.6)}$$

where k denotes the Boltzmann constant.

Let us now sketch the argument of J.C. Maxwell (1860) for finding the velocity distribution of the molecules in a gas. Let x,y,z be Cartesian coordinates in the velocity space and denote by $f(x)dx$ the probability of finding the first coordinate of a molecule in the small interval $(x,x+dx)$. As all directions are equivalent, the density of probability of finding a velocity vector at (x,y,z) is given by $f(x)f(y)f(z)$, and this should be a function of the absolute value $c = (x^2 + y^2 + z^2)^{1/2}$ alone. As Maxwell writes:

$$f(x)f(y)f(z) = \phi(x^2 + y^2 + z^2).$$

By logarithmic differentiation with respect to x

$$\frac{1}{2x}\frac{f'(x)}{f(x)} = \frac{\phi'(x^2+y^2+z^2)}{\phi(x^2+y^2+z^2)}.$$

Differentiation with respect to y or z leads to

$$\frac{1}{2x}\frac{f'(x)}{f(x)} = \text{constant}$$

so that

$$f(x) = C \exp(Ax^2)$$

for appropriate constants C and A. As f should be integrable, $A < 0$, say $A = -\alpha^{-2}$ for some $\alpha > 0$. Moreover $C = \alpha^{-1}\pi^{-1/2}$ because $\int_{\mathbb{R}} f(x)dx = 1$. It is now easy to compute the mean kinetic energy of a molecule of mass m:

$$\frac{3}{2}m\int_{\mathbb{R}} f(x)x^2 dx = \frac{3}{4}m\alpha^2$$

so that $\frac{1}{\alpha^2} = \frac{m}{2kT}$ by comparison with (II.6). Finally, for a simple molecule (monatomic gas), the energy is $E = \frac{1}{2}mc^2$, so that the density of probability of finding a molecule with energy E is given by

$$\left[\frac{m}{2\pi kT}\right]^{3/2} \exp(-E/kT). \qquad \text{(II.7)}$$

For all this, see [Som], Sections 22–23 and [BR].

More generally, consider abstractly a physical system with set of possible states S, and denote by E(s) the energy of the system when it is in the state $s \in S$. The canonical postulate is that the probability of the system being in the state s is proportional to $\exp(-E(s)/kT)$. That this is a sensible postulate can be understood either from general a priori considerations involving additivity with respect to subsystems (see [Gib], Chap. IV), or from the particular case above (II.7). Then, it is clearly quite important to understand and compute Gibbs' partition function, which is the sum-over-states

$$Z = \sum_{s \in S} \exp(-E(s)/kT).$$

As Gibbs has shown, the relevant physical quantities can be expressed in terms of Z. For example the average energy is

$$\frac{\sum_{s \in S} E(s) \exp(-E(s)/kT)}{\sum_{s \in S} \exp(-E(s)/kT)} = kT^2 \frac{\partial}{\partial T} \ln Z.$$

The one dimensional Ising model.

We follow the beginning of Chapter 2 in [Bax], and consider a linear array of n equally spaced atoms

Assume first that each atom can have one of two possible spins. Then a state of the array is a sequence $\sigma = (\sigma_i)_{1 \le i \le n}$ with $\sigma_i \in \{\pm 1\}$. Let us assume moreover that only states with periodic boundary conditions are allowed: $\sigma_1 = \sigma_n$. Assume also that energy comes from interactions of opposite spins at neighboring sites, say

$$E(\sigma) = -J \sum_{j=1}^{n-1} \sigma_j \sigma_{j+1}$$

for some constant J. (One has $J > 0$ because energy in minimal if all spins are alike.) The computational problem is to evaluate

$$Z_n = \sum_{\sigma \in S} \exp\left[K \sum_{j=1}^{n-1} \sigma_j \sigma_{j+1}\right]$$

for large n, where the constant K is J/kT.

Here comes a clever trick. Given a matrix $A = \begin{bmatrix} a_{1,1} & a_{1,2} \\ a_{2,1} & a_{2,2} \end{bmatrix}$ the (k,ℓ)–entry of A^n is

$$\sum_{k_1, \cdots, k_{n-1}} a_{k,k_1} a_{k_1,k_2} \cdots a_{k_{n-1},\ell}.$$

Consequently we set

$$A = \begin{bmatrix} e^K & e^{-K} \\ e^{-K} & e^K \end{bmatrix}.$$

If S_+ denotes the set of states with $\sigma_1 = \sigma_n = +1$ and $Z_{n,+}$ its contribution to Z, one has

$$Z_{n,+} = \sum_{\sigma \in S_{n,+}} \prod_{j=1}^{n-1} \exp(K \sigma_j \sigma_{j+1})$$

which can be recognized as one diagonal entry of A^n. The same holds for S_-, and

$$Z_n = \mathrm{trace}(A^n).$$

It is now quite easy to compute the eigenvalues of A^n, and thus to handle Z_n for large n.

Alternatively, we might assume that each atom has one of p possible spin states, so that S is now the set of sequences $(\sigma_j)_{1 \le j \le n}$ with $\sigma_j \in \{1, \cdots, p\}$ and, say, with $\sigma_1 = \sigma_n$. (Standard notation for the number of spins is q instead of p, but this would conflict here with the notation of Section 2.10.) Assume that neighboring atoms interact in one of only two possible ways, depending on their spins being equal or not. Thus, up to an irrelevant additive constant

$$E(\sigma) = -J \sum_{j=1}^{n-1} \delta(\sigma_j, \sigma_{j+1})$$

where δ is the Kronecker zero–one symbol. Now let A be the p–by–p matrix with e^K on the diagonal and 1 at any other place. Then

$$Z_n = \sum_{\sigma \in S} \prod_{j=1}^{n-1} \exp(K\delta(\sigma_j, \sigma_{j+1})) = \mathrm{trace}(A^n)$$

which is again easy to handle.

E. Ising proposed his model in 1925, and was successful with computations for one dimensional systems. Later, similar ideas have been used for systems with one or two dimensions by several physicists, including L. Onsager (1944) and R.B. Potts (1952). For our purpose, the main point to remember is the trick of using the trace of large powers of an appropriate matrix to evaluate the partition function when the number of atoms is large.

From square lattice Potts' model to $\mathcal{A}_{\beta,n}$.

We follow here part of chapter 12 in [Bax]. In a square lattice, consider a rectangle with m atoms on the base and n on the side

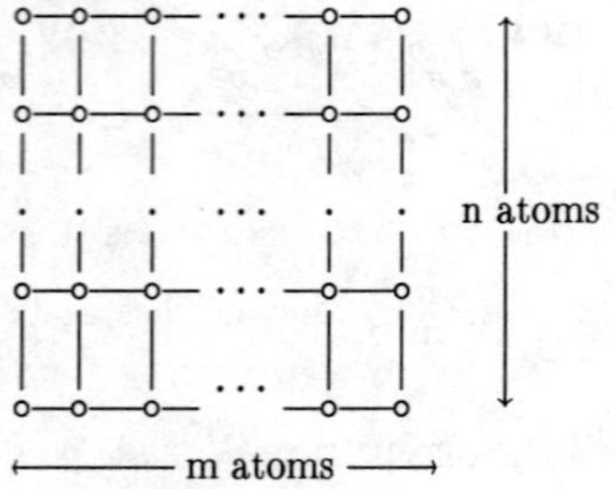

Each vertex is again thought of as one atom with one of p possible spins. A state in S is now a double sequence $s = (\sigma_j^k)_{1\le j\le m, 1\le k\le n}$ with $\sigma_j^k \in \{1,\cdots,p\}$. (No boundary condition.)

We assume that energy comes from either horizontal or vertical nearest neighbor interactions. More precisely, the horizontal contribution of some row with spins $\sigma_1,\cdots,\sigma_m$ to the partition function Z is

$$\exp\{K_1 \sum_{j=1}^{m-1} \delta(\sigma_j,\sigma_{j+1})\} \qquad \text{(II.8)}$$

and the vertical contribution of two adjacent rows with spins $\sigma_1,\cdots,\sigma_m,\tau_1,\cdots,\tau_m$ is

$$\exp\{K_2 \sum_{k=1}^{m} \delta(\sigma_k,\tau_k)\}. \quad \text{(II.9)}$$

Introduce the p^m-by-p^m matrices V,W with entries

$$V_{\sigma,\tau} = \exp\{K_1 \sum_{j=1}^{m-1} \delta(\sigma_j,\sigma_{j+1})\} \prod_{k=1}^{m} \delta(\sigma_k,\tau_k)$$

$$W_{\sigma,\tau} = \exp\{K_2 \sum_{k=1}^{m} \delta(\sigma_k,\tau_k)\}.$$

As $\prod_{k=1}^{m} \delta(\sigma_k,\tau_k) = 0$ if $\sigma \ne \tau$, the matrix V is diagonal and $(VW)_{\sigma,\tau}$ is the product of (II.8) and (II.9).

Consider now a state $s = (\sigma^1,\cdots,\sigma^n) \in S$, where each σ^k denotes a sequence of m spins. Write $\sigma = \sigma^1$ and $\tau = \sigma^n$. Then $((VW)^{n-1}V)_{\sigma,\tau}$

$$= \sum_{\sigma^2,\cdots,\sigma^{n-1}} (VW)_{\sigma,\sigma^2}(VW)_{\sigma^2,\sigma^3} \cdots (VW)_{\sigma^{n-1},\tau}(V)_{\tau,\tau}$$

represents the contribution to Z of the states $s = (\sigma_j^k)_{1\leq j\leq m, 1\leq k\leq n}$ with $\sigma = (\sigma_1^1,\cdots,\sigma_m^1)$ and $\tau = (\sigma_1^n,\cdots,\sigma_m^n)$. Thus Z has again been expressed in terms of iterates of matrices.

But now, the two so-called transfer matrices V and W do *not* commute, and further devices have to be used. One is to define p^m-by-p^m matrices $U_1,\cdots,U_{2m-1}$ by

$$(U_{2i-1})_{\sigma,\tau} = p^{-1/2} \prod_{\substack{1\leq j\leq m\\ j\neq i}} \delta(\sigma_j,\tau_j)$$
$$(U_{2i})_{\sigma,\tau} = p^{-1/2}\delta(\sigma_i,\sigma_{i+1}) \prod_{1\leq k\leq m} \delta(\sigma_k,\tau_k).$$

In particular U_{2i} is diagonal while U_{2i-1} is of the form

$$\mathbb{1}\otimes\mathbb{1}\otimes\cdots\otimes\mathbb{1}\otimes g\otimes\mathbb{1}\otimes\cdots\otimes\mathbb{1} \in \underbrace{M_p(\mathbb{C})\otimes\cdots\otimes M_p(\mathbb{C})}_{m \text{ factors}}$$

where $\mathbb{1}$ denotes the p-by-p identity matrix and where the ith factor on the left is the matrix g with all entries equal to $p^{-1/2}$. The transfer matrices may be written as

$$V = \exp\{p^{-1/2}K_1(U_2 + U_4 + \cdots + U_{2m-2})\}$$
$$W = \prod_{j=1}^{m} \{(e^{K_2}-1)I + p^{1/2}U_{2j-1}\}$$

where I is the p^m-by-p^m identity matrix.

For our point, the crucial fact is that the matrices $U_1,\cdots,U_{2m-1}$ satisfy the relations

$$U_i^2 = p^{1/2}U_i$$
$$U_iU_jU_i = U_i \quad \text{if } |i-j| = 1$$
$$U_iU_j = U_jU_i \quad \text{if } |i-j| \geq 2$$

for $i,j = 1,\cdots,2m-1$. In other words, $p^{-1/2}U_i$ define a representation of $A_{\beta,2m-1}$ for $\beta = p$.

It is readily seen that $R = \prod_{i=1}^{m} U_{2i-1}$ is a p^m-by-p^m matrix with all entries equal to $p^{-m/2}$. For any p^m-by-p^m matrix X, denote by $\Sigma(X)$ the sum of all entries of X, and observe that

$$RXR = p^{-m/2}R\Sigma(X).$$

Then

$$Z = \Sigma((VW)^{m}V)$$

and the computation of the partition function Z is now a problem inside the algebra $\mathcal{A}_{p,2m-1}$. This happens to help, though a full solution is still missing, and constitutes indeed "one of the most tantalizing unsolved models" (see [Bax], page 337).

APPENDIX II.c

More on the algebra $\mathcal{A}_{\beta,k}$ for non–generic β

Consider a field $\mathbb{K}$, a parameter $\beta \in \mathbb{K}^*$, an integer $k \geq 1$, and the corresponding algebra $\mathcal{A}_{\beta,k}$ defined in Section 2.8. If $(P_j)_{j\geq 0}$ are the polynomials of Proposition 2.8.3, we have shown in Proposition 2.8.5 that $\mathcal{A}_{\beta,k}$ is a multi–matrix algebra as soon as $P_j(\beta^{-1}) \neq 0$ for $j \leq k-1$.

Theorem II.10. *If* $P_j(\beta^{-1}) \neq 0$ *for* $j \leq k-1$ *and if* $P_k(\beta^{-1}) = 0$*, then* $\mathcal{A}_{\beta,k+1}$ *is not semi–simple.*

For example $\mathcal{A}_{1,3}$ is not semisimple, as checked in Example 2.9.1, or $\mathcal{A}_{2,4}$ is not semisimple if $\mathbb{K}$ is not of characteristic two.

More precisely, under the hypothesis of the theorem, we show hereafter that $\mathcal{A}_{\beta,k+1}$ contains a one–dimensional two–sided ideal of square zero. Most of the proof is contained in a lemma of H. Wenzl [Wen 1] that we recall below.

As for notations, given two idempotent u and w in some algebra $\mathcal{B}$, we write $u \leq w$ if $uw = u = wu$. This is evidently a partial order on idempotents of $\mathcal{B}$. In case a family $\{u_1,\cdots,u_j\}$ of idempotents in $\mathcal{B}$ has a least upper bound, we denote it by $u_1 \vee u_2 \vee \cdots \vee u_j$.

With β,k as in the theorem and $(P_j)_{j\geq 0}$ as before, we define inductively elements $\delta_1,\cdots,\delta_k \in \mathcal{A}_{\beta,k}$ by

$$\begin{aligned}
\delta_1 &= 1\\
\delta_2 &= 1 - \epsilon_1\\
&\cdot\\
&\cdot\\
&\cdot\\
\delta_{m+1} &= \delta_m - \frac{P_{m-1}(\beta^{-1})}{P_m(\beta^{-1})}\,\delta_m\epsilon_m\delta_m \quad (m \leq k-1).
\end{aligned}$$

Lemma II.11. *One has for* $m \in \{1,\cdots,k\}$:

(i) $(\epsilon_m\delta_m)^2 = \dfrac{P_m(\beta^{-1})}{P_{m-1}(\beta^{-1})}\,\epsilon_m\delta_m.$

(ii) $(\delta_m \epsilon_m)^2 = \dfrac{P_m(\beta^{-1})}{P_{m-1}(\beta^{-1})} \delta_m \epsilon_m$.

Furthermore, if $m \leq k-1$:

(iii) δ_{m+1} *is an idempotent.*

(iv)) $\epsilon_j \delta_{m+1} = \delta_{m+1} \epsilon_j = 0$ for $j = 1, \cdots, m$.

(v) $\delta_{m+1} = 1 - \epsilon_1 \vee \cdots \vee \epsilon_m$.

Proof. Recall that the integer k is fixed in this discussion; we may assume that $k \geq 2$. As the lemma is obvious for $m = 1$, we proceed by induction on m. Thus we assume that $m \geq 2$ and that claims (i) to (v) hold for $1, 2, \cdots, m-1$.

Set $\alpha = (\epsilon_m \delta_m)^2 = \epsilon_m \left(\delta_{m-1} - \dfrac{P_{m-2}(\beta^{-1})}{P_{m-1}(\beta^{-1})} \delta_{m-1} \epsilon_{m-1} \delta_{m-1}\right) \epsilon_m \delta_m$. As δ_{m-1} is a linear combination of monomials in $\epsilon_1, \cdots, \epsilon_{m-2}$, it commutes with ϵ_m. Using (v) for the values $m-1$ and $m-2$, one has $\delta_{m-1} \delta_m = \delta_m$. From these two facts we compute

$$\begin{aligned} \alpha &= \epsilon_m \delta_m - \frac{P_{m-2}(\beta^{-1})}{P_{m-1}(\beta^{-1})} \delta_{m-1} \epsilon_m \epsilon_{m-1} \epsilon_m \delta_m \\ &= \epsilon_m \delta_m - \frac{P_{m-2}(\beta^{-1})}{P_{m-1}(\beta^{-1})} \beta^{-1} \epsilon_m \delta_m = \frac{P_m(\beta^{-1})}{P_{m-1}(\beta^{-1})} \epsilon_m \delta_m. \end{aligned}$$

This shows (i), and (ii) is similar.

From (i) one has $(\delta_m \epsilon_m \delta_m)^2 = \dfrac{P_m(\beta^{-1})}{P_{m-1}(\beta^{-1})} \delta_m \epsilon_m \delta_m$ and (iii) follows.

For $j \in \{1, \cdots, m-1\}$, we know by induction that $\epsilon_j \delta_m = \delta_m \epsilon_j = 0$, so that $\epsilon_j \delta_{m+1} = \delta_{m+1} \epsilon_j = 0$ by definition of δ_{m+1}. Moreover

$$\epsilon_m \delta_{m+1} = \epsilon_m \delta_m - \frac{P_{m-1}(\beta^{-1})}{P_m(\beta^{-1})} (\epsilon_m \delta_m)^2 = 0$$

by (i), and similarly $\delta_{m+1} \epsilon_m = 0$ by (ii). This shows (iv), so that $1 - \delta_{m+1}$ is an upper bound for $\{\epsilon_1, \cdots, \epsilon_m\}$.

Let $\zeta \in \mathcal{A}_{\beta,k+1}$ be an idempotent which is an upper bound for $\{\epsilon_1, \cdots, \epsilon_m\}$. As $1 - \delta_{m+1}$ is a linear combination of (non-trivial) monomials in $\epsilon_1, \cdots, \epsilon_m$, one has

$(1-\delta_{m+1})\zeta = 1-\delta_{m+1} = \zeta(1-\delta_{m+1})$, namely $1-\delta_{m+1} \leq \zeta$. Thus $1-\delta_{m+1}$ is the least upper bound for $\{\epsilon_1,\cdots,\epsilon_m\}$. #

For Theorem II.10, we are interested in $\mathcal{A}_{\beta,k+1}$. Observe however that $\delta_1,\cdots,\delta_k$ are in $\mathcal{A}_{\beta,k}$.

Lemma II.12. *Set* $\gamma = \delta_k\epsilon_k\delta_k \in \mathcal{A} = \mathcal{A}_{\beta,k+1}$. *Then the two-sided ideal* $\mathcal{A}\gamma\mathcal{A}$ *is reduced to* $\mathbb{K}\gamma$, *and is in particular of dimension at most* 1. *Moreover* $\gamma\mathcal{A}\gamma = 0$ *and* $\mathcal{A}\gamma\mathcal{A}$ *has square zero.*

Proof. Let w be a standard monomial in $\mathcal{A}$, as in Proposition 2.8.1. If $w \neq 1$, then $\gamma w = w\gamma = 0$. Indeed, if w starts with ϵ_j, with $1 \leq j \leq k-1$, then $\gamma w = 0$ by claim (iv) of the previous lemma. Also the proof of claims (i) and (ii) shows that

$$\gamma\epsilon_k = (\delta_k\epsilon_k)^2 = \frac{P_k(\beta^{-1})}{P_{k-1}(\beta^{-1})}\,\delta_k\epsilon_k = 0,$$

so $\gamma w = 0$ also for any word beginning with ϵ_k. Similarly $w\gamma = 0$ for all non-trivial words w. By Proposition 2.8.1 and Corollary 2.8.6, this shows that $\mathcal{A}\gamma\mathcal{A} = \mathbb{K}\gamma$.

As $\gamma^2 = \delta_k\epsilon_k\gamma = 0$, it follows that $\gamma w\gamma = 0$ for any standard monomial w, including $w = 1$. This implies $\gamma\mathcal{A}\gamma = 0$. #

To prove Theorem II.10, it is enough to show that $\gamma \neq 0$, namely that the radical of $\mathcal{A}_{\beta,k+1}$ contains the non-zero ideal $\mathcal{A}\gamma\mathcal{A}$ of Lemma II.12.

If $k = 2$, this is straightforward. Indeed $\gamma = (1-\epsilon_1)\epsilon_2(1-\epsilon_1)$ is not zero because $\{1,\epsilon_1,\epsilon_2,\epsilon_1\epsilon_2,\epsilon_2\epsilon_1\}$ is a basis of $\mathcal{A}_{1,3}$. (In example 2.9.1, $\gamma \mapsto 0$ under the map from $\mathcal{A}_{1,3}$ to T, so that the radical of $\mathcal{A}_{1,3}$ is strictly larger than $\mathcal{A}\gamma\mathcal{A}$.) But for $k \geq 3$, we need more lemmas.

Lemma II.13. (i) *For any* $j \geq 2$, *the subspace* $\mathcal{A}_{\beta,j}\epsilon_j\mathcal{A}_{\beta,j}$ *of* $\mathcal{A}_{\beta,j+1}$ *is the linear span of the reduced monomials distinct from* 1 *(see Proposition 2.8.1). In particular*

$$\dim(\mathcal{A}_{\beta,j}\epsilon_j\mathcal{A}_{\beta,j}) = \dim(\mathcal{A}_{\beta,j+1}) - 1.$$

(ii) *The dimension of* $\mathcal{A}_{\beta,j} \otimes_{\mathcal{A}_{\beta,j-1}} \mathcal{A}_{\beta,j}$ *as a vector space over* K *is not more than* $\dim(\mathcal{A}_{\beta,j+1}) - 1$.

Proof. For any $i \in \{1,\cdots,j\}$ one has

$$\epsilon_i = \beta^{j-i}\epsilon_i\epsilon_{i+1}\cdots\epsilon_{j-1}\epsilon_j\epsilon_{j-1}\cdots\epsilon_{i+1}\epsilon_i$$

in $\mathcal{A}_{\beta,j+1}$, and claim (i) follows.

The vector space of (ii) is spanned by elements of the form $w_1 \otimes w_2$ where w_1 and w_2 are standard monomials in $\mathcal{A}_{\beta,j}$.

If w_2 does not involve ϵ_{j-1} then $w_1 \otimes w_2 = w_1w_2 \otimes 1$. Otherwise $w_2 = w_2'\epsilon_{j-1}w_2''$ where $w_2' \in \mathcal{A}_{\beta,j-1}$ ends with ϵ_ℓ for some $\ell \le j-2$ and where $w_2'' = 1$ or $w_2'' = \epsilon_{j-2}\cdots\epsilon_n$ for some $n \in \{1,2,\cdots,j-2\}$. Then $w_1 \otimes w_2 = w_1w_2' \otimes \epsilon_{j-1}\cdots\epsilon_n$. One has $\ell < j-1$ in case $w_2'' = 1$ and $\ell < n$ otherwise, because w_2 is a standard word.

In all cases $w_1 \otimes w_2 = \beta^{-r}\tilde{w}_1 \otimes \tilde{w}_2$ where r is an appropriate integer, and with $\tilde{w}_1\epsilon_j\tilde{w}_2$ a standard monomial in $\mathcal{A}_{\beta,j+1}$. Consequently the vector space of (ii) is spanned by a set containing $\dim(\mathcal{A}_{\beta,j+1}) - 1$ elements. #

Lemma II.14. *For any* $j \ge 2$ *the natural map*

$$\begin{cases} \mathcal{A}_{\beta,j} \otimes_{\mathcal{A}_{\beta,j-1}} \mathcal{A}_{\beta,j} \longrightarrow \mathcal{A}_{\beta,j}\epsilon_j\mathcal{A}_{\beta,j} \\ x \otimes y \longmapsto x\epsilon_j y \end{cases}$$

is a K*-linear isomorphism.*

Proof. The map is clearly well defined (because each of $\epsilon_1,\cdots,\epsilon_{j-2}$ commutes with ϵ_j) and onto. The conclusion follows from the previous lemma. #

Lemma II.15.

(i) *The element* δ_k *is not zero in* $\mathcal{A}_{\beta,k}$.

(ii) *The element* $\delta_k \otimes \delta_k$ *is not zero* $\mathcal{A}_{\beta,k} \otimes_{\mathcal{A}_{\beta,k-1}} \mathcal{A}_{\beta,k}$.

(iii) *The element* γ *of Lemma II.12 is not zero in* $\mathcal{A}_{\beta,k+1}$.

Proof. (i) Denote by B_k' the two-sided ideal generated by $\epsilon_1,\cdots,\epsilon_{k-1}$. The definition of δ_k shows that $1 - \delta_k \in B_k'$. If δ_k were zero, this would imply $B_k' = \mathcal{A}_{\beta,k}$, in contradiction with Lemma 2.8.4.iii, so that $\delta_k \neq 0$. (Observe that δ_k is a central idempotent in $A_{\beta,k}$, so that δ_k is the same as d_k in Lemma 2.8.4).

(ii) By Lemma II.11.iii, the image of $\delta_k \otimes \delta_k$ by the K–linear map

$$\begin{cases} \mathcal{A}_{\beta,k} \otimes_{\mathcal{A}_{\beta,k-1}} \mathcal{A}_{\beta,k} \longrightarrow \mathcal{A}_{\beta,k} \\ x \otimes y \longmapsto xy \end{cases}$$

is δ_k. Thus (ii) follows from (i).

(iii) This follows from (ii) and from Lemma II.14. #

The proof of Theorem II.10 is complete.

In the same way as the braid group on three strings is an epimorphic image of the braid group on four strings, the assignment

$$\epsilon_1 \longmapsto \epsilon_1 \quad \epsilon_2 \longmapsto \epsilon_2 \quad \epsilon_3 \longmapsto \epsilon_1$$

extends to an epimorphism $\mathcal{A}_{1,4} \longrightarrow \mathcal{A}_{1,3}$. It follows that $\mathcal{A}_{1,4}$ is not semi–simple.

However, we do not know in general when $\mathcal{A}_{\beta,j}$ is semi–simple for $j \geq k+2$. It would be interesting to compute the discriminant of $\mathcal{A}_{\beta,j}$.

Following Wenzl [Wen1], we shall record another useful consequence of Lemma II.11.

Theorem II.16. *Consider a real number* $\beta \neq 0$ *and an infinite sequence* $(\epsilon_j)_{j\geq 1}$ *of orthogonal projections on some complex Hilbert space such that the usual relations hold*

$$\epsilon_i^2 = \epsilon_i$$

$$\beta\, \epsilon_i \epsilon_j \epsilon_i = \epsilon_i \qquad \textit{if } |i-j| = 1$$

$$\epsilon_i \epsilon_j = \epsilon_j \epsilon_i \qquad \textit{if } |i-j| \geq 2$$

together with

$$\epsilon_i^* = \epsilon_i .$$

If $\epsilon_1 \neq 0$ *then one has*

$$\textit{either } \beta = 4\cos^2(\pi/q) \textit{ for some integer } q \geq 3 \textit{ or } \beta \geq 4.$$

Observe that $\beta\epsilon_1\epsilon_2\epsilon_1 = \epsilon_1$ obviously implies that $\beta \geq 1$. Before we prove the theorem, we need two more lemmas.

Lemma II.17. *Consider an integer* $k \geq 3$ *with* $P_2(\beta^{-1}) \neq 0, \cdots, P_{k-1}(\beta^{-1}) \neq 0$, *so that* $\delta_1, \cdots, \delta_k$ *are defined and satisfy the relations of Lemma II.11. If* $\delta_{k-1} = \delta_k$ *then* $\epsilon_{k-1} \leq 1-\delta_{k-2}$. *If moreover* $k \geq 4$ *then* $\delta_{k-3}\epsilon_i\epsilon_j = 0$ *whenever* $i,j \geq k-2$ *and* $|i-j| \geq 2$.

Proof. Set $p = \epsilon_{k-1}(\delta_{k-2}-\delta_{k-1})$. By (iv) and (v) of Lemma II.11 and the hypothesis $\delta_{k-1} = \delta_k$, one has

$$\epsilon_{k-1}\delta_{k-1} = \epsilon_{k-1}\delta_k = 0$$
$$\epsilon_{k-1}\delta_{k-2} = \delta_{k-2}\epsilon_{k-1}.$$

Hence p is a projection and $p = \epsilon_{k-1}\delta_{k-2}$. By (i) of the same lemma

$$p = p^*p = \epsilon_{k-1}\left[\frac{P_{k-3}(\beta^{-1})}{P_{k-2}(\beta^{-1})}\delta_{k-2}\epsilon_{k-2}\delta_{k-2}\right]^2\epsilon_{k-1}$$

$$= \frac{P_{k-3}(\beta^{-1})}{P_{k-2}(\beta^{-1})}\epsilon_{k-1}\delta_{k-2}\epsilon_{k-2}\delta_{k-2}\epsilon_{k-1}.$$

As ϵ_{k-1} and δ_{k-2} commute

$$p = \frac{\beta^{-1}P_{k-3}(\beta^{-1})}{P_{k-2}(\beta^{-1})}\epsilon_{k-1}\delta_{k-2} = \left[1 - \frac{P_{k-1}(\beta^{-1})}{P_{k-2}(\beta^{-1})}\right]p.$$

If $p \neq 0$ this implies $P_{k-1}(\beta^{-1}) = 0$ and contradicts the hypothesis. Thus $p = 0$, namely $\epsilon_{k-1}\delta_{k-2} = 0$, and $\epsilon_{k-1} \leq 1 - \delta_{k-2}$.

Assume $k \geq 4$ and set $q = \delta_{k-3}\epsilon_{k-2}\epsilon_k$, which is a projection because δ_{k-3}, ϵ_{k-2} and ϵ_k commute. One has $q = \epsilon_k\epsilon_{k-2}(\delta_{k-3}-\delta_{k-2})$ because

$$\epsilon_k\delta_{k-2} = \beta\epsilon_k\epsilon_{k-1}\delta_{k-2}\epsilon_k = 0$$

by the first part of the present proof. The same computation as above shows that

$$q = \epsilon_k\{\epsilon_{k-2}(\delta_{k-3}-\delta_{k-2})^2\epsilon_{k-2}\}\epsilon_k$$

$$= \left[1 - \frac{P_{k-2}(\beta^{-1})}{P_{k-3}(\beta^{-1})}\right]q$$

and consequently that $q = 0$. This shows the second claim for $i = k-2$ and $j = k$.

For $j \geq k$, set $t = (j-k)/2$ and $v_j = \beta^t \epsilon_k \epsilon_{k+1} \cdots \epsilon_j$. Then $v_j^* \epsilon_k v_j = \epsilon_j$ and v_j commutes with $\delta_{k-3}\epsilon_{k-2}$, so that

$$\delta_{k-3}\epsilon_{k-2}\epsilon_j = v_j^* \delta_{k-3}\epsilon_{k-2}\epsilon_k v_j = 0.$$

For $i \geq k-2$, set similarly $s = (i-k+2)/2$ and $u_i = \beta^s \epsilon_{k-2} \cdots \epsilon_i$, so that $u_i^* \epsilon_{k-2} u_i = \epsilon_i$. If moreover, $i \leq j-2$ then u_i commutes with $\delta_{k-3}\epsilon_j$ and one has

$$\delta_{k-3}\epsilon_i\epsilon_j = u_i^* \delta_{k-3}\epsilon_{k-2}\epsilon_j u_i = 0. \quad \#$$

Lemma II.18. *Consider now an integer* $\ell \geq 2$ *with*

$$4\cos^2[\pi/(\ell+1)] < \beta < 4\cos^2[\pi/(\ell+2)].$$

Then $P_2(\beta^{-1}) \neq 0, \cdots, P_{\ell+1}(\beta^{-1}) \neq 0$, *so that* $\delta_1, \cdots, \delta_{\ell+2}$ *are defined and satisfy the relations of Lemma II.11. Moreover* $\delta_{\ell+1} = \delta_{\ell+2}$.

Proof. The first claim is a restatement of Proposition 2.8.3.(iii), which shows moreover that $P_\ell(\beta^{-1}) > 0$ and $P_{\ell+1}(\beta^{-1}) < 0$. By Lemma II.11.(i) one has

$$0 \leq (\delta_{\ell+1}\epsilon_{\ell+1}\delta_{\ell+1})^2 = \frac{P_{\ell+1}(\beta^{-1})}{P_\ell(\beta^{-1})}\delta_{\ell+1}\epsilon_{\ell+1}\delta_{\ell+1} \leq 0$$

so that $(\epsilon_{\ell+1}\delta_{\ell+1})^*(\epsilon_{\ell+1}\delta_{\ell+1}) = 0$ and $\epsilon_{\ell+1}\delta_{\ell+1} = 0 = \delta_{\ell+1}\epsilon_{\ell+1}$. As $\delta_{\ell+1} = 1 - \epsilon_1 \vee \cdots \vee \epsilon_\ell$, this means that $\epsilon_{\ell+1} \leq \epsilon_1 \vee \cdots \vee \epsilon_\ell$, so that $\delta_{\ell+2} = 1 - \epsilon_1 \vee \cdots \vee \epsilon_{\ell+1} = \delta_{\ell+1}$. #

Proof of Theorem II.16. Assume that $1 < \beta < 4$ and that $\beta \notin \{4\cos^2(\pi/q)\}_{q\geq 4}$. We shall find a contradiction.

Let $\ell \geq 2$ be the integer defined by

$$4\cos^2[\pi/(\ell+1)] < \beta < 4\cos^2[\pi/(\ell+2)].$$

Obviously $1 = \delta_1 \neq \delta_2 = 1-\epsilon_1$, because $\epsilon_1 \neq 0$. Observe also that δ_3 is defined (because $\beta \neq 1$) and that $\delta_2 \neq \delta_3$, otherwise one would have $\epsilon_1 = \epsilon_1 \vee \epsilon_2$ and $\epsilon_1\epsilon_2\epsilon_1 = \epsilon_1$, which is incompatible with $\beta\epsilon_1\epsilon_2\epsilon_1 = \epsilon_1$ and $\beta > 1$. Consequently, by Lemma II.18, there is a first integer k such that

$$3 \le k \le \ell+1 \text{ and } \delta_{k-1} \ne \delta_k = \delta_{k+1}.$$

For $m \ge 1$, set now $\gamma_m = \delta_{k-2}\epsilon_{k-2+m}$, which is a projection because δ_{k-2} and ϵ_{k-2+m} commute. Moreover

$$\begin{aligned} \beta\gamma_m\gamma_n\gamma_m &= \beta\delta_{k-2}\epsilon_{k-2+m}\epsilon_{k-2+n}\epsilon_{k-2+m} = \gamma_m \text{ if } |m-n| = 1 \\ \gamma_m\gamma_n &= \gamma_n\gamma_m \text{ if } |m-n| \ge 2 \end{aligned}$$

for $m,n \ge 1$. Consequently one may apply Lemmas II.11, II.17 and II.18 to the γ_m's, so that

$$\begin{aligned} &\gamma_1 \vee \cdots \vee \gamma_\ell = \gamma_1 \vee \cdots \vee \gamma_{\ell+1} && \text{(by II.18)} \\ &\gamma_{\ell+1} \le \gamma_1 \vee \cdots \vee \gamma_{\ell-1} && \text{(by II.17)} \\ &\gamma_m\gamma_{\ell+1} = \delta_{k-2}\epsilon_{k-2+m}\epsilon_{k-1+\ell} = 0 \end{aligned}$$

for $m = 1,2,\cdots,\ell-1$ (by II.17). The last two relations imply respectively

$$\begin{aligned} \gamma_{\ell+1}(\gamma_1\vee\cdots\vee\gamma_{\ell-1}) &= \gamma_{\ell+1} \\ \gamma_{\ell+1}(\gamma_1\vee\cdots\vee\gamma_{\ell-1}) &= 0 \end{aligned}$$

so that $\gamma_{\ell+1} = 0$.

As $\beta\gamma_m\gamma_{m+1}\gamma_m = \gamma_m$ for $1 \le m \le \ell$, one has also $\gamma_1 = 0 = \delta_{k-2}\epsilon_{k-1}$, namely $\epsilon_{k-1} \le \epsilon_1 \vee \cdots \vee \epsilon_{k-3}$. This implies $\delta_{k-1} = \delta_k$, in contradiction with the definition of k.

#

The careful reader will observe that we have proved more than Theorem II.16, namely:

Consider a real number, $\beta \ne 0$, an integer $N \ge 3$ and a sequence $(\epsilon_1,\cdots,\epsilon_{N-1})$ of nonzero orthogonal projections on some complex Hilbert space which satisfy the same relations as in the theorem. Set $r = \left[\frac{N+3}{2}\right]$. Then one has

$$\begin{aligned} &\text{either} && \beta = 4\cos^2(\pi/q) \text{ for some integer } q \text{ with } 3 \le q < r \\ &\text{or} && \beta \ge 4\cos^2(\pi/r). \end{aligned}$$

APPENDIX III

Hecke groups and other subgroups of PSL(2,ℝ) generated by parabolic pairs

All facts exposed in this Appendix are well known, and most of them may be found in Magnus' survey [Mag2]. Our purpose is twofold: firstly to expose background for the examples in Section 3.3; secondly to offer a self-contained proof of the following result, which we believe is due to Fricke and Klein. It provides spectacular comparison with Kronecker's result (our Theorem 1.1.1) and with the results described in Section 3.4, but we do not know whether it is a superficial curiosity or if there are interesting hidden explanations. We denote by $\begin{bmatrix} a & b \\ c & d \end{bmatrix}$ the class in PSL(2,ℝ) of a matrix $\begin{pmatrix} a & b \\ c & d \end{pmatrix} \in SL(2,\mathbb{R})$.

Theorem III.1. *Let* λ *be a strictly positive real number and let* Γ'_λ *be the subgroup of* PSL(2,ℝ) *generated by* $\begin{bmatrix} 1 & \lambda \\ 0 & 1 \end{bmatrix}$ *and* $\begin{bmatrix} 1 & 0 \\ \lambda & 1 \end{bmatrix}$. *Then* Γ'_λ *is discrete if and only if*

$$\text{either } \lambda = 2\cos\frac{\pi}{q} \text{ for some integer } q \geq 3 \text{ or } \lambda \geq 2.$$

Moreover Γ'_λ *has elements of finite order in the first case and* Γ'_λ *is free on two generators in the second case.*

For comparison with works of Fricke and Klein, see §II.2.11–12 in [FK1] and II.1 in [FK2]. Among the many places where variations and complement on what follows can be found, we shall quote [Eva], [Kna], [Leu], [Mat], [Puz]. [Rus].

Let $G_{\mathbb{C}}$ denote the group PSL(2,ℂ), acting by fractional linear transformations on the Riemann sphere $\tilde{\mathbb{C}} = \mathbb{C} \cup \{\infty\}$. Recall that $\begin{pmatrix} a & b \\ c & d \end{pmatrix}$ is parabolic if it is not one of the two matrices ± 1 and if $(a+d)^2 = 4$, namely if the transformation $z \mapsto (az+b)(cz+d)^{-1}$ of $\tilde{\mathbb{C}}$ has exactly one fixed point. The study of non commuting parabolic pairs reduces to that of pairs as in Theorem III.1, as the next proposition shows.

Proposition III.2. *Let* $\mathcal{A},\mathcal{B}$ *be two parabolic matrices in* SL(2,ℂ) *and let* $\lambda \in \mathbb{C}$ *be such that*

$$4\,\mathrm{tr}(\mathcal{A}\mathcal{B}) = (2+\lambda^2)\mathrm{tr}(\mathcal{A})\mathrm{tr}(\mathcal{B}).$$

Then $\lambda = 0$ *if and only if* $\mathcal{A}$ *and* $\mathcal{B}$ *commute. Moreover, if* $\lambda = 0$, *there exist* $\mathcal{C} \in SL(2,\mathbb{C})$ *and* $z \in \mathbb{C}^*$ *with*

$$\mathcal{C}\mathcal{A}\mathcal{C}^{-1} = \alpha\begin{bmatrix}1 & 1\\ 0 & 1\end{bmatrix} \qquad \mathcal{C}\mathcal{B}\mathcal{C}^{-1} = \beta\begin{bmatrix}1 & z\\ 0 & 1\end{bmatrix}$$
$$\alpha = \frac{1}{2}\,\mathrm{tr}(A) \in \{\pm 1\} \qquad \beta = \frac{1}{2}\,\mathrm{tr}(\beta) \in \{\pm 1\}.$$

On the other hand, if $\lambda \neq 0$, *there exists* $\mathcal{C} \in \mathrm{SL}(2,\mathbb{C})$ *with*

$$\mathcal{C}\mathcal{A}\mathcal{C}^{-1} = \alpha\begin{bmatrix}1 & \lambda\\ 0 & 1\end{bmatrix} \qquad \mathcal{C}\mathcal{B}\mathcal{C}^{-1} = \beta\begin{bmatrix}1 & 0\\ \lambda & 1\end{bmatrix}.$$

Proof. Let A,B be the classes of $\mathcal{A},\mathcal{B}$ modulo $\{\pm 1\}$, and let v,w be the fixed points of A,B in $\tilde{\mathbb{C}}$.

If $v = w$, a conjugation by $\begin{bmatrix}1 & 0\\ -v^{-1} & 1\end{bmatrix}$ reduces the situation to the case $v = w = \infty$, so that one has

$$\mathcal{A} = \alpha\begin{bmatrix}1 & x^{-2}\\ 0 & 1\end{bmatrix} \qquad \mathcal{B} = \beta\begin{bmatrix}1 & y\\ 0 & 1\end{bmatrix}$$

for some $x,y \in \mathbb{C}^*$. After a second conjugation by $\begin{bmatrix}x & 0\\ 0 & x^{-1}\end{bmatrix}$, one obtains matrices as in the theorem. It is now straightforward to check that $4\,\mathrm{tr}(\mathcal{A}\mathcal{B}) = 2\,\mathrm{tr}(\mathcal{A})\mathrm{tr}(\mathcal{B})$ and $\mathcal{A}\mathcal{B} = \mathcal{B}\mathcal{A}$.

If $v \neq w$, a conjugation by the appropriate scalar multiple of $\begin{bmatrix}1 & -v\\ -w^{-1} & 1\end{bmatrix}$ reduces the situation to the case $v = 0$, $w = \infty$, so that one has

$$\mathcal{A} = \alpha\begin{bmatrix}1 & x\\ 0 & 1\end{bmatrix} \qquad \mathcal{B} = \beta\begin{bmatrix}1 & 0\\ y & 1\end{bmatrix}$$

for some $x,y \in \mathbb{C}^*$. If $\mathcal{C} = \begin{bmatrix}z & 0\\ 0 & z^{-1}\end{bmatrix}$, then

$$\mathcal{C}\mathcal{A}\mathcal{C}^{-1} = \alpha\begin{bmatrix}1 & z^2x\\ 0 & 1\end{bmatrix} \qquad \mathcal{C}\mathcal{B}\mathcal{C}^{-1} = \beta\begin{bmatrix}1 & 0\\ z^{-2}y & 1\end{bmatrix}$$

and one may choose z so that $z^2x = z^{-2}y$; thus, we may now assume that $x = y$. A computation shows that $4\,\mathrm{tr}(\mathcal{A}\mathcal{B}) = (2+x^2)\mathrm{tr}(\mathcal{A})\mathrm{tr}(\mathcal{B})$ and that $\mathcal{A}\mathcal{B} \neq \mathcal{B}\mathcal{A}$; in particular $x^2 = \lambda^2$. If $x = \lambda$, the proof is finished; if $x = -\lambda$, it is enough to choose iz instead of z. #

For any $\lambda \in \mathbb{C}^*$, let Γ'_λ be the subgroup of $G_{\mathbb{C}}$ generated by

$$A_\lambda = \begin{bmatrix}1 & \lambda\\ 0 & 1\end{bmatrix} \text{ and } B_\lambda = \begin{bmatrix}1 & 0\\ \lambda & 1\end{bmatrix}$$

and let Γ_λ be the subgroup of $G_{\mathbb{C}}$ generated by A_λ and

$$J = \begin{bmatrix} 0 & 1 \\ -1 & 0 \end{bmatrix}.$$

As $JA_\lambda J = B_\lambda^{-1}$ one has $\Gamma'_\lambda \subset \Gamma_\lambda$. As $J^2 = 1$, the index satisfies $[\Gamma_\lambda : \Gamma'_\lambda] \leq 2$. The cases $[\Gamma_\lambda : \Gamma'_\lambda] = 2$ and $\Gamma_\lambda = \Gamma'_\lambda$ occur both, depending on λ (see below).

Corollary III.3. *For* $\kappa,\lambda \in \mathbb{C}^*$, *the groups* Γ'_κ *and* Γ'_λ *are conjugate in* $G_\mathbb{C}$ *as soon as* $\kappa^4 = \lambda^4$.

Proof. If $\kappa^2 = \lambda^2$, it is clear from the proposition. If $\kappa^2 = -\lambda^2$, it follows from the fact that Γ'_λ is generated by A_λ and B_λ^{-1} and from the equality

$$4\ \mathrm{tr}\left(\begin{bmatrix} 1 & \lambda \\ 0 & 1 \end{bmatrix}\begin{bmatrix} 1 & 0 \\ -\lambda & 1 \end{bmatrix}\right) = (2-\lambda^2)\mathrm{tr}\begin{bmatrix} 1 & \lambda \\ 0 & 1 \end{bmatrix}\mathrm{tr}\begin{bmatrix} 1 & 0 \\ \lambda & 1 \end{bmatrix}. \quad \#$$

It would be nice to know whether the converse is known, but we have not found this in the literature, even under the assumption $\lambda \in \mathbb{R}_+$.

The next proposition shows the claim of Theorem III.1 for $\lambda \geq 2$. We denote by $\{1,J\}$ the group of order 2 generated by J, by $A_\lambda^\mathbb{Z}$ the infinite cyclic group generated by A_λ, and by $H * K$ the free product of two groups H,K.

Proposition III.4. *Let* $\lambda \in \mathbb{C}$ *with* $|\lambda| \geq 2$. *Then*

$$\Gamma_\lambda = A_\lambda^\mathbb{Z} * \{1,J\} \qquad \Gamma'_\lambda = A_\lambda^\mathbb{Z} * B_\lambda^\mathbb{Z}$$
$$[\Gamma_\lambda : \Gamma'_\lambda] = 2$$

and these groups are discrete in $G_\mathbb{C}$.

Proof. We write A for A_λ during the proof. Set

$$S_A = \{z \in \mathbb{C} : |\mathrm{Re}\left[\frac{\overline{\lambda}}{|\lambda|}z\right]| > 1\}$$
$$S_J = \{z \in \mathbb{C} : |z| < 1\}.$$

One has $J(S_A) \subset S_J$ (the 8-shaped domain below) and $A^n(S_J) \subset S_A$ for all $n \in \mathbb{Z}$ with $n \neq 0$. Let W be a non empty reduced word spelled out with the letters A and J. Then $W \neq 1$ in $G_\mathbb{C}$, as results from the four following cases; $n_1,\cdots,n_k$ are integers in $\mathbb{Z}\backslash\{0\}$.

(1) $W = A^{n_1} J A^{n_2} J \cdots J A^{n_k} \Rightarrow W(S_J) \subset S_A \Rightarrow W \neq 1$.

(2) $W = J A^{n_1} J \cdots J A^{n_k} J \Rightarrow JWJ \neq 1$ by (1) $\Rightarrow W \neq 1$.

(3) $W = A^{n_1} J A^{n_2} \cdots J A^{n_k} J \Rightarrow A^{-m} W A^m \neq 1$ by (1) for $m \in \mathbb{Z}$ with $m \neq n_1 \Rightarrow W \neq 1$.

(4) $W = J A^{n_1} J \cdots J A^{n_k} \Rightarrow W \neq 1$ by (3).

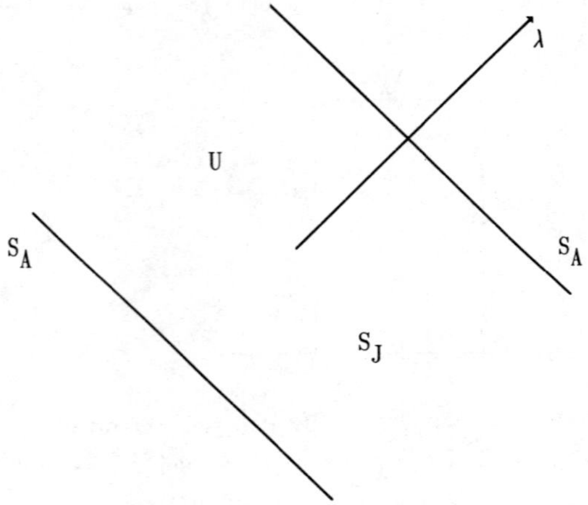

Hence Γ_λ is the free product of $A_\lambda^{\mathbb{Z}}$ and $\{1,J\}$. It follows that Γ'_λ is $A_\lambda^{\mathbb{Z}} * B_\lambda^{\mathbb{Z}}$ and is of index 2 in Γ_λ.

Let U be a non empty open subset of S_J disjoint from $J(S_A)$ and such that $J(U) \cap S_A = \phi$. Then $g(U) \cap U = \phi$ for all $g \in \Gamma_\lambda \setminus \{1\}$, as it is clear for each of the four cases above. Hence Γ_λ has a domain of discontinuity in $\tilde{\mathbb{C}}$, and this implies that Γ_λ is discrete in $G_{\mathbb{C}}$. #

We denote by G the group $PSL(2,\mathbb{R})$ acting on the Poincaré half plane $\mathcal{P}$, and we normalize the Haar measure on G as in §3.3.d.

Proposition III.5. *Consider an integer* $q \geq 3$ *and set* $\lambda = 2\cos(\pi/q)$.

(i) *The group* Γ_λ *is discrete in* G. *It has a fundamental domain in* $\mathcal{P}$ *of finite area* $\pi(1 - \frac{2}{q})$, *so that the volume of* G/Γ_λ *is also* $\pi(1 - \frac{2}{q})$.

(ii) Γ_λ *is the free product of the group generated by* $A_\lambda J$, *which is cyclic of order* q, *and of* $\{1,J\}$. *In particular, all conjugacy classes in* $\Gamma_\lambda \backslash \{1\}$ *are infinite.*

(iii) $\Gamma_\lambda = PSL(2,\mathbb{Z})$ *if* $q = 3$ *and* $\lambda = 1$.

(iv) $\Gamma'_\lambda = \Gamma_\lambda$ *if* q *is odd and* $[\Gamma_\lambda : \Gamma'_\lambda] = 2$ *if* q *is even.*

Proof. Let Δ_λ be the (hyperbolic) triangle in $\mathcal{P}$ with vertices $e^{i\pi/q}$, ∞, $-e^{-i\pi/q}$. The transformation A_λ maps the half-line $(-e^{-i\pi/q},\infty)$ onto $(e^{i\pi/q},\infty)$, and J

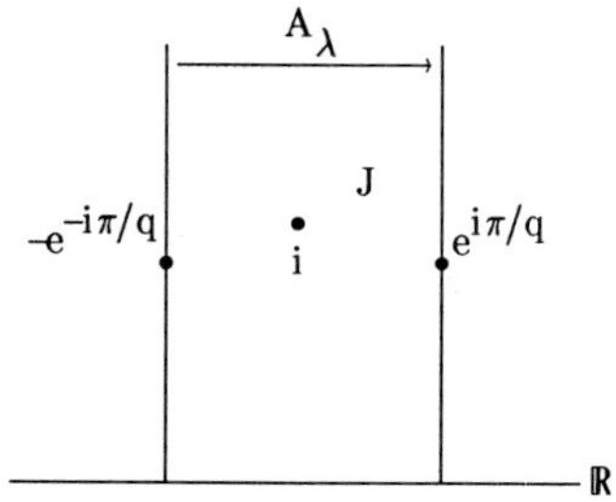

exchanges the edges $(-e^{-i\pi/q},i)$ and $(i,e^{i\pi/q})$. By Poincaré's theorem (see [Mas] or [Rha]), Δ_λ is a fundamental domain for Γ_λ acting on $\mathcal{P}$, and Γ_λ is discrete in G. This and Gauss–Bonet formula show (i). Poincaré's theorem also shows (ii), namely that Γ_λ has a presentation with generators A_λ, J and with relations $(A_\lambda J)^q = 1 = J^2$. Statement (iii) is standard: see for example [Ser2], Chapter VII.

Let us demonstrate (ii) in a different way (see also [Eva]). Set $K_\lambda = A_\lambda J = \begin{bmatrix} -\lambda & 1 \\ -1 & 0 \end{bmatrix}$, which is a rotation of order q around $e^{i\pi/q}$. Consider the action of Γ_λ on the boundary $\hat{\mathbb{R}} = \mathbb{R} \cup \{\infty\}$ of $\mathcal{P}$ in $\hat{\mathbb{C}}$. Then K_λ maps $]0,\lambda^{-1}]$ onto $]-\infty,0]$. Set $S_K =]0,\infty]$ and $S_J =]-\infty,0]$. Then $J(S_K) = S_J$ and $K^r(S_J) \subset S_K$ for $r \in \{1,2,\cdots,q-1\}$. It follows as in the proof of Proposition III.4 that Γ_λ is the free product of $\{1,K,\cdots,K^{q-1}\}$ with $\{1,J\}$. (Digression: let $\mu = e^{i\pi/q}$ and μ^{-1} be the two fixed points of K_λ in $\mathbb{C}$. Then μ and μ^{-1} are given by $\frac{1}{2}(\lambda + (\lambda^2-4)^{1/2})$, as in Section I.2.)

One has $A_\lambda B_\lambda^{-1} = K_\lambda^2$. If q is odd, the group Γ'_λ generated by A_λ and B_λ contains also K_λ and J, so that $\Gamma'_\lambda = \Gamma_\lambda$. If $q = 2p$ is even, then $\Gamma'_\lambda = A_\lambda^{\mathbb{Z}} * \{1,K^2,\cdots,K^{q-2}\} \approx \mathbb{Z} * (\mathbb{Z}/p\mathbb{Z})$ as shown in [LyU]. #

The groups of Proposition III.5 are called Hecke groups [Hec] and have received considerable attention. For some authors, Hecke groups include Γ_λ when $\lambda \in \mathbb{R}$ and $\lambda \geq 2$; see page 333 in [Leh].

As a digression, let us observe that this family of groups shows dramatically that rigidity à la Mostow–Margulis (see [Zim]) does not hold in $G = PSL(2,\mathbb{R})$. Indeed, if $q \to \infty$ so that $\lambda = 2\cos(\pi/q) \to 2$ one has a sequence of groups having fundamental domains with one cusp which "tends" to a group for which any fundamental domain has more than one cusp, and moreover all these groups are pairwise non isomorphic. Also if $\lambda > 2$ tends to 2, one has a family of groups with infinite covolume in G which tends to the group Γ_2 with covolume π.

The "only if" part of Theorem III.1 is yet to be proved. We consider again $\lambda \in \mathbb{C}^*$ rather than $\lambda \in \mathbb{R}^*_+$, as long as it does not complicate the arguments.

<u>Proposition III.6.</u> *Let* $\lambda \in \mathbb{C}$ *with* $0 < |\lambda| < 1$. *Then* Γ_λ *is not discrete in* $G_{\mathbb{C}}$.

<u>Proof.</u> Define inductively $C_n = \begin{bmatrix} a_n & \beta_n \\ \gamma_n & \delta_n \end{bmatrix} \in \Gamma'_\lambda$ by

$$C_2 = ABA^{-1}B^{-1} = \begin{bmatrix} 1+\lambda^2+\lambda^4 & -\lambda^3 \\ \lambda^3 & 1-\lambda^2 \end{bmatrix}$$

$$C_{n+1} = C_n B C_n^{-1} = \begin{bmatrix} 1+\lambda\beta_n\delta_n & -\lambda\beta_n^2 \\ \lambda\delta_n^2 & 1-\lambda\beta_n\delta_n \end{bmatrix} \quad \text{for } n \geq 2.$$

Thus $\beta_n = \lambda^{2^n-1}$, and in particular $|\beta_n| \to 0$ if $n \to \infty$.

Observe that $\delta_{n+1} - 1 = \lambda^{2^n}\delta_n$, so that δ_n is a polynomial in λ, with coefficients in $\{-1,0,1\}$ and with constant term 1. Hence $|\delta_n - 1| < \frac{1}{1-|\lambda|}$ and $|\delta_{n+1} - 1| < \frac{|\lambda|^{2^n}}{1-|\lambda|}$, so that $\delta_n \to 1$ if $n \to \infty$. It follows that $\gamma_n \to \lambda$ and $\alpha_n \to 1$ (because C_n is parabolic for $n \geq 3$) if $n \to \infty$. Thus $C_n \to \begin{bmatrix} 1 & 0 \\ \lambda & 1 \end{bmatrix}$ if $n \to \infty$ and Γ_λ is not discrete in $G_{\mathbb{C}}$. #

<u>Corollary III.7.</u> *Let* $P = \begin{bmatrix} a & b \\ c & d \end{bmatrix} \in G_{\mathbb{C}}$ *and let* $\lambda \in \mathbb{C}$ *with* $0 < |c\lambda| < 1$. *Then the subgroup generated by* P *and* A_λ *is not discrete in* $G_{\mathbb{C}}$.

<u>Proof.</u> As $\operatorname{tr}(APAP^{-1}) = 2 - \lambda^2c^2$, Proposition III.2 shows that the group is conjugated to $P_{i\lambda c}$, and the claim follows from Proposition III.6. #

Proposition III.8. *Consider a real number* $r < \frac{1}{3}$, *and set* $\lambda = 2\cos(\pi r)$. *Assume either that* r *is irrational or that* $r = \frac{p}{q}$ *with* $(p,q) = 1$ *and* $p \geq 2$. *Then* Γ_λ *is not discrete in* G.

Proof. The group Γ_λ contains

$$C_\lambda = JA_\lambda = \begin{bmatrix} 0 & 1 \\ -1 & -2\cos(\pi r) \end{bmatrix}.$$

As the eigenvalues of $\begin{bmatrix} 0 & 1 \\ -1 & -2\cos(\pi r) \end{bmatrix}$ are $-e^{i\pi r}$ and $-e^{-i\pi r}$, this C_λ is a rotation of angle $2\pi r$ around $-e^{-i\pi r}$, in the sense of hyperbolic geometry. If r is irrational, the subgroup of G generated by C_λ is not discrete in G.

If $r = \frac{p}{q}$ with $(p,q) = 1$, there is some power D_λ of C_λ which is the rotation of angle $\frac{2\pi}{q}$ around $-e^{-i\pi r}$. Set $\alpha = \pi/q$, so that D_λ is the rotation of angle 2α around $-e^{-ip\alpha}$. Set also

$$E = \begin{bmatrix} (\sin p\alpha)^{1/2} & -(\cos p\alpha)(\sin p\alpha)^{-1/2} \\ 0 & \sin(p\alpha)^{-1/2} \end{bmatrix}$$

$$R_{2\alpha} = \begin{bmatrix} \cos\alpha & \sin\alpha \\ -\sin\alpha & \cos\alpha \end{bmatrix}.$$

As $E(i) = -e^{-ip\alpha}$ and as $R_{2\alpha}$ is the rotation of angle 2α around i (mark the factor 2!), one has

$$D_\lambda = ER_{2\alpha}E^{-1} = \begin{bmatrix} * & * \\ -\sin\alpha/\sin p\alpha & * \end{bmatrix}.$$

Assume now that $p \geq 2$. One has

$$\left|\frac{\sin\alpha}{\sin p\alpha}\lambda\right| = 2\sin\alpha\frac{\cos p\alpha}{\sin p\alpha} \leq 2\sin\alpha\frac{\cos 2\alpha}{\sin 2\alpha} = \frac{\cos 2\alpha}{\cos\alpha}.$$

As $r < \frac{1}{3}$ one has $q > 6$ and $\alpha < \pi/6$, so that $\frac{\cos 2\alpha}{\cos\alpha} < 1$. Then Corollary III.7 shows that the subgroup Γ_λ generated by D_λ and A_λ is not discrete in G. #

For $\lambda \in \mathbb{C}$ with $\lambda - 2\cos(\pi/q)$ not zero and small enough, it is known that Γ_λ is not discrete in $G_{\mathbb{C}}$. See [Mag2].

REFERENCES

[Arv] W. Arveson: "An invitation to C^*-algebras", Springer 1976.

[Ana] C. Anantharaman-Delaroche: "On Connes' property T for von Neumann algebras", Preprint, Orléans 1986.

[AS] M. Atiyah and W. Schmid: "A geometric construction of the discrete series for semisimple Lie groups", Invent. Math. **42** (1977) 1–62.

[BR] S. Banach and S. Ruziewicz: "Sur les solutions d'une équation fonctionnelle de J. Cl. Maxwell", in S. Banach, Oeuvres I, p. 51–57.

[Bax] R.J.Baxter: "Exactly solved models in statistical mechanics", Academic Press 1982.

[Big] N.L. Biggs: "Algebraic graph theory", Cambridge Univ. Press 1974.

[BW] J. Birman and H. Wenzl: "Braids, link polynomials and a new algebra", Trans. Amer. Math. Soc. (to appear).

[Bor] A. Borel: "Density properties for certain subgroups of semi-simple groups without compact components", Ann. of Math. **72** (1960) 179–188.

[BA2] N. Bourbaki: "Algèbre, chap. 2, algèbre linéaire", 3rd edition, Hermann 1962.

[BA4] N. Bourbaki: "Algèbre", Chap. 4-7, Masson 1981.

[BA8] N. Bourbaki: "Algèbre, chap. 8, modules et anneaux semi-simples", Hermann 1958.

[BAC1] N. Bourbaki: "Algèbre commutative, chap. 1 et 2", Hermann 1961.

[BLie] M. Bourbaki: "Groupes et algèbres de Lie, Chap. 4, 5 et 6", Hermann 1968.

[Boy] D.W. Boyd: "Variations on a theme of Kronecker", Canad. Math. Bull. **21**(2) (1978), 129–133.

[Bra] O. Bratteli: "Inductive limits of finite dimensional C^*-algebras", Trans. Amer. Math. Soc. **171** (1972) 195–234.

[Br] R. Brauer: "On algebras which are connected with the semisimple continuous groups", Ann. of Math. **38** (1937) 854–872.

[Bur] W. Burnside: "Theory of groups of finite order", second ed., Cambridge Univ. Press 1911.

[CGSS] P.J. Cameron, J.M. Goethals, J.J. Seidel and E.E. Shult: "Line graphs, root systems, and elliptic geometry", J. of Algebra **43** (1976), 305–327.

[Car] M. P. do Carmo: "Differential geometry of curves and surfaces", Prentice Hall 1976.

[Che] M. Chein: "Recherche des graphes des matrices de Coxeter hyperboliques d'ordre ≤ 10", Rev. Française Informat. Recherche Opérationelle, no. R-3 (1969), 3–16.

[Con] A. Connes: "Indice des sous-facteurs, algèbres de Hecke et théorie des noeuds", Astérisque **133-134** (Soc. Math. France 1986) 289–308.

[Cox] H.S.M. Coxeter: "The product of the generators of a finite group generated by reflections", Duke Math. J. **18** (1951), 765–782.

[CR] C.W. Curtis and I. Reiner: "Methods of representation theory with applications to finite group and orders, Vol. 1", Wiley–Interscience 1981.

[CD] D. Cvetković and M. Doob: "Developments in the theory of graph spectra", Linear and multilinear algebra **18** (1985) 153–181.

[CDG] D. Cvetković, M. Doob and I. Gutman: "On graphs whose spectral radius does not exceed $(2+\sqrt{5})^{1/2}$", Ars Combinatoria **14** (1982), 225–239.

[CDGT] D. Cvetković, M. Doob, I. Gutman, and A. Torgašev: "Recent results in the theory of graph spectra", North Holland, 1987.

[CDS] D. Cvetković, M. Doob and H. Sachs: "Spectra of graphs", Academic Press, 1980.

[DC*] J. Dixmier: "C^*–algebras", North–Holland 1982.

[DvN] J. Dixmier: "Von Neumann algebras", North–Holland 1981.

[Eff] E.G. Effros: "Dimensions and C^*–algebras", Reg. Conf. Ser. in Math. **46**, Amer. Math. Soc. 1980.

[Eva] R.J. Evans: "A new proof on the free product structure of Hecke groups", Publ. Math. Debrecen **22** (1975), 41–42.

[FH] T. Fack and P. de la Harpe: "Sommes de commutateurs dans les algèbres de von Neumann finies continues", Ann. Inst. Fourier **30**–3 (1980) 49–73.

[Fel] W. Feller: "An introduction to probability theory and its applications", Vol. I, 3rd edition, J. Wiley, 1968.

[Fre] P. Freyd, D. Yetter, J.Hoste, W. Lickorish, K. Millet, and A. Ocneanu: "A new polynomial invariant of knots and links", Bull. Amer. Math. Soc. **12** (1985) 239–246.

[FK] R. Fricke and F. Klein: "Vorlesungen über die Theorie der automorphen Functionen" 1 and 2, 2nd edition, Teubner 1926.

[Fro] F.G. Frobenius: "Über Matrizen aus nicht negativen Elementen", Sitzungberichte 1912, Oeuvres III, p. 546–567.

[Gan] F.R. Gantmacher: "The theory of matrices", Volume 2, Chelsea 1960.

[Gib] J.W. Gibbs: "Elementary principles in statistical mechanics", Yale University Press 1902 and Dover 1960.

[GHM] C. Godsil, D.A. Holton and B. McKay: "The spectrum of a graph", Lecture Notes in Math. **622**, Springer 1979, p. 91–117.

[Gol] M. Goldman: "On subfactors of type II_1", Mich. Math. J. **7** (1960) 167–172.

[GJ] I.P. Goulden and D.M. Jackson: "Combinatorial enumeration", J. Wiley 1983.

[HKW] P. de la Harpe, M. Kervaire and C. Weber: "On the Jones polynomial", l'Enseignement math. **32** (1986) 271–335.

[HW] P. de la Harpe and H. Wenzl: "Opérations sur les rayons spectraux de matrices symétriques entières positives", C.R. Acad. Sci. Paris, Ser. I **305** (1987) 733–736.

[HHSV] M. Hazewinkel, W. Hesselink, D. Sierma and F. D. Veldkamp: "The ubiquity of Coxeter–Dynkin diagrams (an introduction to the A–D–E problem)", Nieuw Arch. Wiskunde, III. Ser. **25** (1977) 257–307.

[Hec] E. Hecke: "Über die Bestimmung Dirichletscher Reihen durch ihre Funktionalgleichung", Math. Ann. **112** (1936), 664–699 = Werke 591–626.

[Hof] A.J. Hoffman: "On limit points of spectral radii of nonnegative symmetric integral matrices", Lecture Notes in Math. **303**, Springer 1972, p. 165–172.

[HS] A.J. Hoffman and J.H. Smith: "On the spectral radii of topologically equivalent graphs", in: "Recent advances in graph theory", Proc. 2d Czechoslovak. Symp., Prague 1974, p. 273–281.

[Jac1] N. Jacobson: "Basic algebra II", Freeman, 1980

[Jac2] N. Jacobson: "Structure theory of simple rings without finiteness assumptions", Trans. Amer. Math. Soc. **57** (1945) 228–245.

[Jo1] V.F.R. Jones: "Index for subfactors", Inventiones Math. **72** (1983) 1–25.

[Jo2] V.F.R. Jones: "Braid groups, Hecke algebras and type II_1 factors", in "Geometric methods in operator algebras (Kyoto, 1983)", Pitman Res. Notes in Math. 123 (1986) 242–273.

[Jo3] V.F.R. Jones: "A polynomial invariant for knots via von Neumann algebras", Bull. Amer. Math. Soc. **12** (1985) 103–112.

[Jo4] V.F.R. Jones: "Index for subrings of rings", Contemp. Math. **43** (Am. Math. Soc. 1985) 181–190.

[Jo5] V.F.R. Jones: "Hecke algebra representations of braid groups and link polynomials", Ann. of Math. (2) **126** (1987) 335–388.

[Jo6] V.F.R. Jones: "Actions of finite groups on the hyperfinite type II_1 factor", Mem. Amer. Math. Soc. **28**, Number 237 (1980).

[Jo7] V.F.R. Jones: "Sur la conjugaison de sous-facteurs de facteurs de type II_1", C.R. Acad. Sci. Paris, Sér. A–B **284** (1977) no.11, A597–598.

[Jo8] V.F.R. Jones: "Subfactors of type II_1 factors and related topics" in: Proc. of the International Congress of Mathematicians, Berkeley, 1986, Amer. Math. Soc. 1987.

[KvN] I. Kaplansky: "Von Neumann's characterization of factors of type II_1", in J. von Neumann's Collected Works III, Pergamon Press 1961, 562–563.

[Kna] A.W. Knapp: "Doubly generated Fuchsian groups", Michigan Math. J. **15** (1968), 289–304.

[Kos] J.L. Koszul: "Lectures on hyperbolic Coxeter groups," University of Notre Dame 1967.

[Kro] L. Kronecker: "Zwei Sätze über Gleichungen mit ganzzahligen Coefficienten", Crelle 1857, Oeuvres I, p. 105–108.

[Lag] J.L. de Lagrange: "Recherches sur la nature et la propagation du son," Oeuvres I. p. 37–148.

[Le 1] D.H. Lehmer: "Factorization of certain cyclotomic functions," Ann. of Math. **34** (1933), 461–479.

[Le 2] D.H. Lehmer: "A note on trigonometric algebraic numbers," Amer. Monthly **40**(3) (1933), 165–166.

[Leh] J. Lehner: "Discontinuous groups and automorphic functions", Math. Surveys VIII, Amer. Math. Soc. 1964.

[Leu] A. Leutbecher: "Über die Heckschen Gruppen $G(\lambda)$", Abh. Math. Sem. Un. Hamburg **31** (1967), 199–205.

[Lin] D.A. Lind: "The entropies of topological Markov shifts and a related class of algebraic integers," Ergod. Th. and Dynam. Sys. **4** (1984), 283–300.

[Lus] G. Lusztig: "Characters of reductive groups over a finite field", Princeton University Press 1984.

[LyU] R.C. Lyndon and J.L. Ullman: "Groups generated by two parabolic linear fractional transformations", Canad. J. Math. **21** (1969), 1388–1403.

[Mag1] W. Magnus: "Non–euclidean tesselations and their groups", Academic Press 1974.

[Mag2] W. Magnus: "Two generator subgroups of $PSL(2,\mathbb{C})$", Nachr. Akad. Wiss. Göttingen, II, Math. Phys. Kl. **7** (1975), 81–94.

[Mas] B. Maskit: "On Poincaré's theorem for fundamental polygons", Adv. in Math. **7** (1971), 219–230.

[Mat] J.P. Matelski: "The classification of discrete 2–generator subgroups of $PSL(2,\mathbb{R})$", Israel J. Math. **42** (1982), 309–317.

[Mxl] G. Maxwell: "Hyperbolic trees", J. of Algebra **54** (1978), 46–49.

[Max] J.C. Maxwell: "Illustrations of the dynamical theory of gases – Part I, on the motions and collisions of perfectly elastic spheres", Philosophical magazine, Ser. 4, **19** (1860) 19–32.

[Mis] M. Misiurewicz: Private communication.

[MvNI] F.J. Murray and J. von Neumann: "On rings of operators", Ann. of Math. **37** (1936) 116–224.

[MvNIV] F.J. Murray and J. von Neumann: "On rings of operators, IV", Ann. of Math. **44** (1943) 716–808.

[NT] Y. Nakagami and M. Takesaki: "Duality for crossed products of von Neumann algebras", Lecture Notes in Math. **731**, Springer 1979.

[Ocn] A. Ocneanu: lectures at Santa Cruz, July 1986 and at Warwick, July 1987, and personal communications.

[Ogg] A. Ogg: "Modular forms and Dirichlet series", Benjamin 1969.

[Pet] M.M. Petrović: "On graphs whose spectral spread does not exceed 4", Publ. de l'Inst. Math. **34**(48) (1983), 169–179.

[PP1] M. Pimsner and S. Popa: "Entropy and index for subfactors", Ann. Sci. École Norm. Sup. (Paris), Ser. 4, **19** (1986) 57–106.

[PP2] M. Pimsner and S. Popa: "Sur les sous–facteurs d'indice fini d'un facteur de type II_1 ayant la propriété T", C.R.Acad. Sci.Paris, Sér. I, **303** (1986) 359–361.

[PP3] M. Pimsner and S. Popa: "Iterating the basic construction", to appear in Trans. Amer. Math. Soc.

[PP4] M. Pimsner and S. Popa: "Finite dimensional approximation of pairs of algebras and obstructions for the index", preprint, 1988.

[Pop1] S. Popa: "Maximal injective subalgebras in factors associated with free groups", Adv. in Math. **50** (1983) 27–48.

[Pop2] S. Popa: "Orthogonal pairs of *-subalgebras in finite von Neumann algebras", J. Op. Theory **9** (1983) 253–268.

[Pop3] S. Popa: "Relative dimension, towers of projections and commuting squares of subfactors", Pac. J. Math., to appear.

[P] R.T. Powers: "Representations of uniformly hyperfinite algebras and their associated Von Neumann rings, Ann. Math. **86** (1967) 138–171.

[Puz] N. Puzitsky: "Two generator discrete free products", Math. Z. **126** (1972), 209–223.

[R] I. Reiner: "Maximal orders", Academic Press 1975.

[Rha] G. De Rham: "Sur les polygones générateurs de groupes Fuchsiens", l'Enseignement math. **17** (1971), 49–61.

[Rbt] A. Robert: "Introduction to the representation theory of compact and locally compact groups", Cambridge U. Press 1983.

[Rob] R.M. Robinson: "Intervals containing infinitely many sets of conjugate algebraic integers", in "Studies in mathematical analysis and related topics: Essays in honor of George Polya", H. Chernoff et al. editors, Stanford Univ. Press 1962, p. 305–315.

[Ros] G. Rosenberger: "All generating pairs of all two-generator Fuchsian groups", Arch. Math. **46** (1986), 198–204.

[Sak] S. Sakai: "C^*-algebras and W^*-algebras", Springer Verlag 1983.

[Sch1] A.J. Schwenk: "Computing the characteristic polynomial of a graph", Lecture Notes in Math. **406**, Springer 1974, p. 153–172.

[Sch2] A.J. Schwenk: "Tight bounds on the spectral radius of asymmetric nonnegative matrices", Linear Alg. Appl. **75** (1986) 257–265.

[Ser1] J.P. Serre: "Représentations linéaires des groupes finis", 3rd edition, Hermann 1978.

[Ser2] J.P. Serre: "Cours d'arithmétique", PUF 1970.

[Sh] J. Shearer: "On the distribution of the maximum eigenvalue of graphs", Lin. Alg. and its Appl. (to appear).

[Slo] P. Slodowy: "Surface singularities and simple algebraic curves", Lecture Notes in Math. **815**, Springer 1980.

[Smi] J.H. Smith: "Some properties of the spectrum of a graph", in: "Combinatorial structures and their applications", Gordon and Breach 1970, p. 403–406.

[Smy] C.J. Smythe: "On the measure of totally real algebraic integers", J. Austral. Math. Soc. (Ser. A) **30** (1980),137–149.

[Som] A. Sommerfeld: "Thermodynamics and statistical mechanics", Academic Press 1964.

[Ste1] R. Steinberg: "Finite reflection groups", Trans. Amer. Math. Soc. **91** (1959), 493–504.

[Ste2] R. Steinberg: "Lectures on Chevalley groups", Lecture Notes, Yale University 1967.

[Str] S. Stratila: "Modular theory in operator algebras", Abacus Press 1981.

[SZ] S. Stratila and L. Zsidó: "Lectures on von Neumann algebras", Abacus Press 1979.

[SV] S. Stratila and D. Voiculescu: "Representations of AF–algebras and of the group U(∞)", Lecture Notes in Math. **468**, Springer Verlag 1975.

[Sun] V.S. Sunder: "A model for AF algebras and a representation of the Jones projections", J. Operator Theory **18** (1987) 289–301.

[Tak] M. Takesaki: "Theory of operator algebras I", Springer Verlag 1979.

[TL] H.N.V. Temperly and E.H. Lieb: "Relations between the "percolation" and "colouring" problem and other graph–theoretical problems associated with regular planar lattices: some exact results for the "percolation" problem", Proc. Roy. Soc. (London) ser A **322** (1971) 251–280.

[Tor] A. Torgasev: "On infinite graphs whose spectrum is uniformly bounded by $\sqrt{2+\sqrt{5}}$ ", Proc. of Fourth Yugoslav Seminar on Graph Theory, Novi Sad. 1983, p. 299–309.

[vN] J. von Neumann: "Zur Algebra der Funktionaloperationen und Theorie der normalen Operatoren", Math. Ann. **102** (1929/30) 370–427.

[Wa] A. J. Wassermann: preprint, 1988.

[We] P.M. Weichsel: "The Kronecker product of graphs", Proc. Amer. Math. Soc. **13** (1962), 47–52.

[Wei] A. Weil: "Algebras with involutions and the classical groups", Journ. Ind. M.S. **24** (1960) 589–623.

[Wen1] H. Wenzl: "On sequences of projections", C.R. Math. Rep. Acad. Sci. Canada **9** (1987) 5–9.

[Wen2] H. Wenzl: "Representations of Hecke algebras and subfactors", Thesis, Univ. of Pennsylvania 1985; and "Hecke algebras of type A and subfactors", Invent. Math. (to appear).

[Wen3] H. Wenzl: "On the structure of Brauer's centralizer algebras", Ann. of Math. (to appear)

[Weyl] H. Weyl: "The classical groups", Princeton Univ. Press 1939.

[Zim] R. J. Zimmer: "Ergodic theory and semisimple groups", Birkhäuser Verlag 1984.

INDEX

Adjacency matrix of a graph: 1.3
$\mathcal{A}_{\beta,k}$: 2.1.7, 2.8, II.b
  and Hecke algebras: 2.1.9,2.11
$\mathcal{A}_{gen,k}$: 2.8
$A_{tr,k}(M_0 \subset M_1)$: 2.1.8, 2.7, 2.8

$B_{\beta,k}$: 2.8, 2.9
Bicommutant theorem: 2.2.3
Bicoloration number of a graph: 1.4
Borel subgroup: 2.10.a
Braiding relations: 2.1
Bratteli diagram: 2.3
Bruhat decomposition: 2.10.a

C^*–algebra: II.a
Catalan numbers: 2.7
Central algebra: 2.1
Chains of multi–matrix algebras: 2.3
Characteristic polynomial of a graph: 1.3
Commutant and bicommutant: 2.1
Commuting square: 4.1, 4,2
Conditional expectation: 2.1, 2.6
  and the fundamental construction: 2.6.4
Connected pair of algebras: 2.1
Coupling constant: 2.2, 3.2
Covolume of a lattice: 3.3.e
Coxeter exponents: 1.4
Coxeter graph: 1.1, 1.4
Coxeter invariant: 4.6
Cusp form: 3.3.e

Depth: 4.1, 4.6
Derived tower: 4.1, 4.6
Dimension of a projection: 3.2
$\dim_M(H)$: 3.2
Discrete series: 3.3.a

$\mathbb{E} = \mathcal{M}(\mathbb{N})$: 1.1, 1.5
E– extension: 2.6.6
e_i – relations: 2.1.6, 2.7.5

Factor: 2.1, 2.2, 3.2
Factor of type II_1: 3.2
Faithful conditional expectation: 2.6
Faithful trace: 2.1, 2.5
Finite depth: 4.1, 4,6
Finite factor: 3.2
Finite index: 3.1, 3.5
Finite representation of a pair: 3.5
Floor of a Bratteli diagram: 2.3
Formal dimension: 3.3.a
Full factor: 3.4

Fundamental construction
  for multi–matrix algebras: 2.1, 2.4
  for finite von Neumann algebras: 3.1, 3.6

Generic β : 2.1, 2.7
Graph:
  labelled: 1.1
  marked: 1.1
  norm: 1.3
  principal: 4.1, 4.6
  spectral radius: 1.3
  spectral spread: 1.4
Hecke groups: 3.1, III
Hecke algebra $H(G, G_0)$: 2.10.a
Hecke algebra $H_{q.k}$: 2.1, 2.10.b, 2.11

Inclusion matrix, index matrix: 2.1, 2.3, 3.1, 3.5
Index: 2.1, 3.1, 3.4, 3.7
  of semi–simple pairs: 2.1.1
  of subfactors: 3.4
Index
  of pairs of finite von Neumann algebras: 3.7.5
Infinite conjugacy class (icc) group: 3.3.b
Involution: II.a
Irreducible subfactor: 3.4
Ising model: II.b

$\mathbb{K} = \{0,2\} \cup \{2\cos(\pi/k)\}_{k\geq 3}$: 1.1
$\mathbb{K}$ = a given field: 2.1
K_0 : 2.3
$\left\{\begin{matrix}k\\i\end{matrix}\right\} = \begin{bmatrix}k\\i\end{bmatrix} - \begin{bmatrix}k\\i-1\end{bmatrix}$: 2.8
Kronecker's theorem: 1.1.1, 1.2.1, 1.2.2

Lattice (in a Lie group) : 3.3.b
Λ_N^M : 2.1, 2.3, 3.1, 3.5
$(\lambda_q)_{q\geq 2}$: 1.1, I.1

Markov relation: 3.1
Markov trace: 2.1, 2.7, 3.1, 3.7
  and index for multi–matrix algebras: 2.1.4, 2.7.3
$Mat_{m,n}(S)$, $Mat_{fin}(S)$: 1.1
Matrix
  adjacency: 1.3
  aperiodic (=primitive) non–negative: 1.3
  of a bicolored graph: 1.3
  equivalent: 1.3
  indecomposable: 1.1, 1.2

index matrix = inclusion matrix: 2.1, 2.3, 3,1, 3,5
irredundant: 1.3
norm of a matrix: 1.1
parabolic: III
pseudo-equivalent: 1.1, 1.3
reducible: 1.3
trace matrix: 3.1, 3.5
transfer matrix: II.b
Modulus of a Markov trace: 2.1, 2.7, 3.1, 3.7
Monomial in $\mathcal{A}_{\beta,k}$: 2.8
Multi-matrix algebra: 2.1

$\mathcal{N}$(S): 1.1
Natural trace: 3.2
von Neumann algebra: 3.2
Normalized trace: 3.2

$P_k \in \mathbb{Z}[\lambda]$: 2.8
Parabolic matrix: III
Partition function: II.b
Path model: 2.3.11
and the fundamental construction: 2.4.6, 2.6.5
and the tower construction: 2.7.6
Perron-Frobenius theory: 1.4
Peterson inner product: 3.3.e
Pimsner-Popa basis: 3.6.4
Popa's theorem: 4.7.3
Positive conditional expectation: II.a
Positive involution: II.a
Positive trace: 2.1, 2.5, II.a
Principal graph: 4.1, 4.6

Rank of a module: 2.1
Reduction by a projection: 2.2, 3.2
Regular subfactor: 3.4
Row vector: 2.1, 2.5

Self-adjoint conditional expectation: II.a
Skau's lemma: 4.4.3
Skolem-Noether theorem: 2.2.6
Square lattice Pott's model: II.b
Story of a Bratteli diagram: 2.3

Temperley-Lieb algebras ($\mathcal{A}_{\beta,k}$): 2.1, 2.7, 2.8, 2.11, II.b
$3 + 3^{1/2}$: 4.5.2
Tower: 2.1, 2.4
Trace: 2.1, 2.5, 3.2, 3.5, II.a
Trace matrix: 3.1, 3.5
Transfer matrix: II.b
$(2 + 5^{1/2})^{1/2}$: I.1
Tunnel construction: 4.7.e

Ultraweak topology: 3.2

Very faithful conditional expectation: 2.6

Weights of a trace: 2.5
Wenzl's representations: 2.10.d
index formula: 4.3

Young diagrams: 2.10.c